SUSTAINABLE AGROECOSYSTEM MANAGEMENT

Integrating Ecology, Economics, and Society

Advances in Agroecology

Series Editor: Clive A. Edwards

Advisory Board

SUSTAINABLE AGROECOSYSTEM MANAGEMENT

Integrating Ecology, Economics, and Society

Edited by

Patrick J. Bohlen and Gar House

CRC Press
Taylor & Francis Group
Boca Raton   London   New York

CRC Press is an imprint of the
Taylor & Francis Group, an **informa** business

CRC Press
Taylor & Francis Group
6000 Broken Sound Parkway NW, Suite 300
Boca Raton, FL 33487-2742

© 2009 by Taylor & Francis Group, LLC
CRC Press is an imprint of Taylor & Francis Group, an Informa business

No claim to original U.S. Government works
Printed in the United States of America on acid-free paper
10 9 8 7 6 5 4 3 2 1

International Standard Book Number-13: 978-1-4200-5214-5 (Hardcover)

Library of Congress Cataloging-in-Publication Data

Sustainable agroecosystem management: Integrating ecology, economics, and society / editors, Patrick J. Bohlen and Gar House. -- 1st ed.
       p. cm. -- (Advances in agroecology ; 14)
    Includes bibliographical references and index.
    ISBN 978-1-4200-5214-5 (alk. paper)
    1. Agricultural ecology. 2. Sustainable agriculture. 3. Sustainable development. I. Bohlen, P. J. (Patrick J.) II. House, Garfield J. III. Title. IV. Series.

S589.7.A474 2009
630--dc22   2008040757

Visit the Taylor & Francis Web site at
http://www.taylorandfrancis.com

and the CRC Press Web site at
http://www.crcpress.com

"Like our meeting, life is then short, so let us give to each our best right now."

—Ben Stinner

Contents

SECTION I Ben Stinner's Contribution to Agroecosystem Science

SECTION II Unifying Concepts and Principles of Sustainable Agroecosystem Management

Preface

This book is dedicated to Ben Stinner, beloved colleague, pioneer in the field of agroecosystem ecology, and former Kellogg Endowed Chair of Ecological Management at the Ohio State University, who died tragically in a car accident in 2004. Ben's pioneering studies comparing nutrient cycling in agricultural fields and natural old field systems remain a classic in the field of agroecosystem ecology. He conducted significant research and led innovative programs that included the ecology and economics of whole-farm systems, arthropod ecology, nutrient cycling, the role of organic matter in soil fertility, and the ecology of Amish farming systems. As outlined in a chapter by his wife, Deborah Stinner, another accomplished agroecologist, Ben's earlier focus on agroecosystems as natural systems expanded to a watershed perspective, which encompassed natural and social science perspectives. This perspective allowed Ben to build collaborative efforts with farmers, researchers, and policy makers that resulted in novel approaches for agroecosystem watershed management. Ben was an inspiring natural leader, ecological innovator, supportive mentor, devoted husband and father, and true friend.

In 1994, Ben Stinner and his colleagues at the University of Georgia, Richard Lowrance and Gar House, coedited a book entitled *Agricultural Ecosystems: Unifying Concepts,* which included contributions from authors who participated in a symposium initially held at the Ecological Society of America annual meeting in State College Pennsylvania in 1992. Five of the authors from that original volume (Coleman, Crossley, Jackson, Pimentel, House) contributed to this new volume, published 24 years later. Since 1984, when agroecology and agroecosystem science were in their infancy, there has been an explosion of interest in these fields, and the book by Lowrance, Stinner, and House remains an important early synthesis. This current book is another synthesis, reflecting the tremendous growth and broad dimensions of agroecology and agroecosystem science.

Central to Ben's work and the ideas presented here is the concept of the agroecosystem, originally applied as an extension of Eugene Odum's ecosystem concept, with the added dimension of a coupling between natural and social systems. We conducted a search of the term "agroecosystem" and its variants in the ISI Web of Knowledge[SM] to assess how the use of this term has grown since the early 1980s. Prior to 1984, there were only 7 references that included this term. From 1980 to 1984, there were 31 references. Thereafter the use of the term increased exponentially, with 154 references in the 1980s, 1078 in the 1990s, and 2144 so far in 2000s (through August 4, 2008). During the same time that the use of the agroecosystem concept has become commonplace, the global agriculture and food system has been completely transformed, with increasing industrialization, specialization, and concentration in developed countries, and persistent lack of development in many poorer parts of the world. As discussed in Chapter 1, the scope and scale of environmental problems associated with agriculture have grown dramatically, including persistent problems with nonpoint-source pollution, contamination with agrochemicals, concentration of animal production and wastes, and depletion of groundwater and soil resources. The scale and complexity of these problems cannot be addressed in a single volume or by a single approach. The goal of this book is not to provide simple prescriptions for the application of ecological principles to agriculture, but rather to emphasize the continued centrality of the ecosystem perspective, and the need for integrated approaches to agroecosystem science and management that combine ecological, economic, and social considerations. It is a fitting tribute to Ben Stinner that many of the authors of this book are social transformers committed to turning the world of ideas into actions that provide enduring solutions to the challenges of sustainable agroecosystem management.

This book includes chapters on a variety of themes that are organized in four sections. An introductory chapter by Bohlen and House is followed two chapters in Part 1 that review Ben Stinner's contribution to agroecosystem science and work with his colleagues and partners at Ohio State University and the Sugar Creek Watershed in northeastern Ohio.

Part 2 includes different perspectives on unifying concepts and themes related to sustainable agroecosystem management. John Ikerd (Chapter 4) offers fundamental commonsense principles for agroecology. Fred Kirschenmann (Chapter 5) reviews challenges for agriculture and illustrates the potential for developing new biological systems. Wes Jackson and his coauthors from the Land Institute (Chapter 6) outline the necessity for developing an agriculture based on perennial cropping systems. Dave Pimentel (Chapter 7) discusses the connection of agriculture to energy and human population growth, and Bland and Bell (Chapter 8) grapple with the definition of systems and boundaries in agroecosystem analysis.

Part 3 focuses on the ecological foundation of agroecosystem management including chapters that address: biological buffering (Phelan, Chapter 9), humus theory and its relation to agroecosystem integrity (Wander, Chapter 10), the role of biodiversity in agroecosystems (Hillel and Rosenzweig, Chapter 11), and biogeochemical changes in the long-term Horseshoe Bend study that Ben Stinner was involved in earlier in his career (Coleman et al., Chapter 12).

Part 4 examines integrated approaches for managing agroecosystems for multiple functions. Westra and Boody (Chapter 13) illustrate how agricultural policies could be used to help develop multifunctional landscapes with no loss of production at no greater cost to the taxpayer. Bohlen and Swain (Chapter 14) use their experience with management of a working cattle ranch to explore a conceptual model that integrates ecological and economic aspects of sustainability. Sassenrath et al. (Chapter 15) summarize their work on integrated agricultural systems in the southeastern United States. Karin Eksvärd and her colleagues from Scandinavia describe their experience with participatory research and stakeholder involvement in sustainable agriculture research (Chapter 16). Finally, Gar House explores the necessity and possibility for retrofitting suburban landscapes to support sustainable agriculture.

Many people have contributed to the production of this volume. We thank the authors for their contributions, especially Deb Stinner, who supported the idea for this book from the outset, and despite her own tremendous loss had the courage to contribute a chapter on the development of Ben's career. Many of the chapters in this volume came from joint symposiums that were held in 2005 at the Ecological Society of America annual meeting in Memphis, Tennessee, and the joint annual meeting of the Agronomy Society of America (ASA), Crop Science Society of America (CSSA), and Soil Science Society of America (SSSA) in Indianapolis, Indiana in 2006. These symposia were supported with a conference grant from the Managed Ecosystems Program of the U.S. Department of Agriculture (USDA), Cooperative State Research, Extension and Education Service (CSREES), and National Research Initiative (NRI). We thank Managed Ecosystems program director, Diana Jerkins, for supporting and attending these symposia. Laurie Drinkwater (Cornell University), Michelle Wander (University of Illinois), and Richard Lowrance (USDA Agricultural Research Service) helped write the grant proposal. The symposium at the ASA-CSSA-SSSA meeting was cosponsored by the Integrated Agricultural Systems Division of the Agronomy Society of ASA the Committee on Sustainable Agriculture Systems (COSA). Special thanks to Integrated Agricultural Systems Chair Paul Porter and COSA Chair Kim Lavel (Center for Rural Affairs), and other COSA members (Heather Darby, Caron Gala, Stefan Seiter) for their support. Karl Glasener, director of science policy for ASA-CSSA-SSSA provided a letter of support from the Tri-Societies for our proposal and supported our efforts throughout the planning process.

In addition to the contributions from the authors, production of a book such as this would not be possible without a supportive editor and publisher. John Sulzycki of CRC Press, Taylor & Francis group has been a strong supporter of this project from its outset. He made astute suggestions for content and organization of this volume and has maintained enthusiasm and support, even as the manuscript deadline disappeared farther over the horizon. The field of agroecology owes John a debt of gratitude for his tremendous contributions to agroecology through supporting publication of the Advances in Agroecology Series and the other significant agroecology books over the years. We thank the Advances in Agroecology Series Editor Clive Edwards for embracing this tribute as part of the series. Finally, no manuscript gets submitted without laborious editing. Julie Mitchell helped format the chapter headings and references, cross-check citations, and prepare the manuscript for submission. We thank Pat Roberson, project coordinator for this book at CRC Press and all the production staff for helping produce this volume.

Patrick Bohlen
Lake Placid, Florida

Gar House
Vista, California

The Editors

Patrick J. Bohlen, PhD, is Director of the MacArthur Agro-ecology Research Center (MAERC), a division of Archbold Biological Station in Florida. MAERC is located at Buck Island Ranch, a 4252-ha cattle ranch dedicated to long-term ecological research, education, and environmental stewardship. Dr. Bohlen earned his BS at the University of Michigan, MS at Miami University, and PhD at The Ohio State University, where Ben Stinner served on his doctoral committee. He is Courtesy Associate Professor in the Soil and Water Science Department at the University of Florida.

Dr. Bohlen's expertise is in the area of soil ecology and ecosystem ecology. For the past decade he has directed interdisciplinary research projects that have focused on (1) controlling nonpoint-source nutrient runoff from cattle pastures, (2) quantifying the effects of management and land use intensity on wetland ecology and biogeochemistry, and (3) developing concepts and procedures for quantifying ecosystem services in working agricultural landscapes. His previous work focused on soil ecology, particularly the influence of earthworms on nutrient turnover and retention in agricultural and forest systems. He is coauthor with Dr. Clive Edwards of *Earthworm Biology and Ecology*. He has served on the editorial boards of *Applied Soil Ecology, Bulletin of Environmental Contamination and Toxicology, Ecology and Ecological Monographs, Frontiers of Ecology and the Environment*, and *Rangeland Ecology and Management*.

As director of a unique ecological research program on a working ranch, Dr. Bohlen has fostered extensive collaborations in research and conservation. He has provided support for a wide array of visiting researchers and investigators, and has collaborated with numerous university faculty. He has mentored more than 20 student research interns and actively supports graduate student research at MAERC. He serves as an advisor to the Agroecology Program at Florida International University, and has worked with ranchers, the Florida Cattleman's Association, state and federal agencies, and conservation and environmental organizations. His efforts with Archbold Director Hilary Swain and Ranch Manager Gene Lollis to support environmentally and economically sustainable cattle ranches in Florida has been recognized by the 2006 Environmental Stewardship Award from the Florida Cattlemen's Association, and the 2007 Florida Agriculture Commissioner's Ag-Environmental Leadership Award. Dr. Bohlen lives on Buck Island Ranch with his wife, Julie Mitchell, and their two children.

Gar House had the honor of enjoying a close friendship with Ben Stinner. While pursuing their PhD degrees together at the University of Georgia's Institute of Ecology, Ben and Gar pioneered research on nutrient cycling within agroecosystems. In 1982, Ben and Gar coorganized one of the first agroecosystems conferences, with participation from ecological luminaries that included Eugene Odum, David Pimentel, and Wes Jackson. Gar has since actively pursued agroecosystem advocacy through his involvement in the research and production of sustainable agriculture within the urban and suburban environment.

As founder and president of Building Sustainable Communities (BSC) (www.buildingsustainablecommunities.org), Gar directs the organization with support from a broadly based, interdisciplinary staff of dedicated individuals. BSC is active in the San Diego County area promoting sustainable practices based on ecosystem concepts and principles. He assists the organization in providing leadership and guidance for local communities and organizations with special emphasis on educational outreach promoting sustainable practices.

Gar also serves as Research Director at the San Pasqual Agroecosystems Research Center (SPARC), where he has initiated a collaborative perennial grain project with plant breeders at the Land Institute in Salina, Kansas. Additional projects at SPARC include investigations of agricultural and land use practices that enhance soil biomass for the purpose of long-term sequestration of atmospheric carbon dioxide.

Contributors

Sofia Arce-Flores
Institute of Ecology
University of Georgia
Athens, Georgia

David W. Archer
USDA-ARS Northern Great Plains Research
  Laboratory
Mandan, North Dakota

Michael M. Bell
Department of Rural Sociology and the
  Agroecology Program
University of Wisconsin–Madison
Madison, Wisconsin

Johanna Björklund
Department of Urban and Rural Development
Swedish University of Agricultural Sciences
Uppsala, Sweden

William L. Bland
Department of Soil Science and the
  Agroecology Program
University of Wisconsin–Madison
Madison, Wisconsin

Patrick J. Bohlen
Director
MacArthur Agro-Ecology Research Center
Lake Placid, Florida

George Boody
Executive Director
Land Stewardship Project
White Bear Lake, Minnesota

David C. Coleman
Institute of Ecology
University of Georgia
Athens, Georgia

Cindy Cox
The Land Institute
Salina, Kansas

Stan Cox
The Land Institute
Salina, Kansas

D. A. Crossley, Jr.
Institute of Ecology
University of Georgia
Athens, Georgia

Lee DeHaan
The Land Institute
Salina, Kansas

Karin Eksvärd
Department of Urban and Rural
  Development
Swedish University of Agricultural Sciences
Uppsala, Sweden

Charles Francis
Department of Agronomy and Horticulture
University of Nebraska–Lincoln
Lincoln, Nebraska

Ulrika Geber
Centre for Sustainable Agriculture
Swedish University of Agricultural Sciences
Uppsala, Sweden

Jerry Glover
The Land Institute
Salina, Kansas

John F. Halloran
USDA-ARS New England Plant, Soil, and
  Water Research Laboratory
University of Maine
Orono, Maine

Jon D. Hanson
USDA-ARS Northern Great Plains Research
  Laboratory
Mandan, North Dakota

John R. Hendrickson
USDA-ARS Northern Great Plains Research
  Laboratory
Mandan, North Dakota

Paul F. Hendrix
Institute of Ecology
University of Georgia
Athens, Georgia

Daniel Hillel
Center for Climate Systems Research
Columbia University
New York, New York

Gar House
Vista, California

Mark D. Hunter
Department of Ecology and Evolution
University of Michigan
Ann Arbor, Michigan

John E. Ikerd
College of Agriculture, Food, and Natural
  Resources
University of Missouri Columbia
Columbia, Missouri

Wes Jackson
President
The Land Institute
Salina, Kansas

Frederick L. Kirschenmann
Distinguished Fellow, Leopold Center
Iowa State University
Ames, Iowa

Geir Lieblein
Norwegian University of Life Sciences
Plant and Environmental Sciences
Ås, Norway

Richard H. Moore
Human and Community Resource
  Development
OARDC/The Ohio State University
Wooster, Ohio

P. Larry Phelan
Department of Entomology
OARDC/The Ohio State University
Wooster, Ohio

David Pimentel
Department of Entomology
Cornell University
Ithaca, New York

Cynthia Rosenzweig
NASA Goddard Institute for Space Studies
New York, New York

Lennart Salomonsson
Department of Urban and Rural Development
Swedish University of Agricultural Sciences
Uppsala, Sweden

Gretchen F. Sassenrath
Plant Physiologist
USDA-ARS Application and Production
  Technology Research Unit
Stoneville, Mississippi

Breana Simmons
Natural Resource Ecology Laboratory
Colorado State University
Ft. Collins, Colorado

Nadarajah Sriskandarajah
Department of Urban and Rural Development
Swedish University of Agricultural Sciences
Uppsala, Sweden

Jeffrey J. Steiner
USDA-ARS George Washington Carver
  Center–Beltsville
Beltsville, Maryland

Deborah H. Stinner
Department of Entomology
OARDC/The Ohio State University
Wooster, Ohio

Karin Svanäng
Centre for Sustainable Agriculture
Swedish University of Agricultural Sciences
Uppsala, Sweden

Hilary M. Swain
Executive Director
Archbold Biological Station
Lake Placid, Florida

David Van Tassel
The Land Institute
Salina, Kansas

Michelle Wander
Department of Natural Resources and
 Environmental Sciences
University of Illinois
Urbana, Illinois

John Westra
Department of Agricultural Economics and
 Agribusiness
Louisiana State University AgCenter
Baton Rouge, Louisiana

Kyle Wickings
Institute of Ecology
University of Georgia
Athens, Georgia

1 Agroecosystem Management for the Twenty-First Century

Sustaining Ecosystems, Economies, and Communities in a Time of Global Change

Patrick J. Bohlen and Gar House

CONTENTS

1.1 INTRODUCTION

Agricultural production and related activities are the foundation of human communities (MacNeill 1992). Yet the current extractive operations and methods of industrial agriculture, dependent on currently inexpensive and readily available energy sources (primarily oil and methane gas), are increasingly vulnerable (Heinberg 2007). In recent years crop yields have flattened, reaching diminishing marginal returns for each added unit of input. Indeed, industrial agriculture appears to be approaching the peak or the downside of the classic, ecological subsidy–stress curve (Odum et al. 1979). Furthermore, on a global scale, industrial agricultural production methods continue to exacerbate soil erosion, water pollution (especially pernicious is hypoxia at river deltas), and negative climate change. As Kirschenmann states in Chapter 5: "the industrialization of agriculture which enabled us to dramatically increase production during the past half-century also is a principal cause of the ecological degradation that now threatens our ability to maintain productivity."

A central hypothesis of agroecology is that the incorporation of internal ecosystem control enhances agricultural production and restores degraded soil and water resources. Restoration and incorporation of internal ecosystem control within agricultural production methods and processes (originating from biological and social sources) provide a conceptual and practical framework for mitigating global resource degradation. A substantial body of agroecological methodology exists, but integration and implementation of these sustainable practices on a large scale is lacking. Although no single approach or set of ideas can fully address the complexities of agroecosystem management, current approaches clearly are failing to achieve desired outcomes of sustainability. This volume is an attempt to provide conceptual underpinnings and examples of integrated approaches for more sustainable agroecosystem management.

Several authors in this book demonstrate the value of the systems approach to problem solving in agriculture (e.g., Kirschenmann, Chapter 5; Jackson et al., Chapter 6; Bland and Bell, Chapter 8). Others explore and identify the underlying ecological, economic, and social principles of sustainable practices (e.g., Ikerd, Chapter 4; Moore, Chapter 3; Sassenrath et al., Chapter 15), including methods for expanding their acceptance, insemination, and accessibility (e.g., Eksvärd et al., Chapter 16). This volume represents the convergent effort of natural and social scientists to employ the ecosystem concept as a working paradigm toward the investigation and building of sustainable agroecosystems. Patterns and processes occurring within sustainable agroecosystems are explored through the four distinct investigative lenses: (1) the physical systems themselves (e.g., Coleman et al., Chapter 12; Phelan, Chapter 9; Wander, Chapter 10), (2) social and economic influence on the structure and operation of agroecosystems (e.g., Bohlen and Swain, Chapter 14; Hillel and Rosenzweig, Chapter 11; Moore, Chapter 3; Sassenrath et al., Chapter 15), (3) the philosophically and ethical underpinnings for pursuing sustainable practices (e.g., Ikerd, Chapter 4; Jackson et al., Chapter 6), and (4) the economic and policy incentives that influence and constrain potential outcomes for multifunctional agroecosystems (e.g., Bohlen and Swain, Chapter 14; Westra and Boody, Chapter 13).

1.2 CHALLENGES FOR AGROECOSYSTEM MANAGEMENT IN THE TWENTY-FIRST CENTURY

The great gains in agricultural productivity in the past century have been accompanied by substantial degradation to global ecosystems (Millennium Ecosystem Assessment 2005), and these strains are projected to increase substantially in the coming decades (Matson et al. 1997; Tilman et al. 2002; Robertson and Swinton 2005). Agriculture has contributed to losses in biodiversity and declining water quality worldwide (Carpenter et al. 1998; Collins and Qualset 1999; Dixon and Gulliver 2001). Agriculture also exacerbates global warming, and currently contributes about 30 percent of the global anthropogenic emission of greenhouse gases (Smith et al. 2007). The impacts of global climate change on agricultural ecosystems are unpredictable, but existing evidence suggests they could be substantial, including a potential increase in the frequency of climatic extremes, and potential significant losses in crop yield in some regions (Rosenzweig and Parry 1994; Dixon and Gulliver 2001; Rosenzweig and Hillel 1995). Addressing these systemic problems will depend on input from diverse scientific disciplines and will require new approaches to research and education in the agricultural sciences. Agroecological approaches that combine the perspectives of ecological and agronomic science, and that integrate the social science and policy perspectives, will be critical to developing sustainable agricultural systems in the face of unprecedented and rapid global change. In short, a new perspective from within the ecosystem paradigm is needed.

The massive gains in agricultural productivity in the twentieth century were due primarily to improvement in crop production technologies, with an emphasis on improved yield, and the development of nitrogenous fertilizers and other industrial inputs that enhance productivity (Dixon and Gulliver 2001; Smil 2001). The research to support these unprecedented gains in agricultural productivity focused principally on intensifying crop and livestock production, usually by means of purchased, fossil fuel–derived inputs. There has been far less research on integrated approaches for sustaining whole system productivity, diversifying agricultural operations, and managing for agricultural systems for multiple purposes, including the ecological integrity of farming systems (Jackson and Jackson 2002; Dixon and Gulliver 2001; Robertson and Swinton 2005). In Chapter 10, Michelle Wander compares and contrasts the systems approach with the reductionists' approach, placing each within its historic perspective, while also discussing the benefits and limitations of each method.

Growing acceptance of the multiple challenges of sustaining food production for a burgeoning human population while preserving natural ecosystems and global biodiversity and sustaining rural communities, has prompted calls for a reorientation of research and policy agendas toward integrated analysis that includes these various perspectives (Giampietro and Pastore 2001; Francis et al. 2004). It is unlikely that major new technological breakthroughs will lead to productivity gains such as those observed in the last century, despite the claims and potential of biotechnology (Altieri 2005). Thus, the most important challenge of managing agroecosystems in the twenty-first century will be to foster the development of production systems that sustain agricultural productivity, support natural functions and diversity, and provide economically viable and socially attractive opportunities for farmers in an increasingly urban world.

Sustaining increased agricultural production while preserving ecological integrity and environmental quality is one of the grand challenges for agriculture for the twenty-first century (Matson et al. 1997; Smil 2000; Robertson and Swinton 2005). Sustaining increased production will doubtless require increasing productivity and yields on existing agricultural land (Tilman et al. 2002). Some have argued that prime agricultural areas should be considered ecological "sacrifice zones," devoted exclusively to agricultural production, so that other natural areas can be preserved (Avery 1995). This view has been countered by others, who have argued that such ecological sacrifice zones represent an unnecessarily high environmental and social cost, and that greater attempts need to be made to incorporate more natural structures and functions within agricultural landscapes (Jackson and Jackson 2002; Francis et al. 2004; Boody et al. 2005). There is a need for more research on the fertile middle ground of these opposing views, in which the advantages of industrial technology are combined with the benefits of more ecological approaches to the management of agricultural landscapes both to mitigate ecosystem degradation and to enhance natural ecosystem processes. Thus, it is critical to develop the conceptual basis and practical strategies for incorporating broader environmental, ecological, and social factors into agroecosystems management and science. In their respective contributions, Kirschenmann (Chapter 5), Jackson et al. (Chapter 6), Bohlen and Swain (Chapter 14), and House (Chapter 17) discuss the need for employing the ecosystem paradigm as a solution framework for building sustainable agricultural ecosystems, and this general theme is touched on throughout this book.

Energy is another critical aspect of agricultural sustainability that presents huge challenges to sustaining agricultural output. Although the industrialized model of agricultural production has been very successful at increasing agricultural yields, it is highly dependent on fossil fuel inputs and thus is vulnerable to increased energy costs and declining energy supplies (Pimentel and Pimentel 1996; Smil 2003; Pimentel, Chapter 7). The sixfold gain in energy yield of food crops since 1900 has been accompanied by an 85-fold increase in energy inputs for production (Smil 2008). Enormous biofuels projects, despite their current vogue, have serious biophysical limitations, especially their very low net energy yield and the low-density energy ethanol contains. The rush to produce ethanol from grain crops caused significant increases in prices of staple grains such as corn, wheat, and rice, causing massive social unrest in many parts of the world in 2008 (Brown 2008). Biofuel production from grain also has the potential to greatly exacerbate the negative environmental impacts of agriculture, especially soil erosion and water pollution, due to increased cultivation and use of fertilizer (Donner and Kucharik 2008). At the same time increased fuel prices in 2008 are driving up the costs of inputs to agricultural production substantially, adding further upward pressure to food prices. Global fertilizer prices rose 200 percent in 2007, due mainly to increases in new demands for food crops and grain for ethanol and other biofuels, increased energy and fuel prices, and higher demand for grain-fed meat in China, India, and Brazil (IFDC 2008).

This grand confluence of energy forces has shocked world food and economic systems indicating that we are reaching a turning point in human history regarding energy and agriculture. Developing systems that are more energy efficient and less dependent on external inputs, but instead rely on internal ecosystem control processes, will be a critical focus of future agricultural systems. However, the reliance on less concentrated forms of energy has the potential to enhance beneficial impacts within agricultural ecosystem, especially biodiversity and nutrient cycling (Allen et al. 2003). Any coherent strategy for managing agroecosystems in the future needs to consider potential impacts of changing energy supplies and cost on agricultural production. Solutions offered throughout this book rely on using the ecosystem concept as a unifying solution framework.

1.3   THE EMERGENCE OF AGROECOLOGY AND THE AGROECOSYSTEM CONCEPT

Agroecology is the integrative study of the interactions among biological, environmental, and management factors in agricultural systems. It has been variously defined as (1) the application of ecological science to the design and management of sustainable agroecosystems, (2) a holistic approach to agriculture, drawing from traditional, alternative, and local, small-scale systems, and (3) an approach that links the ecological and socioeconomic factors to develop agroecosystems that sustain agricultural production, farming communities, and environmental health (Gliessman 1998). Agroecology emerged as a field of study in the 1980s and includes diverse perspectives and various attempts to develop unified concepts and approaches (Lowrance et al. 1984; Altieri 1987; Coleman and Hendrix 1988; Gliessman 1998; Robertson and Paul 1998; Rickerl and Francis 2004a).

The field of agroecology is fundamentally an application of the ecosystem concept within a social science framework. Early developments in agroecosystem ecology drew heavily from ecosystem ecology, which emerged as a discipline in middle of the twentieth century (Odum 1959). Ecosystem ecology emphasizes the links between organisms and their physical environment, and the flow of energy and materials through such linked biophysical systems (Chapin et al. 2002). Consequently, early attempts to unify concepts of agroecosystem ecology focused mainly on the biophysical interactions in agroecosystems (Lowrance et al. 1984). However, ecological and social scientists have long recognized the importance of social and economic factors in agroecosystems and have debated whether agroecosystems are essentially ecological systems with a strong social component or fundamentally socioeconomic systems with a strong ecological component (Lowrance et al. 1984). This conundrum remains largely unresolved, but is it clear that agroecosystems cannot be understood or managed ecologically without including perspectives from both the ecological and social sciences (Rickerl and Francis 2004b). In this volume several chapters (e.g., Moore, Chapter 3; Ikerd, Chapter 4; Kirschenmann, Chapter 6; Bohlen and Swain, Chapter 14; Eksvärd et al., Chapter 16) stress the importance of social engagement to success. Maintaining a viable sustainable agricultural ecosystem requires more than simply physically altering the landscape. As Eksvärd et al. (Chapter 16) and Moore (Chapter 3) indicate, cooperative engagement among social, cultural, political, and especially educational leaders and stakeholders is essential to build and maintain continuity.

An ecological approach to agriculture recognizes that agroecosystems are a consequence of a complex web of interactions between the biophysical environment and social and economic systems. The complexity of these interactions poses significant challenges to developing unified concepts and creating innovative approaches for research and education in the agricultural and ecological sciences. In Chapter 10, Wander discusses the advantages of such comprehensive approaches, suggesting that the reason ecosystem studies are not more prevalent is due in part to their inherent difficulty.

Several programs that have emerged in the United States in recent years reflect recognition that there is a need to deal with the inherent complexity in agroecosystem management and support integrated research approaches. These have included the USDA-NRI Agricultural Systems Program (defunct), Ecosystems Program (defunct), the new Managed Ecosystems Program, the Initiative for Future Food and Agriculture Systems, and programs in other agencies such as the NSF Biocomplexity Program, which emphasize coupling of human and natural systems. The newly formed USDA CSREES national program in Integrated Agricultural Systems is another reflection of this trend toward integrated approaches in agricultural research and education. As these new innovative programs move forward and other new programs emerge, there is a continuing need for synthesis efforts that build the theoretical and applied knowledge base necessary for developing integrated approaches to agroecosystems management. This book represents one effort to synthesize unifying concepts and practical applications of agroecosystem science and management for sustainability.

1.4　THE NEED FOR NEW APPROACHES TO AGRICULTURAL RESEARCH AND EDUCATION

The modern world is characterized by extreme fragmentation and specialization of knowledge, and modern agricultural science is no exception. Yet, the environmental challenges we face require that scientists bring together these disparate fields of knowledge to inform decision making, and to guide public and private action. Synthesizing the various specialized knowledge bases into coherent strategies for addressing these problems is one of the greatest challenges facing agricultural and environmental sciences (Norgaard and Baer 2005). The ecosystem paradigm continues to offer an integrating solution framework.

Modern agriculture is viewed through the multiple lenses of numerous scientific disciplines that support various aspects of agricultural production but no overarching model has been developed to hold these disparate views together. Agroecology and various related approaches have made progress toward accepting a more holistic view of agriculture, but the agricultural sciences in general have not progressed very effectively toward an integrated whole. The educational process in agricultural and other natural resource disciplines needs to address the difficulty in communication among disciplines and especially understanding the interactions between ecological and social systems. Although promising examples exist, the increased emphasis on interdisciplinary studies is largely rhetorical and there are few examples demonstrating effective communication and collaboration across disciplines leading to significant change in outcomes. Successful coupling of natural and social science perspectives can lead to social transformation both in learning and in seeing the world from an ecosystem perspective, with direct social and cultural feedback to agroecosystem management (Moore, Chapter 3).

Several major universities have begun programs in agroecology and agroecosystems management, reflecting the felt need for more integrated approaches in this field. Several chairs in agroecosystem management or sustainable agriculture have emerged at universities in the United States over the past decade, including positions at such major agricultural institutions as the University of California–Davis, Texas A&M, Ohio State University, Iowa State University, and several universities in Canada. Other programs have created unique positions in the past decade, such as the rotating endowed chair in agricultural sciences at the University of Minnesota, to address the need for innovation and progress on new issues facing agriculture science, rural communities, and management of agroecosystems.

In European countries, too, new centers have formed, and traditional agriculture programs restructured, to reflect a change in focus from agricultural production to balancing production with the protection of the environment and conservation of natural diversity. These new centers focus on developing interdisciplinary programs of research and education that help reconnect science to problems facing society. A few examples of these are (1) the Center for Sustainable Agriculture, which was formed within the Swedish University of Agricultural Sciences in 1997, (2) the National Center of Agroecology, which was formed in Switzerland in 1996 by merging the former Research Station for Agronomy with Research Station for Agricultural Chemistry and Environmental Hygiene, and (3) the Nordic Agroecology Program featured in Chapter 16.

These new positions and programs reflect a change in orientation in many agricultural research universities and agencies toward a systems approach that emphasizes interactions at a variety of scales and recognizes the importance of the ecological, social, and economic context of agroecosystem management. This change does not discount the value of reductionist disciplinary research, which drove the massive increases in agricultural production over the past century, and which will be needed to address some of the agricultural challenges in the coming century. However, there is a general awareness that we have been more successful at addressing components of agricultural systems than we have at addressing the complex interactions and multiple challenges of agroecosystem management (Ikerd 1993, 2008; Dixon and Gulliver 2001; Giampietro and Pastore 2001). The desired outcome of a more systems-oriented approach is to train a new generation of agricultural scientists who are strong within their own disciplines, but who are equally capable of interacting across disciplines, and collaborating with diverse social groups that may include policy makers, agricultural producers, and nonprofit environmental or consumer organizations. Scientists trained in such programs would have a broad understanding of the ecological, economic, and social interactions that contribute to sustainable agroecosystem management.

At the same time that agroecology and agricultural sustainability have emerged within the ecological and agricultural disciplines, new programs are emerging in major research universities around the United States and elsewhere in the area of "sustainability science." Sustainability science has emerged from a broader recognition of the combined implications for human society of resource degradation, global change, energy supply, and population growth (Allen et al. 2003). Several chapters in this volume stress the need to include ecologically based agricultural production as a fundamental component and process of sustainability.

The flurry of new activities and programs and the scope and rapidity of change within traditional land grant universities in the United States mirror a paradigmatic shift in approaches to agricultural research and education, and many programs are still struggling with the cultural change engendered by this shift. This shift will require continuing efforts to develop concepts and new practical strategies for agroecosystem management. It is our intent in this volume to foster a new spirit of communication and cooperation among the broad audience of ecological and agricultural scientists.

1.5 THE CRITICAL IMPORTANCE OF ECOSYSTEMS AS SOLUTION FRAMEWORKS

The ecosystem as a scientific concept was introduced by Tansley in 1935. Eugene Odum defined an ecosystem as "any area of nature that includes living organisms and non-living substances that interact to produce an exchange of materials between the living and non-living parts" (Odum 1959). The ecosystem concept is a unique contribution to our understanding of the world not only because it encompasses both biotic and inert components, but also because it critically manifests how all parts interact cooperatively to function as a sustainable unit (i.e., evolved living systems operating within constraints of natural material and energy flows).

Adopting the ecosystem paradigm as a design template has both social and physical benefits for building and maintaining sustainable agroecosystems. The ecosystem concept provides a convergent, inclusive, durable, yet flexible framework. Ecosystems have been operational in nature from the beginning of life on Earth. Hence they are a very successful way of organization for living communities. A substantial body of applied methodology based on ecological principles is currently practiced worldwide (e.g., low-input systems, alternative rotations, organic agriculture, mixed crop–livestock, biointensive, etc.). What remains is to adopt, extend, and implement these sustainable, ecological practices on a broader scale. The fundamental conflict between cropping systems, which are largely based on annual species, and natural ecosystems, which are largely perennial, may force us to rethink the role of perennialism in future agriculture systems (Jackson et al., Chapter 6).

1.6 PRESERVING AND RESTORING ECOSYSTEM SERVICES

Three fundamental land uses necessarily comprise a sustainable human landscape: (1) the natural ecosystem, which provides critical life-support processes or ecosystem services (including nutrient cycling, soil formation, air and water purification, flood control via biologically rich watersheds, etc.), (2) the agricultural production system itself, which provides food energy for both humans and animals, and (3) the anthropogenic environment which supports human industry, commerce, and habitation.

In ecological terms agriculture returns a measure of primary productivity, that is, solar power, to the natural environment, but the major provider of ecosystem services emanates from the preservation, maintenance, and enhancement of natural ecosystems. A healthy environment, that is, robust ecosystem functionality, must be understood as more than a luxury or amenity, but rather as fundamental to our existence.

Organizations worldwide are working to raise environmental awareness and ecosystem literacy to provide an ecological framework and foundation for governments and public policy-making organizations via ecosystem "templates" (Brown 2006). Ecosystems provide a workable model for local regulation and control. Allowing internal ecosystem control to operate as much as possible mitigates many of the most egregious problems of global resource depletion, ecosystem management, and climate change. By implementing local policies rooted in the ecosystem paradigm, specifically those that restore and promote interconnectivity and biotic processes among the various physical components of an agroecosystem, sustainability begins to emerge.

A common theme throughout all chapters in this volume is the central role of cooperation and partnership of concerned stakeholders to the successful development of sustainable or ecologically based agricultural systems. Interaction and communication among growers, farmers, agencies, researchers, concerned citizens, and the like are essential ingredients in the design of sustainable agricultural systems.

The success of ecological solutions to agricultural issues hinges on an open flow of information among all affected groups and individuals. Agroecosystems can no longer be thought of as merely industrial, commodity production systems, that is, food factories, but must be viewed as existing within and interacting with the larger natural environment, which provides life-supporting ecosystem services; the social environment, which supports civic engagement and community; and the economic environment, which provides people the opportunity to produce and sell agricultural goods. All these aspects are part of an inseparable whole that contribute, or detract from, people's quality of life (Ikerd 2008, and Chapter 4).

All authors in this volume emphasize explicitly or implicitly the need for and importance of restructuring, reorganizing, and reconceptualizing agroecosystems to restore and enhance internal ecosystem, that is, biotic, control, and hence ecosystem services. Difficult questions are addressed or implied: (1) Does it make sense to rely on a globalized food production system as we enter an era of energy depletion and constraint? (2) Why is local, sustainable agriculture and thus food security awarded such a low a priority? (3) Are the rigid command and control methods of organization obsolete in today's evolving, network-oriented world, where rapid access to information by all parties is essential?

Perhaps the most important take-home message of the authors is that the current economic paradigm driving our agricultural systems is causing more problems than it is solving. The concept that the deleterious environmental effects of industrial agriculture, that is, soil erosion, water pollution, greenhouse gas emissions, reliance on currently inexpensive and ultimately finite fossil fuel for inputs, processing, and global transportation, can be ignored or "externalized" is no longer valid. Furthermore, although many authors emphasize disciplinary perspectives there is a broader recognition that the economic and social consequences of production systems cannot be separated from their ecological impacts.

Humanity is standing on the precipice of dramatic global changes that will strongly affect agriculture and of which agricultural will be a critical part. Human population growth, energy depletion, climate change, loss of habitat and biodiversity, soil, air, and water pollution—all these implacable global issues and challenges must be addressed, and rapidly, if we are to preserve the world within which we have evolved. The job of agroecologists and agricultural scientists is to work cooperatively to help create agroecosystems that have the flexibility to endure these changes, ameliorate their effects, and initiate a move to a more sustainable culture coming to terms with its limits.

Ecological and physical limits must be recognized, and, critically, the environmentally negative effects of industrial agriculture production must be viewed within the overall societal cost/benefit equation. In brief, the current economic paradigm must be replaced with an integrated ecological–economic–social paradigm based on the ecosystem framework. This new approach is needed if we are to create an environment for the evolution of flexible, dynamic, multifunctional agroecosystems that do more than produce commodities. All human economic activity must be viewed existing within and subject to the larger "budget" and operating principles of our Earth's ecosystems, while addressing the values and meeting the needs of people living within human social systems. The contributions in this volume are a tribute to one agroecologist, Ben Stinner, who devoted his career to bringing that vision closer to reality, and bringing reality closer to that vision.

REFERENCES

Allen, T. F. H., Tainter, J. A., and Hoekstra, T. W. 2003. *Supply-Side Sustainability*. Columbia University Press, New York.

Altieri, M. 1987. *Agroecology: The Scientific Basis for Sustainable Agriculture*. Westview Press, Boulder, CO.

Altieri, M. 2005. The myth of coexistence: Why transgenic crops are not compatible with agroecologically based systems of production. *Bulletin of Science, Technology, & Society*, 25, 361–371.

Avery, D. 1995. *Saving the Planet with Pesticides and Plastics: The Environmental Triumph of High-Yield Farming*. Hudson Institute, Indianapolis.

Boody, G. et al. 2005. Multifunctional agriculture in the United States. *Bioscience*, 55, 27–38.

Brown, L. 2006. Plan Bm2.0: Rescuing a Planet under Stress and a Civilization in Trouble. W. W. Norton, New York.

Brown, L. 2008. World facing huge new challenges on food front: business-as-usual not a viable option. Earth Policy Institute, http://www.earth-policy.org/Updates/2008/Update72.htm (accessed May 30, 2008).

Carpenter, S. R. et al. 1998. Nonpoint pollution of surface waters with phosphorus and nitrogen. *Ecological Applications,* 8, 559–568.

Chapin, F. S., Matson, P. A., and Mooney, H. A. 2002. *Principles of Terrestrial Ecosystem Ecology*. Springer-Verlag, New York.

Coleman, D. C. and Hendrix, P. F. 1988. Agroecosystem processes. In *Concepts of Ecosystem Ecology*, Pomeroy, L. R. and Alberts, J. J., Eds., Springer-Verlag, New York, 149–170.

Collins, W. W. and Qualset, C. O., Eds. 1999. *Biodiversity in Agroecosystems*. CRC Press, Boca Raton, FL.

Dixon, J. and Gulliver, A. 2001. Global farming systems study: challenges and priorities to 2030. Synthesis and overview. FAO and World Bank, Rome and Washington, D.C.

Donner, S. D. and Kucharik, C. J. 2008. Corn-based ethanol production compromises goal of reducing nitrogen export by the Mississippi River. In *Proceedings of the National Academy of Sciences,* 105(11), 4513–4518.

Francis, C. A. et al. 2004. Serving multiple needs with rural landscapes and agricultural systems. In *Agroecosystems Analysis,* Rickerl, D. and Francis, C., Eds., ASA-CSA-SSSA, Madison, WI, 147–166.

Giampietro, M. and Pastore, G. 2001. Operationalizing the concept of sustainability in agriculture: Characterizing agroecosystems on a multi-criteria, multiple scale performance space. In *Agroecology: Ecological Processes in Sustainable Agriculture*, Gliessman, S. R., Ed., CRC Press, Boca Raton, FL, 177–202.

Gliessman, S. R. 1998. *Agroecology: Ecological Processes in Sustainable Agriculture*. CRC Press, Boca Raton, FL, 384.

Gliessman, S. R. 2001a. *Agroecosystem Sustainability: Developing Practical Strategies*. CRC Press, Boca Raton, FL.

Gliessman, S. R. 2001b. The ecological foundations of agroecosystem sustainability. In *Agroecosytem Sustainability: Developing Practical Strategies,* Gliessman, S. R., Ed., CRC Press, Boca Raton, FL, 3–14.

Heinberg, R. 2007. *Peak Everything: Waking Up to the Century of Declines*. New Society Publishers, Gabriola Island, BC.

IFDC. 2008. World fertilizer prices soar as food and fuel economies merge. International Fertilizer Development Center. http://www.ifdc.org/i-wfp021908.pdf (accessed July 23, 2008).

Ikerd, J. E. 1993. The need for a system approach to sustainable agriculture. *Agriculture, Ecosystems, and Environment,* 46, 147–160.

Ikerd, J. E. 2008. *Crisis and Opportunity: Sustainability in American Agriculture*. University of Nebraska Press, Lincoln.

Jackson, D. and Jackson, L. 2002. *The Farm as Natural Habitat: Reconnecting Food Systems with Ecosystems.* Island Press, Washington, D.C.

Lowrance, R., Stinner, B. R., and House, G. J., Eds. 1984. *Agricultural Ecosystems: Unifying Concepts.* John-Wiley & Sons, New York, 233.

Matson, P. A. et al. 1997. Agricultural intensification and ecosystem properties. *Science*, 277, 504–509.

McNeill, W. 1992. *The Rise of the West: A History of the Human Community.* University of Chicago Press, Chicago.

Millennium Ecosystem Assessment. 2005. *Ecosystems and Human Well-Being: Current State and Trends,* Island Press, Washington, D.C.

Norgaard, R. B. and Baer, P. 2005. Collectively seeing complex systems: the nature of the problem. *Bioscience*, 55, 953–960.

Odum, E. 1959. *Fundamentals of Ecology.* Saunders, Philadelphia.

Odum, E. P., Finn, J. T., and Franz, E. H. 1979. Perturbation theory and the subsidy-stress gradient. *Bioscience*, 29, 349–352.

Pimentel, D. and Pimentel, M. 1996. *Food, Energy, and Society.* University Press of Colorado, Boulder.

Rickerl, D. and Francis, C. 2004a. Agroecosystems analysis. American Society of Agronomy, No. 43 in the Agronomy Series, Madison, WI.

Rickerl, D. and Francis, C. 2004b. Multidimensional thinking: A prerequisite to agroecology. In *Agroecosystems Analysis,* Rickerl, D. and Francis, C., Eds., American Society of Agronomy, No. 43 in the Agronomy Series, Madison, WI, 1–18.

Robertson, G. P. and Paul, E. A. 1998. Ecological research in agricultural ecosystems: Contributions to ecosystem science and to the management of agronomic resources. In *Successes, Limitations, and Frontiers in Ecosystem Science,* Pace, M. L. and Groffman, P. M., Eds., Springer-Verlag, New York, 142–164.

Robertson, G. P. and Swinton, S. M. 2005. Reconciling agricultural productivity and environmental integrity: A grand challenge for agriculture. *Frontiers in Ecology and the Environment,* 3, 38–46.

Rosenzweig, C. and Hillel, D. 1995. Potential impact of climate change on agriculture and food supply. Consequences: The Nature and Implications of Environmental Change Volume 1, No. 2. U.S. Global Change Research Information Office, Washington, D.C. http://www.gcrio.org/CONSEQUENCES/summer95/agriculture.html (accessed June 27, 2008).

Rosenzweig, C. and Parry, M. L. 1994. Potential impact of climate change on world food supply. *Nature*, 367, 133–138.

Smil, V. 2000. *Feeding the World: A Challenge for the 21st Century.* MIT Press, Cambridge, MA.

Smil, V. 2001. *Enriching the Earth: Fritz Haber, Carl Bosch, and the Transformation of World Food Production.* MIT Press, Cambridge, MA.

Smil, V. 2003. *Energy at the Crossroads Global Perspectives and Uncertainties.* MIT Press, Cambridge, MA.

Smil, V. 2008. *Energy in Nature and Society.* MIT Press, Cambridge, MA.

Smith, P. et al. 2007. Agriculture. In *Climate Change 2007: Mitigation.* Contribution of Working Group III to the Fourth Assessment Report of the Intergovernmental Panel on Climate Change, Metz, B. et al., Eds., Cambridge University Press, Cambridge.

Tilman, D. et al. 2002. Agricultural sustainability and intensive production practices. *Nature*, 418, 671–677.

Section I

Ben Stinner's Contribution to Agroecosystem Science

2 Evolution of Agroecosystem Management in the Life of Benjamin R. Stinner

A Reflection on His Journey and Legacy

Deborah H. Stinner

CONTENTS

2.1 INTRODUCTION

On November 23, 2004, the sustainable agricultural community and agriculture in general lost a strong and inspiring leader of great intellect and compassion. Ben Stinner was respected and loved by colleagues, students, farmers, friends, family, and virtually everyone with whom he interacted. His life, though shorter than those of us who loved him most would wish, was well-lived and leaves a rich and deep legacy. The evolution of his understanding of what agroecosystem management is and can be has much to teach those of us left to carry on. I had the honor of sharing more than 25 years with Ben, as his wife and colleague, up until his untimely death in an automobile accident in 2004. In this chapter, I trace the historical development of Ben's experience and thinking and reflect on his contribution to the future of agroecology and sustainable agroecosystem management.

2.2 AGROECOSYSTEMS AS EXPERIMENTAL PLOTS

2.2.1 THE EARLY YEARS

Ben and I first met as fellow graduate students at the University of Georgia (UGA) in 1978. He was doing his dissertation work on the new Horseshoe Bend project funded by the Ecosystems Studies Program of the National Science Foundation (NSF) that he, Gar House, and faculty advisors D. A. Crossley, Jr., E. P. Odum, and R. L. Todd had conceived. This project was one of the earliest agro-ecosystem projects funded by the NSF. It represents a milestone in the history of both agricultural and ecosystem science in the United States in that it explicitly linked ecology and agriculture. The project was based on an ecosystem approach that perceived an agronomic unit "as being comprised of interacting components that form a whole which has system-level properties" (Lowrance et al. 1984). Major influences on Ben's thinking and development as an agroecosystem scientist at this time were Eugene Odum with his ecosystem concept, and "Dac" Crossley, Ben's major advisor, with his focus on soil ecology, in particular soil arthropods and nutrient cycling processes. Another key influence that continued until the end of Ben's life was the work Gene Likens and his colleagues conducted on nutrient cycling in forested watersheds at Hubbard Brook Experimental Forest in New Hampshire (Likens et al. 1967, 1969; Likens and Bormann 1977). Finally, there was the synergism between Ben and Gar House, his close friend and colleague in the Horseshoe Bend project, which generated many lively discussions about agroecosystems and the essence of which continue to motivate the contemporary work of colleagues influenced by either or both these men.

At this early stage in Ben's development as an agroecologist, agroecosystems were viewed primarily as experimental plots or fields and "as appropriate systems with which to test hypotheses concerning the effects of perturbation on nutrient retention and loss (Loucks 1977)" (Stinner et al. 1984). Ben's dissertation research at Horseshoe Bend led to numerous publications (Stinner and Crossley 1980, Stinner et al. 1983a, 1983b, 1984). Of particular importance was a landmark publication in *Ecology* in 1984 entitled, "Nutrient budgets and internal cycling of N, P, K, Ca, and Mg in conventional, no-tillage, and old-field ecosystems on the Georgia Piedmont" (Stinner et al. 1984). This research was done during the time when the U.S. Department of Agriculture (USDA) began promoting no-tillage on a large scale. Ben and his colleagues contributed a whole system perspective to investigating effects of tillage on agroecosystems and also brought in the idea of using natural or semi-natural systems as "controls" of sort in agroecological research. It is worthwhile to note that Ben and Gar House's work at Horseshoe Bend was just the beginning of a long distinguished record of important agroecosystem research at that site.

In the same year the *Ecology* paper was published, Ben and UGA colleagues Richard Lowrance and Gar House published the seminal book *Agricultural Ecosystems Unifying Concepts* (Lowrance et al. 1984). The introduction of that book reveals a broadening and foreshadowing of thinking about agroecosystems and understanding of the implications of such broadening:

> Whether agroecosystems are ecological systems under a high degree of socioeconomic control or ... essentially socioeconomic systems with varying levels of ecological control is more than a semantic question. If agroecosystems are essentially socioeconomic systems as Spedding asserts, the study of ecological controls must be done within an existing or projected socioeconomic framework in order to be useful in the development of ecosystem management. (Lowrance et al. 1984, p. 2)

It also reveals beginnings of ideas on sustainability that would become very important in Ben's career:

> Our theme is that sustainability can be accomplished through agroecosystem management—incorporating ecological, social and economic goals into the design of sustainable agroecosystems for specific portions of the landscape. Therefore agricultural management can evolve into agroecosystem management by broadening the traditional agricultural goals of productivity, production and conservation. (Lowrance et al. 1984, p. 2)

2.2.2 The Middle Years

Ben was hired in the Department of Entomology at the Ohio State University (OSU) Ohio Agriculture Research and Development Center April 1, 1982 as a corn entomologist, with responsibilities to conduct research on controlling insect pests in corn with particular emphasis on reduced and no-till maize systems. His no-tillage research experience at UGA and studies on effects of tillage on insect communities helped him get this desirable position at a major land grant university (House and Stinner 1983, Stinner et al. 1984). Insect pest management research at most land grant universities during that time involved a great deal of insecticide evaluations. Although Ben did some of this type of work early on, he often evaluated effects of insecticides on nontarget beneficial arthropods in addition to the target insect pests (Stinner et al. 1984, Brust et al. 1985, Stinner et al. 1986, Hammond and Stinner 1987). From the beginning he was most interested in how the system could be constructed to maximize natural biological control by predators and parasites.

Although an outstanding entomologist in his own right, Ben was also a very good soil scientist and had a special passion for organic matter and soil biological processes in general. Furthermore, from his days as a graduate student and postdoc at UGA he retained his basic conceptual interest in *ecosystem conservancy* as it pertains to agroecosystems and their relative ability to retain nutrients. These ideas were brought together in 1985 in a proposal to the NSF Long Term Ecological Research (LTER) program. System conservancy was the unifying concept and research was proposed that would have evaluated the effects of different agricultural practices on nominal ecosystem function in the Eastern–Midwestern U. S. region. This proposal was Ben's first attempt in Ohio at bringing together a substantial interdisciplinary team to address ecological questions on a watershed scale in agroecosystems. Although the proposal was not funded (funding went to Michigan State and what has become their outstanding agroecological LTER at the Kellogg Biological Station), it developed important ideas that were seeds for a major interdisciplinary agroecosystem management program.

The Sustainable Agriculture Project (SAP) (1987 to 1990) was the next interdisciplinary effort in which Ben was involved. The goal was to establish a replicated experiment that compared different farming systems with varying sustainability factors. Here again, agroecosystems were primarily viewed as experimental plots. The ambitious design sought to examine multiple factors that would simulate alternative management systems. Extreme wet weather conditions early in the experiment created serious management problems that highlighted the difficulty of simulating farm systems in plot experiments. Weed problems encountered during the project led Ben to seek advice from Harold Hartzler, a local long-time organic farmer. This new connection with an organic farmer, which was followed by subsequent connections with many other organic and sustainable farmers in Ohio, proved pivotal in Ben's career, and contributed to his turning away from replicated experimental plot agroecosystems to whole farms.

2.3 AGROECOSYSTEMS AS FARMS

Ben discovered that organic and sustainable farmers shared his passion for soil organic matter and biology and were eager to hear his scientific knowledge about these things. Ben had an extraordinary gift to communicate complex scientific information about soils to farmers. At this point in his career, he largely turned away from replicated experimental plot research and began investigating working farms with a continued goal of learning how sustainable farm management systems influence various agroecosystem processes. This shift led to the first of several grants from the newly established USDA Low Input Sustainable Agriculture (LISA) program, later changed to the Sustainable Agriculture Research and Education program (SARE). In a grant entitled "Ecological, Economic, and Environmental Analyses of Farms under Long-Term LISA Management," a team of us studied three case study organic farms in depth. In addition to measuring ecological and agronomic parameters, we forged our first link with the social sciences by collecting economic data. This involved what we came to call "kitchen table research" in which the researcher sits, usually at the farm household kitchen table, with the farmer and sometimes other family members and collects information on farm management and economics. As ecologists, this new type of "research" proved to be both challenging and very interesting. Of key importance to the evolution of Ben's thinking about agroecosystem management was the direct connection with the human side of agroecosystems as farms.

The next grant (1994 to 1997) "Ecological and Economic Analyses of Whole Farms to Promote Cropping System Diversification" from the North Central SARE Program involved whole farm planning and Allan Savory's holistic management ideas (Savory 1999, Stinner et al. 1997). This project helped us understand the importance of people's quality of life values as major drivers of change and innovation. This was our first participatory on-farm research project (Chambers et al. 1989, Scoones and Thompson 1994) and first experience with meta-level qualitative research methodologies (Creswell 1994, Miles and Huberman 1994, Rubin and Rubin 1995) in which researchers are in the test tube with the "objects of study" as colearners (see Eksvärd et al., Chapter 16). We were forced to learn how to maintain rigorous observation of the process.

From 1995 to 1998, Ben and I teamed up with Richard Moore, a cultural anthropologist, and Fred Hitzhusen, a resource economist on another SARE-funded grant, "Integrating Quality of Life, Economic, and Environmental Issues: Agroecosystem Analysis of Amish Farming." Although our research questions were similar in this project to those in our first LISA project with non-Amish farmers, working within the context of a different culture helped us realize how important the social system is, with its ideology, technology, social structure, and other cultural factors, in determining the structure and function of the classically defined agroecosystem (Marten 1986). We also were influenced by the human ecology perspective that concerns *how* farms function, and provides the holism needed to comprehend interactions (exchanges of energy, materials, and information) within and between agroecosystems and human social systems (Marten and Saltman 1986; Stinner et al. 1989, 1992; Moore et al. 1999).

2.4   AGROECOSYSTEMS AS WATERSHEDS—LINKING ECOLOGICAL, SOCIAL, ECONOMIC ASPECTS OF AGROECOSYSTEMS MANAGEMENT

In 1996, as the Amish project was getting under way, a small interdisciplinary group of faculty and graduate students at OSU from both the Columbus and Wooster campuses who had interests in systems thinking began meeting monthly for brown bag lunch discussions. The goal was to bridge interdisciplinary communication and to develop group coherency and trust as a basis for collaborative inquiry concerning agroecosystems. The focus of the original group was on the Amish project mentioned above. Having an anthropologist (Richard Moore, see Chapter 3 this volume) in the group was critical to a major paradigm shift for the natural scientists, including Ben. This paradigm shift was the realization that it was not sufficient to look only at biophysical parameters and processes to understand agroecosystems, but that social attributes and function were at least equally important. This cognition was not easy or comfortable for the natural scientists to accept.

The core interdisciplinary group expanded and began meeting weekly with Ben's facilitation. The group discussed papers by Conway on properties and analysis of agroecosystems (Conway 1985, 1986, 1987). The question of scale became very important. Although the group had strong conceptual abilities, there was a fundamental commitment to putting ideas to use to solve real problems. To such end a local sustainable dairy farmer (Joe Hartzler, son of Harold Hartzler mentioned earlier) whose farm had been part of earlier case studies was invited to join the weekly discussions to help ground the group. In response to sustainability issues with respect to projected profitability, his extended family was in the process of building a dairy and retail store to process and sell their chemical milk, pasteurized but not homogenized, in glass bottles. The group took on the issues and this farm family's response as a case study that extended beyond the farm gate (Allaire et al. 2001).

After much discussion about the appropriate scale for the new agroecosystem group to analyze agroecosystems, there was consensus that we should be looking at the watershed scale and the group began searching for case studies. This marked yet another shift in the agroecosystem concept in Ben's mind—from farms to watersheds. Watersheds allow for incorporation of interactions among agricultural and urban systems and of important political and socioeconomic influences that are external and yet extremely important to farm systems. Production, economic, social, and environmental issues had to be considered at multiple scales: field, farm, community, and watershed. Geographical Information Systems became essential as a research tool. A case study of the local Apple Creek Watershed was initiated to help the group focus its methods and thought processes. This collaboration eventually led to development of an "Agroecosystem Health Index" using an analytical hierarchy process (Figure 2.1) (Vadrevu et al. 2008).

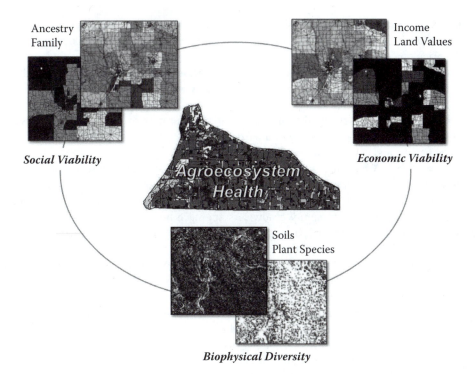

FIGURE 2.1   A depiction of agroecosystem health analyzed using spatially referenced hierarchical process, using social, economic, and biophysical aspects.

At the same time that the early brown bag lunch meetings were going on, the OSU College of Food, Agriculture, and Environmental Science (FAES) was working with a Project Reinvent grant from the Kellogg Foundation that was intended to catalyze major changes in the way the college was doing business by creating a more open atmosphere with respect to stakeholder input and sustainability. One of the many positive outcomes of Project Reinvent was the creation of three interdisciplinary programs, the first of which became known as the Agroecosystems Management Program (AMP). Subsequently, discussions among Kellogg, leaders of the Innovative Farmers of Ohio, a sustainable farmers' organization, Ben, and OSU administrators resulted in the Kellogg Foundation endowing the AMP program and the College of Food, Agriculture, and Environmental Science endowing a Chair of Agroecosystem Management. It took more than 2 years of searching to find the right person to fill this chair. From the beginning, Ben refused to apply in spite of considerable pressure from his colleagues. However, after an extensive international search, he finally agreed and was easily selected for the position.

Major concepts/models that guided Ben's thinking about agroecosystems management were outlined in the lecture he presented when he interviewed for the Kellogg Chair position in 1999. These included: emergent properties, linking academic and experiential knowledge, material cycles and energy flow, linking ecological, economic, and social currencies, scale-neutral solutions, and complexity at many levels. Ben continued to use the case study approach as an important medium with which to explore how these concepts played out in the real world. Under Ben's facilitation, the group developed the following process for examining fundamental conceptual and pragmatic issues in agroecosystem management:

- Learn how to trace problems and issues, and the associated complexity, back to their fundamental causes and systemic nature, in order to ask the *why* questions first and the *how* questions later.

- Focus on issues that are widely shared, reach out into the larger landscape and community, and offer opportunities and alternatives for economic, social, and ecological viability in agriculture.
- Facilitate a local agroecosystems team/learning group that has long-term sustainability with ample points of participation for university and nonuniversity partners.

A key practical lesson that emerged from this period was learning to meet people wherever they are in their thinking and understanding of a particular issue and then gradually bring in the larger system perspective and connections to effect change.

The Sugar Creek Watershed, located in Wayne County, Ohio and close to OARDC (Ohio Agriculture Research and Development Center), with its significant water quality issues, was chosen as an in-depth case study. Richard Moore, the cultural anthropologist on the AMP team, took the leadership in this local community-headwaters approach project which has grown immensely since it was first conceived. There is a separate contribution in this volume that details the history and development of the project, including Ben's involvement.

In the final months of his life Ben presented several talks to diverse groups that showed how his thinking about agroecosystem management was continuing to expand. His focus shifted from sustainable agriculture per se to sustainability in general. In this context, he became very enthusiastic about the concept of an "entrepreneurial ecosystem" that he learned about from economists (James Curry, pers. commun. 2004; Neck et al. 2004; Cohen 2006) and the role that agriculture can play in local and regional economic development and sustainability. Through this new and fertile cross-disciplinary lens, Ben saw the importance of helping others see more of the innate complexity in agroecosystems and the importance of bringing that perspective into integrated research, education, and extension activities of land grant universities.

Under Ben's leadership, a large interdisciplinary group of scientists working in the Sugar Creek Watershed developed two NSF proposals to the Biocomplexity in the Environment program. Both focused on biocomplexity issues in the Sugar Creek Watershed. The first was a funded planning grant and the second was a full proposal submitted just before Ben was killed. The proposed research in the full proposal extended from the hierarchical level of molecular biology to the level of regional agroecosystems. Ben understood the importance of the both the "hawk" and "mouse" perspectives, which provide the high overview and close detail needed to fully understand an issue and to make constructive changes. He also saw the need and was committed to helping mentor future leaders who could combine an understanding of the big picture, that is, the system, with disciplinary knowledge and ability to work in teams. He was convinced that this combination is what it would take to manage agroecosystems wisely and solve the challenges facing humanity in the twenty-first century.

Although Ben was an outstanding academic scholar, he was not afraid, and indeed loved, to engage with nonacademic stakeholders in deep and meaningful ways to help create healthier agroecosystems and sustainable communities in general. He did not stay in the Ivory Tower. He brought his concepts out into the living, breathing world of real agroecosystems and in the process deeply touched countless people's lives. This is his challenge to those of us left to carry on his legacy. Ben's vision lives on in the continuing work of the OSU AMP, through the Ben Stinner Endowment for Healthy Agroecosystems and Sustainable Communities, and in the continuing contributions of the many committed and talented people he influenced throughout his life.

REFERENCES

Allaire, F. et al. 2001. Discovery learning of the processes needed to develop jointly a farming enterprise and its community. *Journal of Sustainable Agriculture,* 19, 65–83.

Brust, G. E., Stinner, B. R., and McCartney, D. A. 1985. Tillage and soil insecticide effects on predator-black cutworm (Lepidoptera: Noctuidae) interactions in corn agroecosystems. *Journal of Economic Entomology,* 78, 1389–1392.

Chambers, R., Pacey, A., and Thrupp, L. A., Eds. 1989. *Farmer First, Farmer Innovation, and Agricultural Research.* Intermediate Technology Publications, London, 218.

Cohen, B. 2006. Sustainable valley entrepreneurial ecosystems. *Business Strategy and the Environment,* 15, 1–14.

Conway, G. R. 1985. Agroecosystem analysis. *Agricultural Administration,* 20, 31–55.

Conway, G. R. 1986. *Agroecosystem Analysis for Research and Development.* Winrock International, Bangkok.

Conway, G. R. 1987. The properties of agroecosystems. *Agricultural Administration,* 2, 95–117.

Creswell, J. W. 1994. *Research Design Qualitative and Quantitative Approaches.* Sage, Thousand Oaks, CA, 228.

Curry, J. 2004. Personal communication.

Hammond, R. B. and Stinner, B. R. 1987. Soybean foliage insects in conservation tillage systems: Effects of tillage, previous cropping history, and soil insecticide application. *Environmental Entomology,* 16, 524–531.

House, G. J. and Stinner, B. R. 1983. Arthropods in no-tillage agroecosystems. In *Nematodes in Soil Ecosystems,* Freckman, D. W., Ed., University of Texas Press, Austin, 14–28.

Likens, G. E. and Bormann, F. H. 1977. *Biogeochemistry of a Forested Ecosystem.* Springer-Verlag, New York.

Likens, G. E. et al. 1967. The calcium, magnesium, potassium, and sodium budgets for a small forested ecosystem. *Ecology,* 48, 772–285.

Likens, G. E., Bormann, F. H., and Johnson, N. M. 1969. Nitrification: importance to nutrient losses from a cutover forested ecosystem. *Science,* 16, 1205–1206.

Loucks, O. L. 1977. Emergence of research on agro-ecosystems. *Annual Review of Ecology and Systematics,* 8, 171–192.

Lowrance, R., Stinner, B. R., and House, G. J. 1984, *Agricultural Ecosystems Unifying Concepts.* John Wiley & Sons, New York, 233.

Marten, G. G., Ed. 1986. *Traditional Agriculture in Southeast Asia.* Westview Press, Boulder, CO.

Marten, G. G. and Saltman, D. M. 1986. The human ecology perspective. In *Traditional Agriculture in Southeast Asia,* Marten, G. G., Ed., Westview Press, Boulder, CO, 20–53.

Miles, M. B. and Huberman, A. M. 1994. *Qualitative Data Analysis.* Sage, Thousand Oaks, CA. 338.

Moore, R. H., Stinner, D. H., and Kline, E. 1999. Honoring creation and tending the garden: Amish views of biodiversity. In *Cultural and Spiritual Values of Biodiversity,* Possey, D. A. and Dutfield, G., Eds., Intermediate Technology Publications, for United Nations Environmental Programme, London, 305–309.

Neck, H. et al. 2004. An entrepreneurial system view of new venture creation. *Journal of Small Business Management,* 42(2), 190–208.

Rubin, H. J. and Rubin, I. S. 1995. *Qualitative Interviews: The Art of Hearing Data.* Sage, Thousand Oaks, CA, 302.

Savory, A. 1999. *Holistic Management, a New Framework for Decision Making.* Island Press, Washington, D.C., 616.

Scoones, I. and Thompson, J., Eds., 1994. *Beyond Farmer First: Rural People's Knowledge, Agricultural Research, and Extension Practice.* Intermediate Technology, London, 301.

Stinner, B. R. and Crossley, Jr., D. A. 1980. Nutrient cycling in till and no-till systems: an experimental approach to agroecosystems analysis. In *Soil Biology as Related to Land Use,* D. Dindal, Ed., Environmental Protection Agency, Washington, D.C., 280–288.

Stinner, B. R., Hoyt, G. D., and Todd, R. L. 1983a. Changes in soil chemical properties following a 12 year fallow: A two year comparison of conventional tillage and no-tillage agroecosystems. *Soil and Tillage Research,* 3, 277–290.

Stinner, B. R., Odum, E. P., and Crossley, Jr., D. A. 1983b. Nutrient uptake by vegetation in relation to ecosystem processes in conventional, no-tillage, and old-field systems. *Agriculture, Ecosystems, and Environment,* 10, 1–13.

Stinner, B. R. et al. 1984. Nutrient budgets and internal cycling of N, P, K, Ca, and Mg in conventional tillage, no-tillage, and old-field ecosystems on the Georgia Piedmont. *Ecology,* 65, 354–369.

Stinner, B. R., Krueger, H. R., and McCartney, D. A. 1986. Insecticide and tillage effects on pest and non-pest arthropods in corn agroecosystems. *Agriculture, Ecosystems, and Environment,* 15, 11–21.

Stinner, D. H., Paoletti, M. G., and Stinner, B. R., 1989. In search of traditional farm wisdom for a more sustainable agriculture: A study of Amish farming and society. *Agriculture, Ecosystems, and Environment,* 27, 77–90.

Stinner, D. H., Glick, I., and Stinner, B. R. 1992. Forage legumes and cultural sustainability: lessons from history. *Agriculture, Ecosystems, and Environment,* 40, 233–248.

Stinner, D. H., Stinner, B. R., and Martsolf, E. 1997. Biodiversity as an organizing principle in agroecosystem management: Case studies of holistic resource management practitioners in the USA. *Agriculture, Ecosystems, and Environment, 62*, 199–213.

Vadrevu, K. P. et al. 2008. Case study of an integrated framework for quantifying agroecosystem health. *Ecosystems, 11*, 283–306.

3 Ecological Integration of the Social and Natural Sciences in the Sugar Creek Method

Richard H. Moore

CONTENTS

3.1 INTRODUCTION

Twenty-first century environmental problems ranging from global warming to pollution of the Earth's ecosystems all have one thing in common: a necessary linkage of social and natural ecological systems involving human mismanagement of ecosystem complexity. The Sugar Creek Project is an attempt to find new principles to integrate and bring balance back to the physical, biological, social, and economic aspects of agricultural systems. The research is based on a reconfiguration of the position of the researcher so that farmers and researchers can work together to link hierarchical scales of analysis at the field, farm, community, and watershed levels. This chapter examines four components of the Sugar Creek Method that integrate the social and natural sciences. The first is a clear definition of the values of stewardship and social responsibility of the farmers as they relate to the values of the researchers and how this affects the scientific method. In the Sugar Creek case, this led to a new approach to water quality sampling methodology, namely, use of a year-round high density sampling approach to examine the water quality of headwater streams. The second is to categorize modes of intensification within the same watershed. We found different modes of intensification in different subwatersheds of Sugar Creek relating to ethnic differences as well as different farming strategies. Positive and negative feedback loops within environmental biocomplexity were identified and led to an analysis locating leverage points to start to correct environmental degradation. Third, we examined fragmented landscapes and people's social organization and ideas about the landscape along three streams. Heterogeneous landscape patterns, land use, and land tenure are based on abstract cultural rules which affect levels of biodiversity at different hierarchical scales and contribute to the relative degree of the system to be resilient and buffer system perturbations. The fourth principle of integration was to create new social and economic value through connecting social and natural systems.

In the formation of the Agroecosystems Management Program (AMP) at Ohio State University (OSU) led by Ben Stinner, to whom this volume is dedicated, we consciously tried to create a team that consisted of social, natural, and physical scientists. Inspired by National Science Foundation

(NSF) Long-Term Ecological Research (LTER) programs such as Hubbard Brook (Likens and Bormann 1995), we wanted to create a long-term ecological site. However, when we examined the list of NSF LTER sites, we realized that most have either no or very little integrated social science component. Thus, we wanted to add the social science dimension so we would be able to promote more ecological approaches in farming. We tried from the start to avoid using social science as a "plug-in" in the sense that it was sometimes used in U.S. Department of Agriculture (USDA) NRI grants to do an impact assessment or educational component add-on to natural or physical science research. So from the start we tried to move the social, natural, and physical sciences together toward solving ecological problems. For our part as social scientists we introduced Green and Shapiro's (1994) critique of rational choice theory to Ben Stinner who tried to frame the theoretical significance of this within the context of self-organization as mentioned in the works of ecologists such as A. J. Lotka, I. Prigogine, and the Odum brothers and incompatibility of reductionist thinking with the emergent properties of ecosystems. As we tried to bridge the social, natural, and physical sciences, we created a teamwork approach with the farmers in Upper Sugar Creek. Most of our grants were sparked by our interactions with the farming community. When writing grants, we ask for their input and support from the farmers, and share our results with them.

3.2   SUGAR CREEK METHOD CONCEPT 1: FARMERS AND THE RESEARCHERS WORK AS A TEAM

Sugar Creek is located in north central Ohio in Wayne and Holmes Counties, the leading dairy counties in Ohio. The watershed is in the headwaters of the Muskingum Watershed, Ohio's largest watershed flowing to the Ohio River, and therefore bears directly on the hypoxia issue in the Gulf of Mexico. The watershed consists of seven subwatersheds: Upper Sugar Creek, Lower Sugar Creek, North Fork, Middle Fork, South Fork, Indian Trail/Walnut Creek, and East Branch. A participatory approach with farmers in the watershed is used to jointly explore the water quality and related issues. It also helps to increase awareness of the major impairments and to examine the influences of culture on ecological function in these contrasting subwatersheds, with cultures ranging from German descent non-Amish (so-called "English") to Amish. Most important, however, was the discovery that the farmers and researchers could mutually lead each other to making new theoretical and practical advances in watershed ecology. This was building on Robert Rhoades' "Farmer back to Farmer" (Rhoades 1982) approach and taking it to the next level.

Labeled as Ohio's second most polluted watershed, second only to the Cuyahoga River which burned prior to the establishment of the Clean Water Act in 1973, the Sugar Creek Watershed provided an opportunity for our AMP to attempt to link social and natural systems at a site near to the OSU experiment station at Wooster. The project was started in 2000 by two social scientists (Richard Moore and Mark Weaver) in close collaboration with Ben Stinner. The starting date coincided with the Ohio Environmental Protection Agency (EPA) announcement of Ohio's second Total Maximum Daily Load (TMDL) plan which happened to be Sugar Creek located near our experiment station where we were centered. Eventually this project has grown to involve more than 30 faculty researchers and currently funds 14 graduate students in various water quality projects ranging from headwaters biocomplexity, to pathogen transport, to water quality trading.

The Sugar Creek Project currently conducts water quality sampling at 105 sites biweekly all year except when the stream is frozen. The density of sampling is approximately one site per 1 to 2 square miles. This is greater than three times the rate which Ohio EPA used to collect data during the summer sampling season to prepare for its TMDL. Our dense sampling regimen happened as a result of a transformative event that occurred during the second meeting of the Upper Sugar Creek Farmer Partners in 2000. Transformative events can be either social or natural but ultimately have to result in a cultural redefinition of the event in terms of social relationships and cognitive framing. Important aspects of transformative events might include the different organization forms or networks that foster or inhibit coalition formation, the formulation of a coherent set of beliefs and core values to guide the group's actions, new spatial and temporal axes for framing the event, or the complex web of issues surrounding the role of technical and scientific information in the group. Clemens (2007), McAdam et al. (2001), Morris (2004), and Moyer et al. (2001) define transformative events as turning points in the history of a social movement. The founders of the Sugar Creek research project struggled for theoretical ways to bridge the social and natural sciences. Ben Stinner and I each taught agroecology but in different colleges of the university and both of us liked crossing that bridge. As a social scientist I had explored the works of E. P. Odum and Howard Odum as a graduate student and for his part Ben, who worked in the natural and physical sciences, was fond of reading social scientists such as Habermas (1987, 1996) and Kemmis (1988). These theorists favored the construction of autonomous self-organized grassroots groups conducting participatory and collaborative action research. Accordingly, we used the advocacy coalition framework (Sabatier and Jenkins-Smith 1999; Sabatier et al. 2005; Weaver et al. 2005) as a basis for forming stakeholder teams which validate claims and further discourse ethics rooted in a moral point of view linking theory and practice (see Chapter 16 for another perspective on participatory approaches in sustainable agriculture research).

In the case of the Upper Sugar Creek Farmer Partners, the farmer group had been self-selected. In late 2000, a "learning circle" in the Upper Sugar Creek subwatershed was initiated by one farmer, who invited three other farmers to assist in forming a watershed group. These three individuals in turn invited additional participants. Unlike many watershed groups which tried to be inclusive by inviting diverse population segments, the Upper Sugar Creek Farmer Partners wanted to include only farmers willing to take concrete measures to improve ecology on their own farms. One of the first steps they took in the fall of 2000 was to request that we researchers use a geographic information system (GIS) to map parcels of adjacent landowners to whom they would then send a letter stating that they (the Farmer Partners) were simply setting out to improve the water quality on their own farms. It was only after their initial conservation successes that the project expanded to the rest of the community.

In the autumn of 2000 the original 12 farmers teamed with the university AMP research group to investigate the degree and sources of water impairment in Upper Sugar Creek. The key transformative turning event occurred at the farmers' second meeting where a debate occurred. The debate centered on a moral dilemma. In the first meeting the farmers had approached the researchers to conduct water quality sampling on their lands to ascertain whether or not the Ohio EPA data collected for the TMDL was correct. There was a high degree of distrust with the Ohio EPA and some had a strong belief that these regulators had a flawed methodology. Acting out of a sense of social responsibility to do good stewardship for the land, the farmers wanted to know if they were actually polluting, or if there were other sources of the pollution such as new homes that had been linked to their drainage tile lines, or if the Ohio EPA data might be wrong because it sampled only a few times and part of that was during a drought. The transformative debate came when they had to decide what to do if, after 1 year of intensive sampling, they found Ohio EPA data to be wrong. Would they disband having the data in hand to defend themselves in court—or stay together and try and do something good for the watershed anyway? They decided that they would try to improve the water quality regardless. Our role in the university was to provide a neutral test of the Ohio EPA data. This moral decision by the farmers was the transformative event which set the stage for conservation measures once the data confirmed the Ohio EPA study.

There were three main effects of this transformative event. First, it helped solidify the social cohesion of the farmer group and at the same time established a team concept of working together with the researchers. From May 2001 to the present, the research team conducted biweekly water quality testing of various parameters (e.g., nitrates, pH, phosphorus, dissolved oxygen, temperature, fecal coliform bacteria) at a catchment density of 1 to 2 square miles per testing site for 21 sites, one of the highest sampling densities in the United States. In the years to follow, the number increased to 105 sites, and continuous flow monitoring and sampling for several sites were introduced in 2008. Second, it morally obligated farmer action if and when the data showed that pollution was

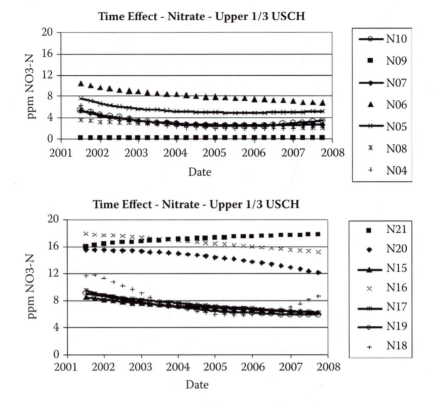

FIGURE 3.1   Nitrate decrease in Upper Sugar Creek Streams 2000–2007.

associated with their farms; as a result of this accepted moral obligation, once we had clear evidence that the farms were a major polluting factor in the watershed, the farmers signed up for the Conservation Reserve Program (CRP) together in 2001 forming more than 8 miles of contiguous stream buffer. Third, the only way to interpret the results was by finding the actual source of the pollution because one site per 1 to 2 square miles normally puts one in the vicinity of one to two farms. We noticed that a few sites were continually "hot spots" and the only way to find the source was to investigate what was happening on that particular headwater stream of less than 1 square mile drainage. To address this hotspot phenomenon we adopted a headwaters approach, the foundation for our first NSF biocomplexity grant. In 2001, the Ohio EPA announced a Primary Headwaters Initiative (http://www.epa.state.oh.us/dsw/wqs/headwaters/) with the same catchment density as our university team. Collaboration between the Ohio EPA headwaters initiative focusing on aquatic ecology and the farmer/researcher team focusing on terrestrial ecology started in 2002 based on mutually trying to understand the role of headwaters in watershed ecology.

Changing outcomes through this process takes time. Figure 3.1 shows the site by site progression of lowering nitrates. Site N19 is the mouth of Upper Sugar Creek and shows the sum of our efforts decreasing the nitrates gradually from 9 milligrams per liter (mg/L) in 2001 to 6 mg/L in 2007. This fine-grained approach also provides information to show us that sites N17 and N21 have been rising in the last several years so added efforts are needed there.

3.3   SUGAR CREEK METHOD CONCEPT 2: ECOLOGICAL PROCESS AND MODES OF INTENSIFICATION

The Sugar Creek Watershed has significant ethnic diversity including the largest settlement of Amish for any watershed in the United States. The dairy farms range in size from large conventional

CAFOs (confined animal feeding operations) with 3000 cows to Amish with only 20 to 30 cows. There are also other farms focusing on corn and soybeans without livestock and others choosing new mixes of livestock and crops. However, we have found intensification of agriculture to have occurred in most cases, although the degree of environmental degradation varies between intensification types. Intensification is not unique to the Sugar Creek Watershed; it is a process that has occurred worldwide and, with it, hypoxia zones have formed in most bays and oceans throughout the world during the last 50 years (Stramma et al. 2008).

The intensification of agriculture has been the dominant anthropogenic influence on a large portion of the world's land surface (Vitousek et al. 1997), and has drastically changed the structure and function of terrestrial and aquatic ecosystems. This historic and ongoing transformation involves ~230 million ha in the United States (NRC 1989). It affects biogeochemical, population, community, and ecosystems process at many levels, powerfully reshaping landscapes (Matson et al. 1997; Watzin and McIntosh 1999). In North America, especially since the 1970s, agriculture has been progressing to large-scale, low-diversity farming practices that simplified ecosystems through monocropping. Far from being self-regulating and in equilibrium (Ashby 1963; Jarvis 1999), these agroecosystems depend on high inputs of nonrenewable resources to maintain their productivity (Bentley 1985; Watzin and McIntosh 1999) and are characterized by loss of biocomplexity on multiple scales including biodiversity and the complexity of ecological structure and function within agricultural fields to the landscape scale (van Mansvelt et al. 1998; Altieri 1999; van Elsen 2000; Torsvik et al. 2002; Stinner 2004; Hole et al. 2005). These simplified agroecosystems are much less nutrient conservative than their natural counterparts (Stinner et al. 1984; House and Stinner 1984) and are major sources of nutrient loading, impairment, and biodiversity loss in aquatic systems (Peterson et al. 2001; EPA 2000). However, while ecological complexity of modern agroecosystems is low, social complexity can be high. Complex interactions of human values, beliefs, and social capital and institutions determine the ecological structure and function of agroecosystems (Martens 1986; Stinner et al. 1992; Williams et al. 2000). Therefore, efforts to restore ecological function of agricultural watersheds will be successful only to the degree that they fully integrate human and environmental dimensions of these systems (Moore and Weaver 2004; Weaver and Moore 2004; Lansing 2004).

The issue of intensification of food production is central to understanding human life on this planet, especially considering that humans have spent 99 percent of their existence as hunter-gatherer-fishers with a very light ecological footprint. Accordingly, in the social sciences there has been considerable debate regarding the drivers of the plant and animal plant domestication process as well as the resulting problem of population increase and carrying capacity. In response to Malthus's (1826) theory that population increase would outrun the food supply leading to mass depopulation, Boserup (1965) proposed an alternative theory whereby agricultural intensification would occur when the population increased under environmental constraints. Boserup's argument rested on three points: (1) environmental limits on the type of agriculture are relatively "elastic" (broad and flexible), meaning that there are a number of possible strategies within these environmental limits that have impact on productivity and carrying capacity, (2) increased production usually leads to decreased yields per unit labor but increased yield/unit land; intensification will lead to shorter fallow cycles, different plowing techniques, more fertilizer application, more labor or mechanization, or engineering such as irrigation or drainage lines, and (3) efforts to minimize labor lead to the decision to intensify; population growth will lead to land scarcity which in turn leads to technological or economic innovations to achieve higher yield/unit accompanied by higher labor costs. New economic opportunities such as new markets or external constraints such as taxes can also force intensification. Furthermore, decreased population or decreased external constraints may lead to intensification (Netting 1977). To date, there have been a number of attempts (Cohen 1989; Conelly 1992; Brookfield 2001) to understand the process of intensification. The key metrics have been fallow length, productivity (yield), efficiency of labor, population density, technological strategy, soil fertilization, land tenure, economic systems, and sociopolitical complexity.

Few studies have connected the magnitude or types of intensification with nutrient loading, which is a necessary link between the human and natural systems. A search on AGRICOLA, BIOSIS Previews 1980 to present, and the ISI Web of Science using the terms "nutrient loading" and "intensification" resulted in only seven articles that linked the causality between nutrients and intensification methods. Only one of the articles (Mattikalli 1996) entailed watershed-scale analysis. Moreover, no studies to our knowledge have attempted to connect the concept of the intensification with the degradation of headwater streams which comprise more than 80 percent of our nation's waterways.

Despite considerable support for best management practices (BMPs) designed to ameliorate negative impacts of agriculture, serious problems persist at multiple scales. One such example is the hypoxia zone in the Gulf of Mexico, which can be attributed largely to excessive nutrient loadings in the headwaters of the Ohio and Mississippi Rivers (USDA-CEAP 2004). Many watershed restoration programs, however, ignore critical historical and cultural factors that affect the behavior of watershed residents (Leach and Pelky 2001; Leach et al. 2002; Moerke and Lamberti 2004) and tend to emphasize descriptive accounts of impairments rather than complex causal mechanisms (Watzin and McIntosh 1999). Only through the study of headwaters can we truly address the problem of nonpoint-source pollution. These small streams at the local and farm level make it possible to research "nonpoint-source pollution" as "point source" and then scale up to larger watersheds and river basins. Then, and only then, can local solutions embodying human and natural systems be applied to solving the specific sources. However, interdisciplinary analyses that integrate social and environmental components of these policies on ecosystem functioning and health are lacking and there is a clear need for a new interdisciplinary model that integrates social and environmental sciences with innovative quantitative and qualitative research methodologies and education in a biocomplexity framework (Covich 2000; Michener et al. 2001; Cottingham 2002).

Based on our research in the Sugar Creek Watershed, we note that there are different modes of agricultural intensification in the watershed, each having coupled social, natural, and physical biocomplexity of nutrient and carbon cycling. Each mode results in different levels of nitrogen, phosphorus, and biological transport of pathogens as well as social organization. Only by linking social and biological intensification at multiple scales (headwaters to entire watershed) can we understand the process of intensification. These watersheds are model systems for the Ohio River Basin as they capture the underlying diversity of geology, biology, land use, and social systems characteristic of the Ohio River Basin that are affecting the quality and quantity of freshwater resources.

Basically the three types of intensification we have observed are (1) corn–soybean rotation land intensification through renting, (2) small-scale dairy intensification through the introduction of milking machines, and (3) large-scale dairy intensification through concentrated feeding and herd expansion. Each of these modes of intensification has its own drivers and respective types of environmental degradation, and they are located in different subwatersheds of Sugar Creek, making for an excellent spatially based comparison.

The corn–soybean rotation common to the corn belt, is found to a lesser degree in the Sugar Creek Watershed which lies just east of the corn belt. Typically, in this farm strategy we find a high degree of rented land in order to expand the acreage of management. In most cases, one of the largest hurdles is the equal heir pattern of farm inheritance that divides up the farm on the death of the owners. Because livestock are not part of this operation, it is customary to use chemical fertilizers rather than manure. Higher levels of nitrate pollution are typical of this type of intensification and can be addressed through conservation measures such as late spring nitrate tests to reduce the level of chemical fertilizer on corn or stream buffer zones, although tile drainage usually bypasses these buffers. There is also the social issue that farmers who rent are more reluctant to implement Farm Bill conservation measures if these measures require the approval or payout to the farm owner. A last observation is that when the scale of operation increases, there is a tendency to simplify the ecological processes and reduce labor costs; so, for example, commonly weed control is simplified through the use of Roundup Ready transgenic soybeans.

Small-scale dairy intensification in the Amish community was triggered in the mid 1990s when the Old Order church groups decided to allow milking machines. Prior to that time there was a high value on families milking about 25 cows by hand on 80 acres. Pioneering work on the ecological benefits of Amish self-sufficient farming was published by Ben Stinner (Stinner et al. 1989, 1992, 1997). When the milking machines were allowed, it took much less time to milk. Usually, a year or two after the introduction of milking machines, steel bulk tanks were introduced so that the milk could be piped directly into one holding container which cooled the milk that could be picked up by a milk tank truck every 2 or 3 days replacing the need for 10 gallon milk cans which had been picked up each day. Most farms increased their herd size about by 10 to 20 percent. Because this necessitated more corn and hay to be grown to feed the cows, rotational grazing using solar-powered electric fences and more permanent pasture grasses has been introduced on most farms resulting in improved pasture yield, milk quality, and herd and soil health. New plows have been invented to be able to plow the deeper roots of these grasses.

Although the level of environmental degradation has not been empirically tested, the transition to milking machines and rotational grazing has been accompanied by major shifts in social organization. Contrasting to corn–soybeans rotation discussed above, where the average age of the farmer increased, we see many Amish young families with children starting to farm. There are 90 to 100 new organic Amish dairy family farms that have appeared during the last 10 years in the Killbuck/Sugar Creek Watershed area.

Like all cases of intensification, coupling the social aspects with the natural science aspects is crucial. In most cases there is a social driver, and in the Amish case the population doubles every 26 years, bringing with it high demand and high prices for land and the pressure to subdivide small plots off to nonsucceeding heirs. This challenges the Amish intent to pass on the family farm intact to one heir and increases the cow/land ratio of intensification and likelihood of environmental degradation.

A third type of intensification in Sugar Creek is the large-scale dairy intensification through concentrated feeding and herd expansion. Although there is only one large CAFO of 1200 dairy cows (a CAFO is defined as having more than 700 adult cows), other farms in the East Branch subwatershed of Sugar Creek also have high herd numbers qualifying them as medium CAFOs. These farms are characterized by high farm inputs for hay, grain, and feed. According to the Pew Commission on Industrial Farm Animal Production (2008), the main problems associated with large CAFOs are "the increase in the pool of antibiotic-resistant bacteria because of the overuse of antibiotics; air quality problems; the contamination of rivers, streams, and coastal waters with concentrated animal waste; animal welfare problems, mainly as a result of the extremely close quarters in which the animals are housed; and significant shifts in the social structure and economy." According to the Ohio EPA Bacterial TMDL on Sugar Creek (OEPA 2007), the East Branch, where these CAFOs in Sugar Creek are concentrated, scored as one of the highest subwatersheds for high fecal coliform counts. The East Branch scored more than 10,000/100 ml accompanied with low stream substrate and channel scores using the Qualitative Headwaters Evaluation Index (QHEI).

Figure 3.2 shows a comparison of the field patterns of these two types of intensification. In the large-dairy area, we can see a clustering of the corn next to the stream. Stream proximity of the corn increases the likelihood that nitrate and phosphorus pollution will occur through the drain tiles despite that a buffer project was implemented. However, there was slightly more hay grown than in the adjacent Amish subwatershed where there is more rotationally grazed pasture. This case supports Altieri's (2008) statement that small farms are a planetary asset. We can also see that the field size is smaller in the Amish case where typically the agricultural plots are rotated through a 4-year rotation of corn, spelts, oats, and hay.

We also see a stark contrast in the social dynamics of these two subwatersheds. In the Amish stream there are 21 farm families, whereas in the large-dairy area there are only a few farms. In the Amish case, these families have a high degree of cooperative labor (Long 2004), for example, in oat threshing, and are organized into small church districts of fewer than 40 households. The Amish are much less mechanized and utilize household labor to a greater degree.

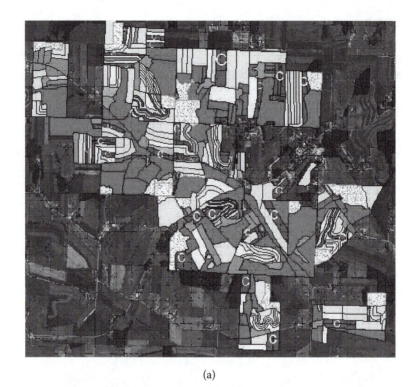

(a)

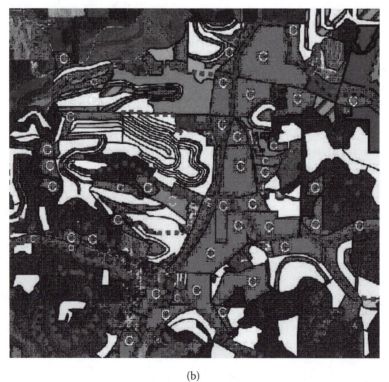

(b)

FIGURE 3.2   Field patterns of Amish and large dairies in Sugar Creek. (a) Amish fragmented field pattern with 4-year rotation of corn, C, spelt, oats, and hay. (b) Non-Amish large dairy farm contiguous corn fields. (From Joannon 2007.)

In terms of water quality, including nutrients, carbon, and bacteria, phosphorus, and nitrogen levels in the Upper Sugar Creek are strongly correlated with sedimentation from cropland soil erosion and intensity of nitrogen inputs, especially seasonal manure applications (Goebel et al. 2002; Holmes 2004). There is a significant difference in the source of fertilizer applications in the two cases as the Amish spread manure with a horse-drawn manure spreader while the non-Amish large dairy farms have liquid manure pit lagoons. The latter has necessitated hauling the manure to other locations, a costly enterprise. This difference also has led to experimentation by these large dairies to sell carbon credits to power companies by burning methane (American Electric Power 2007).

In sum, the three cases of intensification observed in Sugar Creek could only be understood by using both natural science and social science approaches. Although the landscapes in the South Fork and East Branch are highly simplified in structure compared with the ecosystems they replaced, when coupled with the human communities that have shaped them, the underlying complexity and its influence on ecological conditions in the watershed become apparent. Stakeholders experience *transformational learning* when introduced to stream ecology by analogy to soil food webs and nutrient cycling processes in soils. This is particularly true of the case of the cattle exclusion from the streams where the somatic cell count method became embedded into the dairy rules of a new organic co-op the Amish farmer formed. Perhaps the most fundamental change has been in core values. In many parts of Sugar Creek the farming community is starting to change its view of streams: from conduits for removing excess water from their fields to being valued parts of their ecosystem.

3.4 SUGAR CREEK INTEGRATING CONCEPT 3: COUPLING SOCIAL AND NATURAL SYSTEMS IN FRAGMENTED LANDSCAPES

Our research on fragmented landscape patterns has focused on several small streams, each of which represented different landscape patterns. During 2003 to 2005 our team was funded by an NSF biocomplexity grant where we focused on water quality and fragmented landscapes of three headwaters streams in a HUC 14 digit watershed that had an area of 26 square miles. Our goal was to bridge the social and natural sciences through this project. The sites shown on each stream in Figure 3.3 were selected according to the following criteria: (1) land use types, (2) order of the types along the stream, (3) representation of both farming and residential areas, (4) inclusion of Upper Sugar Creek farmer partner locations, (5) the degree of owner farming versus leasing land, and (6) social stratification gradient. Our objective in the research was twofold. The first objective was to benchmark biocomplexity of major terrestrial and aquatic components of the existing landscape and farmers' knowledge and valuation of biocomplexity. In this objective, we built on our benchmarking and modeling activities to achieve an increased understanding of the following: (1) the social organization of humans and physical attributes at headwater, subbasin, and whole watershed scales; this included the settlement pattern, residential permanence, kin and nonkin networks, land rights (including the wide spectrum of ownership and use) between people living and working on the land, and farming strategies which influence land use decisions, (2) impact of current land use on source–sink relationships in carbon and nitrogen, and terrestrial and aquatic food webs, and (3) impacts of changes on watershed ecology if new management systems were adopted.

Our second objective focused on how to initiate restoration and ecological monitoring of representative headwater streams as a landscape level experiment. With this objective, we developed an experimental approach to the restoration of small headwater tributaries of Sugar Creek. These headwater streams, while largely unnamed or mapped, comprise more than 115,000 miles in Ohio compared to 21,000 miles of named, larger streams. The Ohio EPA classifies these tributaries as draining 0.5 ha or less of land and generally having a depth of less than 40 cm. The primary headwaters are characterized by having perennial flow, high diversity of invertebrates, amphibians, and in some cases fish, and high nutrient assimilation capacity. In addition these headwaters play a proportionately large role in the overall watershed ecology by maintaining flow during drought conditions. When riparian vegetation is intact adjacent to these streams, the carbon input from

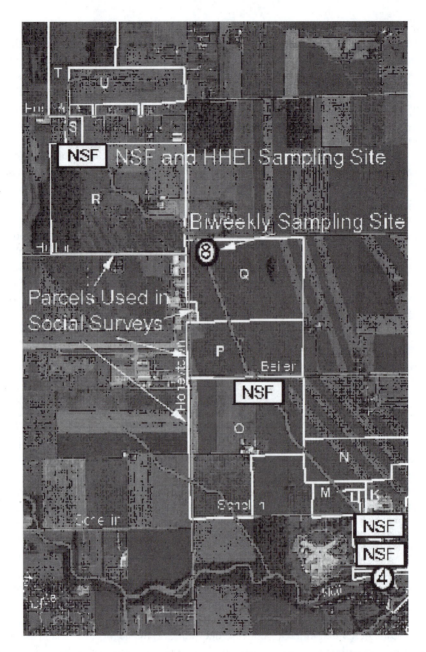

FIGURE 3.3   Combined biological and social science sampling on Upper Sugar Creek primary headwater stream 20.

senescent leaves and woody material provides the necessary energy and nutrients for maintenance of well-developed food webs, especially for key indicator species, such as salamanders, which have complex life cycles shared between terrestrial and aquatic phases (Davic 2003, 2004). Thus, these first-order streams are foci where terrestrial and aquatic ecosystem processes are relatively tightly coupled. Also, it has been shown that these headwaters are fairly resilient, in that when restoration is undertaken, biodiversity and carbon and nutrient cycling processes increase within relatively short time spans (Davic and Anderson 2002).

Our plan was to (1) establish a monitoring system and collect baseline data sets focused on key indicators of aquatic and terrestrial (primarily soil) food webs, and nitrogen and carbon cycling processes, (2) document landowner attitudes about and their understanding of the ecological roles of these headwater areas, and (3) initiate restoration efforts using the protocol developed by the Ohio EPA Surface Waters Division (Davic et al. 2001).

The significance of riparian cover for small stream is underscored by Sugar Creek's classification by Ohio EPA as a warm water aquatic habitat. However, the fact that there are multiple cool springs on the majority of Sugar Creek headwater streams necessitates a consideration for reclassification as potential cool water habitat, a point that Ben Stinner and Richard Moore emphasized after we started comparing (1) stream segments of the same stream, and (2) historical 1874 plat maps showing spring locations. Because the headwater streams are narrow and shallow, the presence or absence of canopy rapidly changes the stream temperature and aquatic makeup. Fittingly, on the same day Ben passed away he gave a lecture to the Upper Sugar Creek Farmer Partners. The main topic of discussion was about connecting patches of canopy on the same stream shown in Figure 3.4. The question arose whether placing canopy upstream or downstream on a headwaters stream would

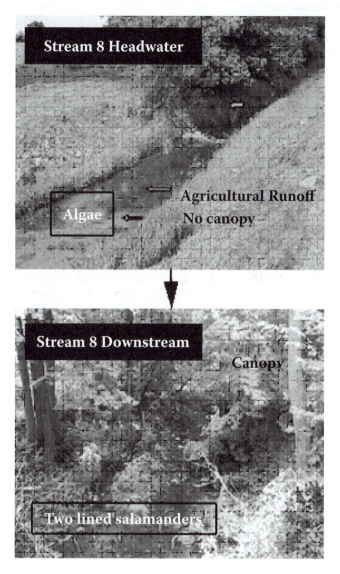

FIGURE 3.4 Upper Sugar Creek primary headwater habitat stream 8: A case of transformation of stream ecology.

have more impact on stream structure and function; a question no doubt inspired by Eugene Odum's work on the structure and function of ecosystems (Odum 1962) which was a common reference for Ben's work and our collaborative work in the watershed. The stream in Figure 3.4 was of particular interest because it started out in a degraded biological state as a result of being fed by an agricultural drainage ditch without canopy but later recovered toward the end of stream where a riparian canopy had been established.

Scaling of our social research was based on the idea (Nicholis and Prigogine 1977; Adams 1988) that a system will tend toward self-regulation as information is provided by the parts of the system (in this case subwatershed residents and participatory teams) and transformative events necessitate the emergence of a new structural hierarchy of social organization such as an organization to coordinate the various activities. Complexity exists because many of the processes, especially human managed ones, are nonlinear. For example, history and symbolic culture can constrain reorganizations in the relationships of power and production. As Rappaport (1968) noted, "adaptive structures are structured sets of processes, and regulatory hierarchies, whether or not they are embodied in particular organs or institutions, are found in all biological and social systems." The relations between subsystems and how they are regulated at different levels of structural complexity in social–ecological systems are ordered along several continua such as: (1) the ability to detect deviation of variables from reference values and ensure that corrective measures are engaged, (2) resilience (Holling 1998) and flexibility leading to appropriately timed and measured responses to system perturbations, (3) division of labor or functions between subsystems so that there is a balance between centralization and decentralization and proper degree of specialization to facilitate structural stability, (4) a hierarchical distribution of organization such that coherence and order are maintained to build infrastructure, and (5) establishment of a value system such that there is a clear demarcation and complementarity between sacred and profane (Rappaport 1968).

The biological findings of the NSF biocomplexity research (Moore et al. 2006) showed that in-stream particulate organic matter (POM) and dissolved organic carbon (DOC) concentrations (both of which are related to riparian vegetation in these heterotrophic stream systems) varied considerably, with the highest values observed along stream segments with wooded riparian areas and areas with adjacent agricultural and urban development. These results suggest that understanding the complexity associated with riparian areas and adjacent land use are important factors that can guide habitat restoration in the Upper Sugar Creek. Two-lined salamander populations were closely associated with riparian canopy over the stream. The HMFEI (Headwater Macroinvertebrate Field Evaluation Index) was most associated with discharge, while a number of two-lined salamanders were positively associated with HHEI (Headwater Habitat Evaluation Index), QHEI (Qualitative Headwater Evaluation Index), and canopy cover, and negatively associated with percent silt. The *Escherichia coli* results showed that the number of *E. coli* on the same headwater stream varied from one testing site to the next. There were different types of *E. coli* present on different days in the same stream at the same sampling site, suggesting that the *E. coli* did not travel far downstream and it is not the same strain of *E. coli* that is persistently contaminating the water.

The social science component of the NSF research focused on the survey of all residences, farm and nonfarm, living on property adjacent to the stream. Flyers were distributed to all residences along the stream. One side of the flyer had a letter of introduction and the other side representative landscape photos from that stream. A formal letter of introduction was included when the household was approached to sign up for an in-depth interview. These interviews used exploratory questions about landscape choices and responses were transcribed and analyzed for key words and concepts. Permission to conduct research on each site was obtained from the landowner. In all, 8 farms and 24 nonfarm residents living adjacent to three primary headwater streams in Upper Sugar Creek were interviewed in July 2004. These were combined with the data from a previous survey of 225 households in the same subwatershed. The survey was designed to measure if individuals with a stronger sense of place (articulated in terms of family and residential history) and community (measured in terms of social capital) tended to exhibit a stronger commitment to remediation or restoration of the

Sugar Creek. Consistent with our approach of advocacy coalition framework (ACF) we were especially interested in the potential for remediation on these small streams. Accordingly, we hypothesized that individuals are more likely to support remediation and restoration of the Sugar Creek if they perceive this goal as congruent with their core (moral and aesthetic) beliefs and values and that individuals would be more likely to support remediation and restoration of the Sugar Creek if they had a clear sense of how to realize these goals.

The results of the study showed that mutual reciprocity was higher in areas where there were more farming residences. Farm residents living along all three streams scored high in participating in community activities, especially in churches and schools. Likewise, there was a strong desire to participate in water quality issues. Ag–town conflict corresponded to areas where there was the highest degree of economic and occupational differences and newcomers. Residential permanence was associated with a high degree of labor exchange, reciprocity, relatedness, and networking. The longer people had lived along a stream, the less they wanted the landscape to include multifamily housing or lots with large houses. People living in this rural community had a high preference for farmland and open spaces. Dairy farmers favored biodiversity more than soybean farmers and were more willing to take steps to improve the stream quality. Farmers were more concerned about EPA regulation and drainage issues than nonfarmers. Nonfarmers were more concerned about the loss of biodiversity, agricultural pollution, and children's health. An unexpected result was a measurable increase in social capital and social complexity related to social self-organization as biodiversity increased. Heterogeneous landscape patterns, land use, and land tenure were associated with higher levels of biodiversity.

3.5   SUGAR CREEK INTEGRATING CONCEPT 4: CREATING VALUE

The Sugar Creek Project has moved forward by creating value in a number of ways. These include improving the bottom line for farmers while valuing ecosystem services, creating innovative non-market based trading systems, and fostering emergent properties of ecosystems. The opportunities for creating value are unlimited for several reasons. First, as Robert Costanza (1976) has pointed out, the economic systems in the world have undervalued the services of the world's ecosystems. Second, as pointed out by H. T. Odum in his work on energy, we can see that the "currency exchange rate" for energy and money (represented by a diamond on his charts) on any two points on his energy flow charts is different. That is to say, while the energetics as a metric are constant, the value for that energy is dependent on the cultural value that humans place on it in the framework of supply and demand. Third, degraded ecological systems are usually simplified systems with ample opportunities for creating feedback loops from the energy and nutrient sinks. These feedback loops can be generated in accord with ecological concepts such as resilience, system buffers, hierarchy, patches, multiple scales, emergent properties, and the tendency of systems to self-organize. These feedback loops must be linked into environmental policy and human decision making to succeed. Conservation practices are inherently social phenomena.

During the first meeting between the researchers and farmers in the fall of 2000, it became clear that the farmers wanted to be socially responsible about the pollution causing problems with water quality but that they also had the idea that fixing the problem would be too costly. They wanted to find solutions that improved their farm operation bottom line, as came across to us most clearly when we discussed the possibility of creating buffers along Sugar Creek. Conservation Reserve Program (CRP) buffers paid farmers by the acre with rates that were slightly below the county average for land rental rates. When the farmers decided whether or not to put in buffers, they were mainly concerned about two things: the per acre return on the project and whether the buffer could be mowed so that it would maintain a neat appearance. As a result of this, the group went through several stages, one of which centered on dreaming about a hunting buffer where they could charge admission for hunting on the CRP buffer and also make money raising pheasants and turkeys for hunting so as to make the buffer profitable. This layering of land use functions has not occurred to date but the fact that more than 8 miles of contiguous buffer is in place makes the possibility for layering land use functions to include a future hunting buffer, nature trail, or bike path. This process transforms individual property rights to include customary use (usufruct) rights or a type of commons because constructing the contiguous buffer makes it socially difficult to "pull out" of the group because other segments of the buffer are dependent on everyone's participation.

The second case where the bottom line was improved was on our work fencing cows out of the stream near Kidron, Ohio. Many of these farmers found that fencing the cows out of the stream has resulted in lower somatic cell counts in their milk resulting in higher milk premiums when they sell their milk at the dairy. Going from 365,000 somatic cells per milliliter of milk to 165,000 allowed a lead Amish farmer to receive a 75¢ per hundredweight (cwt) premium. Our data indicate fencing cattle away from streams will reduce bacterial abundance dramatically within 6 months. The idea of a sliding scale for somatic cell counts milk premiums was incorporated into the rules of the Amish co-op Green Field Farms, which is marketing organic milk, eggs, cheese, butter, and vegetables. Two of the leaders of this cooperative, which has grown to 80 Amish farms, originated from the area where we worked with farmers to fence cows out of the stream.

Our water quality trading system targets the cost effectiveness of having farmers implement conservation measures paid for by local industry or county waste-water treatment plants. In these cases new regulations increased the costs associated with lowering the nitrogen and phosphorus outflow from the treatment plants. In other words, the costs associated with lowering the phosphorus outflow concentration from a factory from 200 to 10 mg/L might be equal to the cost of lowering the same plant from 10 to 1 mg/L. In this situation trading plans make better sense for reaching the last few milligrams per liter.

The Alpine Nutrient Trading Plan goal which officially started in 2007 was developed to help industry and the farming community work together to solve a pollution nutrient loading problem in local streams. Alpine Cheese Company is using the program to gain credit for a portion of its pollution remediation responsibility of its National Pollutant Discharge Elimination System (NPDES) 5-year permit by providing funds to pay farmers to improve on-farm conservation. The net result is that the stream will have less pollution than if the factory solved its pollution problem by itself. Alpine Cheese Company was faced with a large cost to remediate phosphorus.

To comply with new Ohio EPA regulations and simultaneously promote economic development, the Sugar Creek Project helped Alpine Cheese draft a nutrient-trading permit to lower phosphorus pollution coming out of its factory pipe into the stream. The phosphorus TMDL effluent limit for point sources set for Sugar Creek is 1 mg/L. Under the NPDES permit, Alpine Cheese is allowed 3.2 mg/L or a maximum of 1.7 kg/day load which is regulated through the Sugar Creek TMDL policy. The Alpine nutrient trading plan pays farmers' costs for adopting conservation practices to reduce the phosphorus that would otherwise pollute the watershed along with a premium for each pound of phosphorus removed each year after they install conservation measures. In exchange, the cheese company receives phosphorus credits to use to reduce its phosphorus contribution to environmental standards. The plan will reduce up to three times more phosphorus than if the company had met EPA standards by itself, without farmer cooperation. Equally, Alpine Cheese will be able to expand and buy more milk from local producers.

Lowering the pollution levels is economically beneficial for both the factory and the farmers. Alpine Cheese will be able to expand and create 12 new local jobs and use 250,000 pounds more milk per day, up from the estimated 650,000 pounds of milk per day it used in 2004. All the milk used will be Ohio-produced Grade A Class III milk, valued at $17 to $20/cwt to the producer. This increase in local milk usage amounts to the equivalent production of 125 small dairy farms, each with 40 cows. Alpine Cheese is the only U.S. manufacturer of Jarlsberg, a specialty cheese from Norway's TINE Dairy for which the manufacturer has the exclusive U.S. contract. Jarlsberg cheese will be produced at the rate of 50,000 pounds per day, each pound with a retail value of $6 to $10.

The local Soil and Water Conservation District (SWCD) administers the plan and works directly with the farmers and serves as broker between the farmers and the factory. The university serves as a neutral mediator, conducting monitoring, research, planning, writing the plan, and fostering cooperation between agencies. The university, working with the local SWCD, targets the best sources of phosphorus. For each farm, the cost of phosphorus per pound for each conservation measure is calculated along with a cap per farm so as to include the full array of ecologically sound conservation measures including those that do not produce quite so much phosphorus per pound. As equal partners, the local SWCD, industry, and the university share in proceeds from the sale of extra credits that will be generated during the plan.

Perhaps the most important aspect of the Alpine water quality trading program is that it makes cultural sense. Every community has its own character so nutrient-trading programs must fit into what each local community wants with respect to community development and watershed vision. The Alpine plan makes sense to everyone in this community because the pollution is removed from the cheese and the dairy farms. There is also low risk because the cheese factory is legally responsible and in 5 years its permit must be renewed. Most local residents ask why such approaches were not adopted sooner and there is waiting list of farms wanting to participate. In the Alpine plan, a cultural solution emerged to establish a new ecosystem management approach. It was one that could not have been conceived and executed by policy makers and regulatory agencies, but rather evolved from the grassroots when the factory owner and Ohio EPA reached an impasse.

Local knowledge appears to be key for these types of emergent properties. For example, the Kayapo Indians used local knowledge to create deceptively natural ecosystems by "transplanting" wild local varieties (endemism) including termite mounds into wooded concentrations of useful plants (Posey 2000). Historically these ecosystems were considered "natural" by researchers but Posey discovered this was an anthropogenic landscape (Posey 1997, 2000) only after living with the Kayapo for more than 7 years. For the Kayapo, systematic ecological knowledge is "embedded" (Ellen 2000, p. 180); the Sugar Creek case has a long way to go to achieve, for example, the level of indigenous ecological knowledge (IEK) that Woodley (2008) describes in the Solomon Islands as necessary for creating an emergent system. Such systems require not just knowledge as context and practice—as in the Sugar Creek case—but also knowledge as belief (Woodley and Rappaport 1967).

Context portrays the confines of learning due to history (settlement patterns), demographic factors as well as biophysical features of place. Knowledge as *practice* portrays meaningful action, through physical interaction and experiential learning. Knowledge as *belief* portrays the influence that spirituality and values have on how people act within their ecosystem (Woodley 2008).

A cognized model of the ecosystem must include ideological aspects to create a complete understanding of or realization of an emergent system. In the anthropological literature we have excellent cases of this such as the Tsembaga Maring ritual regulation in New Guinea (Rappaport 1968), the Balinese water temples described by Lansing (2006), or the Japanese rice god in my own work (Moore 1990).

This chapter has given examples from the Sugar Creek Watershed that illustrate ways to bridge and integrate the social and natural sciences in agroecological work. Ben Stinner was gifted at bridging the social and natural sciences and his legacy lives on in the Sugar Creek Project. The Millennium Ecosystem Assessment (2005) has shown that the world's ecosystems are being degraded at an unprecedented rate. The cases of intensification described in this chapter have been occurring at a world scale. The hope is that this urgent situation will act as a transformative event to trigger solutions based on combined social and natural science responses.

REFERENCES

Adams, R. N. 1988. *The Eighth Day: Social Evolution as the Self-Organization of Energy.* University of Texas Press, Austin.

Altieri, M. A. 1999. The agroecological dimensions of biodiversity in traditional farming systems. In *Human Values of Biodiversity,* Posey, D., Dutfield, G., and Plenderleith, K. Eds., Intermediate Technologies, London, 305–309.

Altieri, M. 2008. Small farms as a planetary ecological asset: Five key reasons why we should support the revitalization of small farms in the Global South. Food First web site. http://www.foodfirst.org/en/node/2115 (accessed July 14, 2008).

American Electric Power. 2007. AEP to support largest agricultural carbon offset program in U.S.; program will capture and destroy methane from 200 farms. http://www.aep.com/newsroom/newsreleases/?id=1375 (accessed July 27, 2008).

Ashby, W. R. 1963. *An Introduction to Cybernetics*. John Wiley & Sons, New York.

Berkes, F. and Folke, C., Eds. 1998. *Linking Social and Ecological Systems: Management Practices and Social Mechanisms for Building Resilience*. Cambridge University Press, Cambridge.

Boserup, E. 1965. *The Conditions of Agricultural Change: Decision-Making in a Costa Rica Community*. Rutgers University Press, New Brunswick, NJ.

Brookfield, H. 2001. *Exploring Agrodiversity*. Columbia University Press, New York.

Buttel, F. 2002. Environmental sociology and the sociology of natural resources: Institutional stories and intellectual legacies. *Society and Natural Resources,* 15, 205–211.

Campbell, J. 2008. Impacts of collaborative watershed management policies on the adoption of agricultural best management practices. MS thesis, School of Environment and Natural Resources, Ohio State University, Columbus.

Clemens, E. S. 2007. Toward a historicized sociology: theorizing events, processes, and emergence. *Annual Review of Sociology,* 33, 527–549.

Cohen, M. N. 1989. *Health and the Rise of Civilization*. Yale University Press, New Haven, CT.

Conelly, W. T. 1992. Agricultural intensification in a Philippine frontier community: impact on labor efficiency and farm diversity. *Human Ecology: An Interdisciplinary Journal,* 20, 203–223.

Costanza, R. et al. 1997. The value of the world's ecosystem services and natural capital. *Nature,* 387, 253–260.

Cottingham, K. L. 2002. Tackling biocomplexity: The role of people, tools, and scale. *Bioscience,* 50, 793–799.

Covich, A. 2000. Biocomplexity and the future: The need to unite disciplines. *Bioscience,* 50, 1035.

Davic, R. D. 2003. Linking keystone species and functional groups: a new operational definition of the keystone species concept. *Conservation Ecology* 7(1), r11. http://www.consecol.org/vol7/iss1/resp11/ (accessed July 14, 2008).

Davic, R. D. and Anderson, P. 2002. Ohio EPA Primary Headwater Habitat Initiative Data Compendium, 1999–2000: Habitat, Chemistry, and Stream Morphology Data. Ohio Environmental Protection Agency, Division of Surface Water, Columbus.

Davic, R. D. and Welsh, H. H., Jr. 2004. On the ecological roles of salamanders. *Annual Review of Ecology Evolution, and Systematic,* 35, 405–434.

Dietz, T., Rosa, E., and York, R. 2007. Driving the human ecological footprint. *Frontiers in Ecology and the Environment,* 5(1), 13–18.

EPA. 2000. 2000 National Water Quality Inventory. U.S. EPA. Office of Water, Washington, D.C.

Escobar, A. 1998. Whose knowledge, whose nature? Biodiversity, conservation, and the political ecology of social movements. *Journal of Political Ecology,* 5:53–82.

FAOSTAT. 2004. Food and Agriculture Organization of the United Nations, Statistical Databases, http://faostat.fao.org (accessed July 14, 2008).

Goebel, P. C. et al. 2002. Effect of landscape properties on water quality in an agricultural watershed of north-central Ohio. Poster presented at the 2002 Annual Meeting of the Ecological Society of America and Society for Ecological Restoration, Tucson, AZ.

Green, D. and Shapiro, R. 1994. *Pathologies of Rational Choice Theory: A Critique of Applications in Political Science*. Yale University Press, New Haven, CT.

Habermas, J. 1987. *The Theory of Communicative Action,* Vol. 2: *Lifeworld and System: A Critique of Functionalist Reason*. Beacon Press, Boston.

Habermas, J. 1996. *Between Facts and Norms: Contributions to a Discourse Theory of Law and Democracy*. Polity Press, Cambridge.

Hole, D. G. et al. 2005. Does organic farming benefit biodiversity? *Biological Conservation,* 122, 113–130.

Holmes, K. M. 2004. Landscape factors influencing water quality and the development of reference conditions for riparian restoration in the headwaters of a northeast Ohio watershed. M.S. thesis, School of Natural Resources, Ohio State University, Columbus.

Jarvis, S. C. 1999. Nitrogen dynamics in natural and agricultural ecosystems. In *Managing Risk of Nitrates to Humans and the Environment,* Wilson, W. S., Ball, A. S., and Hinton, R. H., Eds. Royal Society of Chemistry, Cambridge, 2–20.

Joannon, A. 2007. Comparison of the dairy farms of the South Fork and East Branch of Sugar Creek Watershed in Ohio. Powerpoint presentation at OARDC, May. (Visiting Scholar from INRA in Rennes, France.)

Kemmis, S. 1988. Action research in retrospect and prospect, In *The Action Research Reader,* 3rd ed., Kemmis, S. and McTaggart, R., Eds. Deakin University Press, Geelong, Victoria, 27–39.

Lansing, S. 2006. Perfect Order: Recognizing Complexity in Bali. Princeton University Press, Princeton, NJ.

Leach, W. D. and Pelky, N. W. 2001. Making watershed partnerships work: a review of the empirical literature. *Journal of Water Resources Planning and Management,* ASCE, 127, 378–385.

Leach, W. D., Pelky, N. W., and Sabatier, P. A. 2002. Stakeholder partnerships as collaborative policymaking: Evaluating criteria applied to watershed management in California and Washington. *Journal of Policy Analysis and Management,* 21, 645–670.

Likens, G. E. and Bormann, F. H. 2003. *Biogeochemistry of a Forested Ecosystem.* Springer-Verlag, New York.

Long, S. 2003. The complexity of labor exchange among Amish farm households in Holmes County, Ohio. PhD dissertation, Department of Anthropology, Ohio State University, Columbus.

Malanson, G. P. 1999. Considering complexity. *Annals of the Association of American Geographers,* 89, 746–753.

Malthus, T. R. 1826. *An Essay on the Principle of Population*, 6th ed. John Murray, London.

Martens, G. G. 1986. *Traditional Agriculture in Southeast Asia.* Westview Press, Boulder, CO.

Matson, P. A. et al. 1997. Agricultural intensification and ecosystems properties. *Science,* 277, 504–509.

Mattkalli, N. M. 1996. Estimation of surface water quality changes in response to land use change: Application of the export coefficient model using remote sensing and geographical information system. *Journal of Environmental Management,* 48(3), 263–282.

McAdam, D., Tarrow, S., and Tilly, C. 2001. *Dynamics of Contention.* Cambridge University Press, Cambridge.

Michener, W. K. et al. 2001. Defining and unraveling biocomplexity. *BioScience,* 51, 1018–1023.

Millennium Ecosystem Assessment. 2005. *Ecosystems and Human Well-Being: Current State and Trends.* Island Press, Washington, D.C.

Moerke, A. H. and Lamberti, G. A. 2004. Restoring stream ecosystems: Lessons from a Midwestern state. *Restoration Ecology,* 12, 327–334.

Moore, R. 1990. *Japanese Agriculture: Patterns of Rural Development.* Westview Press, Boulder, CO.

Moore, R. .2001. Honoring creation and tending the garden: Amish views of biodiversity. In *Human Values of Biodiversity,* Posey, D. and Dutfield, G., Eds. Special volume for the UN Environmental Programme Global Assessment of Indigenous People. Intermediate Technologies.

Moore, R. 2002. Jizokuteki Seitaikei Hozen no shiten yori FukaKachi Fukei wo Kangaeru [Considering value-added landscapes from a viewpoint of sustainable ecological preservation]. *Norin Tokei Chosa,* 52(2), 50–61.

Moore, R., Stinner, B., Williams, L., Goebel, C., Stinner, D., LeJeune, J., and Taylor, R. 2006 Final Report: Impact of Economics-Driven Land-Use Decisions on Watershed Health. NSF grant BE/CNH 0308464.

Moyer, B. et al. 2001. *Doing Democracy: The MAP Model for Organizing Social Movements.* New Society, Gabriola Island, BC.

Netting, R. 1993. Smallholders, Householders; Farm Families and the Ecology of Intensive, Sustainable Agriculture. Stanford University Press, Stanford, CA.

Nicholis, G. and Prigogine, I. 1977. *Self-Organization of Nonequilibrium Systems.* John Wiley & Sons, New York.

Odum, E. 1962. Relationships between structure and function in the ecosystem. *Japanese Journal of Ecology,* 12, 108–118.

Ohio Environmental Protection Agency. 2007. Total Maximum Daily Loads for Bacteria in the Sugar Creek Watershed. Final Report, March 19.

Parker, J. S. 2006. Land tenure in the Sugar Creek watershed: A contextual analysis of land tenure and social networks, intergenerational farm succession, and conservation use among farmers of Wayne County, Ohio. PhD dissertation, Department of Anthropology, Ohio State University, Columbus.

Parker, J., Moore, R., and Weaver, M. 2007. Land tenure as a variable in community based watershed projects: Some lessons from the Sugar Creek Watershed, Wayne and Holmes Counties, Ohio. *Society and Natural Resources,* 20, 815–833.

Peterson, B. J. et al. 2001. Control of nitrogen export from watersheds by headwater streams. *Science,* 292, 86–89.

Peterson, G., Allen, C., and Holling, C. S. 1998. Ecological resilience, biodiversity, and scale. *Ecosystems,* 1, 6–18.

Pew Commission on Industrial Farm Animal Agriculture. 2008. Putting Meat on the Table: Industrial Farm Production in America. http://www.pewtrusts.org/uploadedFiles/wwwpewtrustsorg/Reports/Industrial_Agriculture/PCIFAP_FINAL.pdf (accessed on July 22, 2008).

Posey, D. A. 1997. Ecological consequences of Kayapó Indian presence in Amazonia: Anthropogenic resources and traditional resource rights. In *A Economia da Sustentabilidade: Princípios, Desafios, Aplicacões,* Cavalcanti, C. Ed., Fundacao Joaquim Nabuco, Recife, Brazil.

Posey, D. 2000. Introduction: Culture and nature—The inextricable link. In *Cultural and Spiritual Values of Biodiversity: A Complementary Contribution to the Global Biodiversity Assessment,* Posey, D. and Dutfield, G., Eds. Special Volume for the UN Environmental Programme Global Assessment of Indigenous People. Intermediate Technologies, London, 3–18.

Rappaport, R. 1968. Pigs for the Ancestors: Ritual in the Ecology of a New Guinea People. Yale University Press, New Haven, CT.

Rhoades, R. and Booth, R. H. 1982. Farmer back to farmer: A model for generating acceptable agricultural technology. *Agricultural Administration,* 11, 127–137.

Sabatier P. A. and Jenkins-Smith, H. C. 1999. The advocacy coalition framework, an assessment. In *Theories of the Policy Process,* Sabatier, P. A. and Jenkins-Smith, H. C., Eds. Westview Press, Boulder, CO, 117–166.

Sabatier, P. A. et al. 2005. *Swimming Upstream: Collaborative Approaches to Watershed Management,* MIT Press, Cambridge, MA.

Stramma, L. et al. 2008. Expanding oxygen-minimum zones in the tropical oceans. *Science,* 320, 655–658.

Stinner, B. R. and House, G. J. 1990. Anthropods and other invertebrates in conservation-tillage agriculture. *Annual Review of Entomology,* 35, 299–318.

Stinner, B. R. et al. 1997. Earthworm effects on crop and weed biomass and N content in organic and inorganic fertilized agroecosystems. *Soil Biology and Biochemistry,* 29, 361–368.

Stinner, B. R., Krishna Prasad, V., and Gulbahar, G. 2004. The Great Lakes ecosystem: Regional sustainability and local examples. *Proceedings, Land Use at the Urban-Rural Interface in the Great Lakes Region,* Sandusky, OH, May 4–5.

Stinner, D. H. 2004. Biodiversity and agriculture. In *Encyclopedia of Plant and Crop Science,* Goodman, R. M., Ed. Marcel Dekker, New York, 1–4.

Stinner, D. H., Paoletti, M. G., and Stinner, B. R. 1989. Amish agriculture and implications for sustainable agriculture. *Agriculture, Ecosystems, and Environment,* 27, 77–90.

Stinner, D. H., Glick, I., and Stinner, B. R. 1992. Forage legumes and cultural sustainability: lessons from history. *Agriculture, Ecosystems, and Environment,* 40, 233–248.

Stinner, D. H., Stinner, B. R., and Martsolf, E. 1997. Biodiversity as an organizing principle in agroecosystem management: Case studies of Holistic Resource Management practitioners in the USA. *Agriculture, Ecosystems, and Environment,* 62, 199–213.

Torsvik, V. L., Øvreås, T., and Thingstad, F. 2002. Prokaryotic diversity—Magnitude, dynamics, and controlling factors. *Science,* 296, 1064–1066.

van Elsen, T. 2000. Species diversity as a task for organic agriculture in Europe. *Agriculture Ecosystems and Environment,* 77, 101–109.

van Mansvelt, J. D., Stobbelaar, D. J., and Hendriks, K. 1998. Comparison of landscape features in organic and conventional farming systems. *Landscape and Urban Planning,* 41, 209–227.

Vitousek, P. M. et al. 1997. Human domination of Earth's ecosystems. *Science,* 277, 494–499.

Watzin, M. C., and McIntosh, A. W. 1999. Aquatic ecosystems in agricultural landscapes: A review of ecological indicators and achievable outcomes. *Journal of Soil and Water Conservation,* 54, 636–644.

Weaver, M., Moore, R., and Parker, J. 2005. Understanding grassroots stakeholders and grassroots stakeholder groups: The view from the grassroots in the Upper Sugar Creek. Paper presented at the Annual Meeting of the Political Science Association, Washington, D.C.

Willems, E. et al. 2000. Landscape and land cover index. In *From Land Cover to Landscape Diversity in the European Union.* Report of the European Commission.

Woodley, E. 2008. Local and indigenous ecological knowledge as an emergent property of a complex system: A case study in the Solomon Islands. http://www.millenniumassessment.org/documents/bridging/papers/woodley.ellen.pdf (accessed July 17, 2008).

Section II

Unifying Concepts and Principles of Sustainable Agroecosystem Management

4 Rethinking the First Principles of Agroecology
Ecological, Social, and Economic

John E. Ikerd

CONTENTS

4.1 INTRODUCTION

Agroecology is generally defined as the application of ecological science in the design and management of sustainable agroecosystems (Gliessman 2008). Agroecology integrates agriculture and ecology for the specific purpose of facilitating the scholarly study of agricultural sustainability.

The concept of agroecology was first used by scientists in the late 1970s, in questioning the ecological, social, and economic sustainability of Green Revolution-era agroecosystems. A 1982 symposium at the annual meetings of the Ecological Association of America was among the first efforts to bring scientists from the various disciplines associated with agroecology together in a national academic forum (Lowrance et al. 1984). Agroecology provided a logical, conceptual framework for integrating the disciplines of ecology, economics, and sociology for the purpose of enhancing research and education related to the sustainability of agriculture.

After more than 20 years, however, relatively few scientists have chosen to identify themselves as agroecologists, and those who have are disproportionately from the physical rather than social sciences. The case for agroecology as a framework for interdisciplinary scientific inquiry seems compelling; yet it seems to have gained few advocates. Perhaps it is time to ask why, and to reexamine the first principles of agroecology, in the hopes of finding ways of better communicating the importance of integrating ecology and agriculture in addressing issues of agricultural sustainability.

4.2 SUSTAINABLE AGRICULTURE

The explicit purpose of integrating the science of ecology and agriculture is to enhance the sustainability of agriculture. Sustainable agriculture is defined in different ways by different people, seemingly, at times, to accommodate individual scientific, political, or economic agendas. However, there is no serious disagreement regarding the basic principles of sustainability, at least not among those who have seriously studied the issue. Ben Stinner, for example, had a vision of a sustainable agriculture as agroecosystems that preserve environmental quality, sustain healthy social connections among people, and efficiently recycle natural and social capital, rather than rely on commercial inputs (Ben Stinner Endowment 2006). Stephen Gliessman defines sustainable agriculture as a whole-systems approach to production that balances environmental soundness, social equity, and economic viability (Gliessman 2007). Miguel Altieri views sustainable agriculture as a useful concept, in spite of conflicting definitions and interpretations, because it captures a set of growing concerns about agriculture, which have resulted from the coevolution of socioeconomic and natural systems (Altieri 1995). The authentic proponents of sustainability all agree, a sustainable agriculture must be ecologically sound, socially responsible, and economically viable.

However, little consideration is given to questions of *why* a sustainable agriculture must be ecologically, socially, and economically sustainable. Even among those who stress the logical linkage among ecology, sociology, and economics, few seem to address the question *why* sustainability should be an important priority for human society. Much of the continuing resistance to the concept of sustainability obviously arises from a lack of concern for the future. Many people apparently feel that they are expected to take care of themselves, so those of future generations should expect to do likewise. Others seem to share the belief that the pursuit of short run, individual self-interest is the best means of ensuring the long run well-being of society in general, as proclaimed by neoclassical economists. Others boast that human ingenuity is capable of solving any ecological or social problems we might create and finding an alternative for any resource we might use up. But these are simply beliefs, with little, if any, basis in fact. Differences in priorities afforded sustainability do not arise from differences in information or in intellect among its opponents or advocates, but instead from differences in fundamental beliefs.

4.3   FIRST PRINCIPLES

The advocates of agroecology will not gain widespread support until they can answer the questions of *why* designing and managing *sustainable* agroecosystems should be a critical priority and *why* the ecological, social, and economic dimensions of agriculture must be integrated to ensure agricultural sustainability. Most scientists today avoid such questions because the answers ultimately depend on basic beliefs or *first principles*. Wikipedia defines first principles as "a set of basic, foundational propositions or assumptions that cannot be deduced from any other proposition or assumption."

Since first principles are the most fundamental and general conceptions of thought, action, and reality, they are inherently philosophical rather than scientific in nature. Philosophy differs from science in that its questions cannot be answered empirically—by observation or experiment. Ironically, many scientists proudly accept the title of "doctor of philosophy," but are reluctant to address most important philosophical questions of their academic disciplines.

First principles are sometimes called laws of nature. "Laws of nature are the 'principles' which govern the natural phenomena of the world. That is, the natural world 'obeys' the Laws of Nature," as described by Swartz (2006). Philosophers refer to this as necessitarian theory, in that such principles are considered to be necessary for nature to fulfill its purpose. An alternative theory defines laws of nature as statements or descriptions of the regularities in the world; the way the world works, period, denying any specific purpose for the principles of nature. Regardless of necessity, first principles represent the ultimate truths or *pure knowledge* from which all other truth is derived.

The ethics or morality of our actions—whether they are right or wrong, good or bad—also is determined by first principles (Hamilton 1829). In the case of ethics, first principles are called natural law rather than laws of nature. According to natural law, the moral standards that govern appropriate human behavior can be traced to the basic nature of human beings, to a supreme being, or to the nature of the cosmos in general (Wikipedia). Regardless, natural law exists independently of any given religion, culture, society, political order, or nation-state. Natural laws apply to all people of all times. Belief in the existence of natural law is expressed, explicitly or implicitly, in such historic documents as the Magna Carta and American Declaration of Independence, where rights are described as being *inherent* or *self-evident.*

Plato argued that one could never gain *pure knowledge* through observation because anything that can be observed is always changing whereas pure knowledge is inherently unchanging (Baird 2008). He argued that we can observe examples of the *form* of pure knowledge, and we can visualize ideas of this true *form* in our minds. But we can never actually observe true *form*, or pure knowledge, because it exists only in the abstract. On the other hand, Plato argued that when reason is *properly used* the resulting intellectual insights are certain, universal, and eternal.

By using Plato's terminology, first principles constitute the form or architecture of pure knowledge. Thus, we can see evidence of the existence of first principles in the world around us, but our observations have meaning only insofar as we have some intuitive understanding of the underlying, unchanging principles that guide the ever-changing phenomena we observe. Our understanding of first principles requires reliance on our intellectual insights, or more precisely, on our common sense.

4.3.1 Common Sense

As Thomas Reid, nineteenth-century philosopher wrote, "All knowledge and science must be built upon principles that are self-evident; and of such principles every man who has common sense is competent to judge" (Reid 1863). These self-evident principles provide a starting point, and lacking a starting point, all logic and reasoning eventually become circular and thus useless. For example, first principles of algebra, called axioms or laws, are the foundation for all mathematical proofs. One such axiom is, *a* times *b* equals *b* times *a*. This may seem obvious, but such is the nature of first principles. First principles are common sense, of which every thoughtful person is competent to judge. Without the first principles or axioms of algebra, however, proof of any mathematical proposition would be impossible.

Relying on common sense does not imply rejection of science as a means of understanding the nature of things; it is just that all science must be rooted in common sense. Thomas Huxley, a noted English botanist, once wrote, "It is plain all truth, in the long run, is only common sense clarified" (Huxley 2004). When Albert Einstein wrote, "Common sense is the collection of prejudices acquired by age eighteen," he obviously was referring to prejudices, customs, or conventional wisdom rather than Reid's philosophical concept of common sense. Einstein also wrote, "The whole of science is nothing more than a refinement of everyday thinking." Science can be used to clarify and refine our common sense, but not to replace it.

Common sense admittedly has become an overused, often abused colloquialism, but the concept has deep philosophical roots. Eighteenth-century philosophy of common sense, sometimes called Scottish philosophy, arose in response to John Locke's "doctrine of ideas." George Berkeley's related theory of "pure idealism" attempted to explain reality solely in terms of ideas. On the other hand, David Hume argued that if reality existed only as ideas, there was no logical basis for assuming the existence of any mental substance capable of receiving ideas, the mind being nothing more than a succession of experiences. Between these two propositions, both ideas and reality disappeared, leaving nothing, and thus degenerating into complete skepticism.

In an effort to resolve this dilemma, Thomas Reid set out to vindicate common sense, meaning the natural judgment of common people, as the ultimate judge of reality. He concluded that ideas and the mind are both real, simply because people know they exist. He argued that the ultimate understanding of reality can be found only in human consciousness or human knowledge of reality, and thus reality neither needs to be proved nor can be proved because human understanding of reality must provide the grounds for all proof. Other Scottish philosophers, including Thomas Brook, William Hamilton, and James Mackintosh, added refinements to Reid's philosophy of common sense and extended it to deal with direct knowledge of human *morality* as well as *reality*. According to this eighteenth-century philosophy, common sense is our inner sense or intelligent insight regarding the basic nature of first principles—both laws of nature and natural law—by which people must test the truth of knowledge and the morality of actions.

Deep ecology, first advocated by Norwegian philosopher Arne Naess, is rooted in philosophical thinking very similar to that of the Scottish philosophers (Devall and Sessions 1985). Naess argued that the "shallow" environmental movement was concerned primarily with social welfare issues such as pollution and depletion of natural resources, while the "deep" ecology movement was more concerned with the deeper philosophical questions of how humans *should* relate to their natural environment. He suggested that most Western philosophers hold an outdated view of humans as separate from each other and from their natural environment, whereas a deeper understanding reveals that humans are not truly separate beings, but instead, are integrally interconnected with each other and with the world around them. Equally important, he believed there are right and wrong ways for humans to relate to the world around them.

Naess considers human ecology to be a "genuine part" of "general ecology" and, thus, considers human relationships to nature to be a genuine part of deep ecology. He wrote, "for each species of living being there is a corresponding ecology" (Naess 1988). As a means of differentiating between shallow and deep ecology, Naess contrasted typical slogans of shallow environmentalism, such as "natural diversity is valued as a resource for humans," with alternative slogans of deep ecology, "natural diversity has its own intrinsic value." However, he carefully points out that deep ecology actually *questions* both sets of slogans and provides no unique set of right or wrong answers or conclusions. Deep ecology frames the questions, but the answers ultimately must arise from the common sense of ordinary people. So the first principles of ecology must be derived not from ecological science, but instead from ecological philosophy.

4.4   FIRST PRINCIPLES OF AGROECOLOGY

The first principles of ecology and of agriculture logically provide the first principles for agroecology. The first principle of agroecology is the first principle of agriculture: *life has purpose*. Agriculture, by its basic nature, is a *purposeful* human activity. The basic purpose of agriculture is to shift the ecological balance of nature in favor of humans relative to other species. If there is no purpose for life, there is no purpose for human life, and thus no purpose for agriculture. Without purpose, agriculture would be a senseless activity.

Most people probably never question whether life has purpose, just as they never question whether a times b equals b times a, but scientists do. Most scientists are philosophical materialists, at least in the practice of their professions. In his 1919 classic book, *Modern Science and Materialism*, Hugh Elliott writes, "The age of science is necessarily an age of materialism; ours is a scientific age, and it may be said with truth that we are all materialists now" (Elliott 1972). Elliott emphasized three primary assumptions of materialism. The first is the uniformity of law: When the conditions at any moment in time are precisely the same as those prevailing at some earlier moment, the results also will be identical to the earlier results. Thus, science can link effects with their causes.

The second assumption is a denial of "teleology" or purpose. He writes, "Scientific materialism warmly denies that there exists any such thing as purpose. It asserts that all events are due to the interaction of matter and motion, acting by blind necessity in accordance with those invariable sequences to which we have given the name laws" (Elliott 1972). Elliott refers to the human species as a "mere incident of the universal redistribution of matter and motion." Then, 85 years later, physicist Brian Green wrote, "Newton and Einstein agree, you can, in principle, use the laws of physics to predict everything about the universe arbitrarily far into the future or figure out what it was like arbitrarily far in the past" (Green 2004). Quantum physics casts some doubt on the precise predictability of events, but does nothing to suggest that events unfold for any particular purpose. Modern science treats the unfolding of a human life as nothing more or less than the natural consequences of physical actions and reactions, without any particular purpose or meaning.

The third assumption of materialism is the denial of any form or existence other than that having some kind of palpable material characteristics and quality, which "stands in direct opposition to a belief in any of those existences which are vaguely classed as 'spiritual.'" Among those things, he included not only gods and souls, but also such imaginary entities as intellect, will, feelings, insofar as they are supposed to be different from material processes.

Throughout human history, most people have believed that life has purpose and meaning. Aristotle used the word *telos* to refer to the ultimate goal, final end, or purpose of life in his classic works, *Physics* and *Metaphysics* (Clayton 2006). He suggested that one could not fully understand or describe anything without referring to its purpose. For example, the purpose of a knife is to cut. However, Aristotle was most interested in the purpose of people. He believed the purpose of human life was happiness, that all people were meant to be happy. He further believed that human happiness required a life of virtue, that a person who was not living a moral life could not actually be happy, no matter what he or she might think at the time. Someone who chose to do the right thing because it was the right thing to do was living a life that *flourished*; he suggested, in that such a person was using his or her human capacities to the fullest by living according to his or her purpose. Since each person is confronted with a unique set of life's choices, each person has a unique path to follow in the pursuit of happiness.

Most people today seem to agree with Aristotle rather than the "scientific materialists." Most people do not consider their choices and actions to be predetermined acts of blind necessity or the inevitable consequences of ongoing interaction of matter and motion. They believe they have some degree of autonomy in their choices, that they can affect the future by choosing one course of action rather than another. They may not believe they can change everything but their actions suggest that they believe they can change some things. These beliefs are expressed in the social norms and customs of every civilized society and in the constitutions, laws, and regulations of every government in the world. Most people believe their actions have meaning, that their decisions can be right or wrong and good or bad. Thus, they obviously believe that life has purpose. Lacking purpose, right or wrong and good or bad are indistinguishable.

Human societies clearly define what they consider to be acceptable and unacceptable behavior, and they assign the associated consequences. It matters if some people choose to kill, steal, and rape. Such things are not consistent with the purpose of human life; they do not further human well-being and happiness. It matters if people keep their promises, if they show compassion for other people and respect for other living things. These things are consistent with the purpose of human life; they promote happiness and well-being. The common sense of ordinary people is that life has purpose and the ultimate purpose of human life is the pursuit of happiness.

If life has purpose, it might logically follow that the purposeful activity of agriculture is a legitimate human pursuit. However, the rightness or legitimacy of agriculture is not solely determined by its generic purpose, but also by *why* humans attempt to tip the ecological balance in their favor and, consequently, *how far* they are willing to tip it. Many thoughtful people, including some deep ecologists, question the rightness of agriculture, primarily because it shows an explicit preference for humans over other species. They question whether it is right and good for humans to pursue the interest of their species at the expense of other species. Since humans appear to pursue their self-interests by their very nature, either natural law or the laws of nature, the question becomes whether humans as a species, and thus human life, is inherently good or evil.

However, humans act no differently from other species with regard to self-interests, in that all species act in their individual and collective self-interest. All species, including humans, constantly degrade their natural environment by depleting resources upon which other species also must rely (Lewontin 2000). Many organisms live by consuming the dead carcasses of other organisms, and many species, including humans, do not wait for members of other species to die. So the fact that agriculture is inherently anthropocentric does not necessarily mean that agriculture is ethically wrong. It simply means that in pursuing their self-interests, humans are no different from other species.

However, there still is no consensus among ordinary people regarding the rightness of agriculture. Some people suggest that we should abandon agriculture altogether and return to hunting and gathering, to function at an equal level with other species, while larger numbers suggest that we should abandon animal agriculture, choosing a vegetarian diet to punctuate their ethical position. But, far larger numbers of people question the legitimacy of today's industrial paradigm of agriculture, which seems to show little if any respect for any other living species or even for the future of humanity.

Other species appear to be limited in their pursuit of self-interest and thus are unable to do lasting damage or eliminate other species entirely. Humans, on the other hand, clearly are capable of exploiting other species to the level of extinction and might even be capable of destroying all other life on Earth. Thus, a consensus concerning the good or evil of agriculture seems to rest ultimately with the question of how far humans are willing to tip the ecological balance.

At this philosophical juncture, the integration of ecology with agriculture becomes particularly insightful. The first principle of ecology is that *all of life is interconnected*. Evolutionary scientists, such as Alfred Wallace and Charles Darwin, first pointed the way to a new understanding of biological communities as being inherently systemic and interconnected. Ecology later emerged as a subdiscipline of biology in which species were studied within the context of their physical environment. A century later, deep ecology went further in proclaiming that not only biological communities but also all local and global communities, biological, human, nonhuman—in the past, as well as in the present—are interconnected (Engaged Buddhism, Manzanita Village 2006). Although some ecologists might disagree about the relevance of connections among past, present, and future, ecologists in general agree that all life is interconnected. Although ordinary people may disagree about the relative importance of connections, the general consensus or common sense of people seems to be in agreement with this first principle of ecology. Thus, the first principle of ecology is the second principle of agroecology: *all life is interconnected*.

The third principle of agroecology comes from both agriculture and ecology: *life is good*. If life is evil, or even neutral, neither agriculture nor ecology makes sense. It would make no sense to be concerned with the health, vitality, or survival of living communities, species, or ecosystems if the continuation of life on Earth is not inherently good. Obviously, death of individuals is an inevitable and natural aspect of life, but communities, species, and ecosystems are capable of renewal and regeneration, and thus, life is capable of sustaining life. While individuals, communities, and species of living organisms may appear to pursue their self-interests within their larger ecosystems, individuals naturally function in ways that enhance the long run sustainability of life in general. Nature, including both laws of nature and natural law, is biased in favor of life.

This natural bias is enough to convince many people that life is good. Many other thoughtful, logical, and reasonable people simply reject the assumptions of scientific materialism. They believe that people have a free will to act, that life has purpose, and life is spiritual. They believe that intellect, will, and feelings are more than material processes. They believe in an intangible, unknowable higher order of things, within which all aspects of reality, including all life, have purpose and meaning. And, they believe that life was meant to be good. Very few people believe that reality and life are inherently evil, and those who do are generally labeled as sociopaths. It does not matter whether the principle of goodness arises from the natural bias or the nature of the goodness of some higher order; both arise as matters of faith. Such is the nature of first principles; they cannot be proved, but require no proof. They exist because people know they exist. Without first principles, life simply makes no sense.

The question of the rightness or goodness of specific kinds of agriculture then can be derived from the first principles of agroecology. A purposeful agriculture that is good for all life, including life across generations, is good. An agriculture that diminishes life, including quality of life, is bad, even if its negative consequence is not purposeful. An agriculture that enhances life is right and an agriculture that diminishes life is wrong. Aldo Leopold expressed much the same conclusion when he wrote, "A thing is right when it tends to preserve the integrity, stability, and beauty of the biotic community. It is wrong when it tends otherwise" (Leopold 1966). He proposed a "land ethic" that would lead us to "examine each question in terms of what is ethically and esthetically right, as well as what is economically expedient." An agriculture that is right and good must be rooted in the common sense principles of agroecology.

The ecological, social, and economic principles of agroecology must be interpreted within the context of the first principles of agroecology: *purpose, connectedness, and goodness.* In other words, the ecology of agroecology is ecology with a purpose. The sociology of agroecology is a sociology that embraces human ecology—the connectedness of humans with the other living and nonliving elements of their natural environment. And the economics of agroecology is an economics of goodness that facilitates purposely-positive relationships among people and between people and their natural environment. The first principles of agroecology are unifying principles, which integrate ecology, sociology, and economics in the pursuit of agricultural sustainability.

4.4.1 ECOLOGICAL PRINCIPLES OF AGROECOLOGY

The fundamental principles of ecology include holism, diversity, and interdependence. An ecological whole is more than the simple sum of its parts; the relationships among those parts matter. As relationships change, either spatially or sequentially, the essence of the whole is changed. Living organisms are inherently holistic; they cannot be dissected into their individual parts or processes without destroying their essence, their life.

Diversity among distinct elements, across both space and time, is essential in sustaining all living processes. Distinct cells, organs, organisms, communities, and ecosystems are defined by selective boundaries. These boundaries, whether in cell membranes, connective tissue, skin, social relationships, or natural topography must be semipermeable or selective in nature. When this selectivity is lost, diversity disappears, and life is no longer sustainable. Biological diversity provides the potential for renewal, productivity, resistance, resilience, and regeneration, and thus for the sustainability of life.

Interdependent relationships are necessary to transform the potential of holism and diversity into positive ecological reality. Relationships among diverse elements within wholes can be independent, dependent, or interdependent. Independence implies complete isolation, which is incompatible with life. Even partial isolation limits the positive potential of relationships. Dependence relationships are inherently exploitive, as the life of the parasite is inextricably linked to the life of the host, and thus, either exploits or becomes exploited. Interdependent relationships are mutually beneficial. *Interdependent* relationships among diverse elements are necessary for renewal and regeneration of resistant, resilient, productive wholes.

The social aspects of *sustainable* agroecosystems must reflect these same ecological principles. The essence of human families, communities, and cultures must be something more than the simple collections of their individual members. The capability of any human organization depends as much on the nature of relationships among its members and on the capabilities of the individuals involved. Diversity among individuals within and among families, communities, and cultures creates the potential for the renewal and regenerations of a resistant, resilient, productive human society. Realization of this potential requires mutually beneficial relationships across selective social boundaries, relationships of choice rather than relationships of necessity in the case of human relationships. Human relationships must reflect the principles of biological communities.

The economic aspects of *sustainable* agroecosystems likewise must reflect these same ecological principles. Sustainable enterprises, entrepreneurs, and organizations must function as interrelated components of economies as wholes. The sustainable economy is far more than the simple summation of individual economic enterprises, proprietorships, and corporations. Diversity within farming systems, business organizations, and economies provides potential stability, resilience, productivity, and economic viability. However, the potential for economic sustainability can be realized only through mutually beneficial relationships among people and between people and natural resources. Economic extraction and exploitation are not sustainable.

4.4.2 Social Principles of Agroecology

The same line of reasoning is valid for the social principles of agroecology. For example, trust, kindness, and courage are basic principles of social relationships. True social principles transcend religion, philosophy, race, nationality, and culture. Different groups of people obviously have different values, but the same principles are common to all groups. The Institute for Global Ethics, for example, has conducted surveys, interviews, and focus groups with diverse groups of people around the world, asking, "What do you think are the core moral and ethical values held in the highest regard in your community?"(Kidder 2005). Responses varied widely, as would be expected, but five core values consistently ranked high in virtually every inquiry: honesty, fairness, responsibility, compassion, and respect.

The core values of honesty, fairness, and responsibility together define the social principle of trust or trustworthiness. People who are trustworthy must not only be honest and truthful, but also must be fair and equitable in their treatment of others, and must be willing to accept and fulfill their responsibilities. Violation of these core values—dishonesty, inequities, and irresponsibility—destroys trust, and threatens relationships. Trust is essential in maintaining a positive sense of social connectedness or social capital. As relationships grow in trust, they grow stronger—they build "social capital." When trust is lost, they grow weaker—social capital is depleted. When the social capital is lost, relationships are no longer sustainable.

The core values of respect and compassion define the principle of kindness. Kind people are willing and able to visualize themselves in the place of other people and, then, to treat the other person as they would have liked to be treated, if they were the other person. They are empathetic. Kindness is ultimately rooted in respect, in respecting others, as they would like to be respected by others. Kindness goes beyond being trustworthy, at times requiring people to be more than fair, less than *brutally* honest, and to do more than their share. Relationships of kindness are not exploitative or destructive; they are sustainable.

Trust and kindness are necessary but accomplish little without action. It takes courage to be trustworthy and kind. The principle of courage requires self-confidence, commitment, and discipline. It takes courage to trust other people and to stay committed to relationships through times of inevitable misunderstanding and disappointment. People must have confidence in themselves or they will not be willing to confide in others, but they cannot allow self-confidence to compromise their kindness and trust. They must have the discipline to persevere in relationships that are *right*, the courage to abandon those that have become irretrievably *wrong*, and the wisdom to know when to do which. Sustainable relationships require *moral* courage.

In agroecosystems, the principles of relationships between humans and their environment, meaning human ecology, must be derived from the principles of relationships among humans. People today hold very different values concerning the rightness or wrongness of relationships between humans and nature. Older societies, including Native American, gave a great deal of thought to such relationships and quite likely held many core values in common. Modern industrial societies, however, have abandoned these ancient values, labeling them as primitive superstitions. Today, many people see nature as purely material, a realm over which humans have absolute dominion and the right to do whatever they choose. Others see nature as an inviolable sacred trust, a realm into which humans have no right to intrude. Society is left with no common values or principles to guide human relationships with the other living and nonliving things of nature.

However, laws of nature and natural laws cannot be in conflict with each other. Thus, if relationships of humans with nature are consistent with the principles of human relationships, they cannot be in conflict with the unknown principles governing human relationships with nature. Alternatively, if human relationships with nature violate the principles of relationships among humans, they cannot be consistent with the unknown principles of ecological relationships. So, the principles of right relationships of humans with their natural environment may be derived from the principles of right relationships among people. Thus, relationships between people and *nature* should reflect trust, kindness, and courage with respect to the effects those relationships may have on the natural environment of other *people*, including people of both present and future generations.

For example, people who degrade the land, deplete nonrenewable resources, pollute the natural environment, destroy biological diversity, or simply ignore the needs of future generations are not acting with trust and kindness toward other people. They are not being fair, responsible, compassionate, respectful, or even honest, in their relationships with other people, including people of past and future generations.

When people exceed the natural regenerative capacity of nature, they invariable diminish the quality of life, and may even threaten the life of other people. On the other hand, when they respect the ability of natural ecosystems to assimilate and recycle wastes, they rarely, if ever, create health or environmental risks either for themselves or for others. The carrying capacity of Earth is limited. When people ignore this fact, they are violating one or more ecological principles that are consistent with the social principles of agroecology. Finally, people must find the courage to act on their convictions; crimes against nature are essentially crimes against other human beings, and must be treated as such. Those who show no respect for the things of nature, show no respect for other people, for humanity, or for life.

In agroecosystems, economic relationships must reflect the same social principles of trust, kindness, and courage. Economic relationships in capitalistic market economies must be strictly impartial, meaning strictly impersonal, in order for economies to function efficiently. Thus, those who manage sustainable agroecosystems must have the courage to challenge the conventional wisdom of neoclassical capitalism. Their economic relationships must be trusting relationships—honest, fair, and responsible—going beyond minimum legal requirements. They must act with kindness—treating the less informed and less economically astute, as they would like to be treated, if they were less informed and less astute. They must be willing to cooperate and to share in the costs, benefits, and responsibilities of joint economic endeavors, rather than rely on competitiveness to ensure fairness and equity. Those who manage sustainable agroecosystems must realize that economic viability must be built on trust, kindness, and courage.

4.4.3 ECONOMIC PRINCIPLES OF AGROECOLOGY

Finally, the fundamental principles of economics also must be reflected in sustainable agroecosystems. An economy actually produces nothing; all economic capital either is extracted from nature or is provided by humans, arising from either natural or social capital. Economies simply facilitate individual, material relationships among people, and between people and their natural environment. No economy would be necessary if people were self-sufficient, deriving their total well-being from nature, or could barter with each other directly to meet their needs. But, the potential material gains from specialization and impersonal trade are important to the well-being of society. Thus, the costs of self-sufficiency and barter are unnecessarily high. Specialization and trade need not be extractive and exploitative. In addition, relationships among people and between people and their environment inevitably reflect the basic principles of economics, even in the most primitive of barter economies.

The fundamental principles of economics include value, productivity, and sovereignty. Economic value is determined by scarcity, the quantity of something available relative to how much of something else people are willing and able to give up to get it. Economic value differs from intrinsic value in that the economy may not place much value on things of great intrinsic value, such as air and water. Most people can get all of the air and water they want without having to give up anything else to get them, thus they have little economic value. Clean air and clean water take on economic value only when pollution or depletion makes them scarce. Scarcity exists only in situations where we have to make choices among alternatives. Money is a common measure of scarcity, because money can be traded for many things. If we choose to trade or spend our money for one thing, we cannot spend it for another. So if we can get all we want of something without spending money for it, it is not scarce and, thus, has no economic value. As more of a thing is made available, it becomes less scarce, and the value of each additional unit diminishes. Thus, only those things that are scarce have economic value, and the greater the scarcity, the greater the value.

Productivity may be defined as the creation of value. Productivity arises from the combination of different productive resources, the most basic of which are land, labor, capital, and management. Different combinations of resources can be used to achieve the same level of production and different levels of production can be achieved by varying the amount of any given resource. So production also is about choices—choosing how much of each resource to use. The productivity capacities of resources are always limited by their natural environment. Thus, as production is increased, beyond some point, each additional unit of production requires more resources. In a market economy, the *marginal* costs of production rise, as production is increased.

As production of a thing is increased, the costs of producing an additional unit rise and the value of an additional unit (its scarcity) declines. Eventually production reaches a point where additional cost to the producer is just equal to the additional value to the buyer. This is how market value is determined, by the economic laws of supply and demand.

The economic principle of sovereignty receives less attention than the principles of value and productivity, but is no less important. Without sovereignty, without the freedom to choose, a market economy cannot function for even the individual, material benefit of people. Buyers must be free to choose. They must have adequate information about alternative choices, and be free from coercion or persuasion. Producers must be free to choose. They must have access to markets, without unreasonable requirements for entry or intimidation from competitors. When choices are restricted, when people are not free to choose, market economies simply do not function for even the individual, material well-being of society.

In sustainable agroecosystems, economic principles must be expressed in ways consistent with the ecological and social principles of agroecosystems. Economic relationships that are extractive of nature or exploitative of people are simply not sustainable, as indicated previously. The ecological and social principles of agroecosystems, likewise, must function in ways that reflect fundamental economic principles.

For example, the ecological value of specific kinds of ecosystems becomes more valuable in maintaining overall ecological integrity as they become more scarce, meaning as there are few if any like them left in a given region or the world. Individual species within ecosystems may also become more ecologically valuable as they become rare or endangered. As a single species becomes more dominant in a particular ecosystem, it diminishes the integrity of the ecosystem. As agroecosystems are managed more like natural ecosystems—more like hunting and gathering—beyond some point, their production of things of particular value to humans diminishes. Thus, the principles of value and productivity relate to decisions affecting nature as well as economic decisions.

Economic principles also are reflected in social relationships. Everyone needs positive relationships with other people, but some need more and closer relationships than do others. At some point, however, as the number of meaningful relationships becomes fewer, the remaining relationships become more valuable. As people establish more relationships, at some point, additional friends or acquaintances become less important. Relationships also have costs, in terms of time, energy, emotional capital, and even money. And, as relationships are added, at some point, the cost of each additional relationship increases, particularly in relation to its value. Finally, the principle of sovereignty is particularly important in social relationships. People must be sovereign, free to choose, if they are to sustain interdependent relationships of mutual benefit. Relationships formed and maintained through domination, intimidation, or coercion are not sustainable. The basic principles of economics clearly are relevant and important in sustaining social relationships.

Healthy agroecosystems are systems with integrity. Integrity may be defined as wholeness, completeness, soundness, and strength. Ecological integrity requires wholeness, diversity, and interdependence. Social integrity requires trust, kindness, and courage. Economic integrity requires value, productivity, and sovereignty. Lacking in any one of the three, the agroecosystem is not sustainable. Sustainable agroecosystems must have ecological, social, and economic integrity.

4.5 THE CHALLENGE

Agroecology integrates the principles of ecological, social, and economic aspects of reality for the explicit purpose of sustaining the inherent goodness of all life. Agroecology is firmly rooted in first principles: *life has purpose, life is interconnected*, and *life is good*. For some people, the common sense of rightness in integrating agriculture and ecology to develop sustainable agroecosystems may be sufficient. Others obviously have not yet accepted the necessity for agroecology or for agricultural sustainability.

Perhaps better ways can be found for defining or explaining these first principles of agroecology. Certainly much still remains to be done in exploring the implications of unifying the core principles of ecology, sociology, and economics through agroecology. The principles of sociology and economics have much to contribute in managing natural ecosystems. Ecological and social principles provide an ethical and just context for sustainable economic decisions. And communities and societies are stronger when they are built on the principles of ecology and economics. Rethinking the first principles of agroecology seems the logical place to start.

Perhaps if agroecologists can find the courage to venture outside of their narrow disciplinary boundaries, they will be able to better understand and communicate the importance of agroecology not only to agriculture but also to human society. The fundamental principles of ecology, sociology, and economics are all pretty basic, and are clearly within the intellectual grasp of competent scientists in any of the three disciplines. One need not claim to be an economist to deal effectively with basic economic principles, nor do the basic principles of ecology or sociology need be the exclusive realm of ecologists and sociologists. The true value of agroecology can be realized only when the basic principles of the three disciplines are integrated for the purpose of developing a sustainable agriculture. If agricultural scientists can find the courage to function as true agroecologists—guided by the principles of agroecology—perhaps, their numbers and their contributions to society will be far greater in the future than in the past.

REFERENCES

Altieri, M. 1995. Agroecology in action. University of California, Berkeley. http://www.cnr.berkeley.edu/~agroeco3/principles_and_strategies.html (accessed June 16, 2008).

Baird, R. M. 2008. Plato. Microsoft Encarta online encyclopedia. http://encarta.msn.com/text_761568769__1/Plato.html (accessed June 16, 2008).

Clayton, E. 2006. Aristotle (384–322 BCE): Politics. *Internet Encyclopedia of Philosophy*. http://www.iep.utm.edu/a/aris-pol.htm#H5 (accessed June 16, 2008).

Devall, W. and Sessions, G. 1985. *Deep Ecology, Living as If Nature Mattered*. Peregrine Smith Books, Salt Lake City, 63–77.

Einstein, A. 1936. Quoted by E. T. Bell in *Mathematics, Queen, and Servant of the Sciences*. Math Association of America, Washington, D.C.

Elliott, H. 1972. Materialism. In *Readings in Philosophy*, Randall, Jr., J. H., Buchler, J., and Shirk, E., Eds. Harper & Row, New York, 307–310.

Gliessman, S. 1998–2008. Agroecology Research Group. http://www.agroecology.org/ (accessed June 16, 2008).

Green, B. 2004. *The Fabric of the Cosmos: Space, Time, and the Texture of Reality*. Random House, New York, 79.

Hamilton, W. 1829. *Essays in Edinburgh Review*. Edinburgh Review, Edinburgh, 32.

Huxley, T. 1869 [1967]. *On a Piece of Chalk*. Scribners, New York.

Huxley, T. 2004. On the study of biology: 1876 lecture. In *Science and Education, Essays by Thomas Huxley*. The Project Gutenberg eBook History of Science and Education [online]. http://www.gutenberg.org/dirs/etext04/8sced10h.htm#X (accessed June 16, 2008).

Ikerd, J. 2005. *Sustainable Capitalism: A Matter of Common Sense*. Kumarian Press, Bloomfield, CT.

Kidder, R. 2005. *Moral Courage: Taking Action When Your Values Are Put to the Test.* Harper Collins, New York.

Leopold, A. 1966. *A Sand County Almanac.* Ballantine Books, New York, 262.

Lewontin, R. 2000. *The Triple Helix: Gene, Organism, and Environment.* Harvard University Press, Cambridge, MA, 55.

Lowrance, R., Stinner, B. R., and House, G. J., Eds. 1984. *Agricultural Ecosystems: Unifying Concepts.* Wiley, New York.

Manzanita Village. 2006. Engaged Buddhism. *Ecology and Deep Ecology.* http://www.manzanitavillage.org/retreats/fr_engaged_buddhism.html (accessed June 16, 2008).

Naess, A. 1988. Identification as a source of deep ecological attitudes. In *Deep Ecology,* Tobias, M., Ed., Avant Book, San Marcos, TX, 256.

Ohio State University. 2006. Ben Stinner Endowment for Healthy Agroecosystems and Sustainable Communities. http://www.oardc.ohio-state.edu/entomology/news.asp?strID=504 (accessed June 16, 2008).

Reid, T. 1863. *Works of Thomas Reid.* Hamilton, W., Ed. Thoemmes Continuum Press, Bristol, 422.

Swartz, N. 2006. Laws of nature. *Internet Encyclopedia of Philosophy.* http://www.iep.utm.edu/l/lawofnat.htm (accessed June 16, 2008).

5 Potential for a New Generation of Biodiversity in Agroecosystems*

Frederick L. Kirschenmann

CONTENTS

5.1 INTRODUCTION

The roots of modern industrial agriculture are embedded in the historic publication of Justus von Liebig's *Chemistry in Its Applications to Agriculture and Physiology* (1843). Von Liebig argued that we could sustain agricultural productivity without the complexities of mixed farming practices and the laborious task of manuring soils. All we had to do was to substitute synthetic inputs for these labor-intensive practices. Such synthetic inputs, he argued, could achieve the same results much more effectively, substantially simplify farming operations, and make farming more efficient and productive.

Liebig's assertion has been proved true. The ability to substitute synthetic fertilizers for nutrient cycling enabled farmers to specialize in the production of a handful of commodities, abandon most mixed farming practices, and dramatically reduce labor input. In a single paragraph, David Keller and Charlie Brummer (2002) succinctly describe the kind of modern farming operations that have resulted from Liebig's vision.

> Modern agriculture has become highly industrialized in order to reliably produce the largest amount of plant and animal product possible while minimizing labor inputs. Through the incorporation of numerous components manufactured externally to the farm, including fertilizers, pesticides, and technology, the modern system manipulates the land to make it amenable to industrial processes. Typically, crops are produced as large-hectarage monocultures consisting of a single genotype planted across an entire field. Most farms using modern agriculture methods cultivate only a few crops grown in simple rotations such as wheat–fallow or maize–soybean. Similarly, most animals are grown in feedlots or climate-controlled buildings in order to closely monitor feed efficiency and to guarantee uniform meat, egg, or milk products. Cycling nutrients is not a major consideration of most industrial agricultural systems because the addition of externally derived fertilizers is cheaper and simpler than collecting, storing, and using manure.

---

* Parts of this chapter appeared in a paper in *Agronomy Journal* 99, 373–376, 2007.

From the perspective of producing maximum quantities of food and fiber with the least amount of labor, this industrialized system has been spectacularly successful. The system, however, appears increasingly vulnerable as we enter the twenty-first century. Many of the assumptions—long taken for granted—that bolster the foundations of industrial agriculture are now being challenged.

In the industrial system it is generally assumed that:

- Production efficiency can best be achieved through specialization, simplification, and concentration.
- Therapeutic intervention is the most effective way to control undesirable events.
- Technological innovation will always be able to overcome production challenges.
- Control management is the most effective way to achieve production results.
- Cheap energy to fuel this energy-intensive system will always be available.

But in the early twenty-first century most, if not all, of these assumptions are open to question.

5.2   CHALLENGES FACING INDUSTRIAL AGRICULTURE AS WE ENTER THE TWENTY-FIRST CENTURY

The world is experiencing a major energy transformation which is bound to have a profound effect on our industrialized farming systems. At the same time that the global demand for fossil fuels is skyrocketing, the global production capacity of oil and natural gas either has peaked or will do so shortly (Heinberg 2003). Oil and natural gas constitute two-thirds of our hydrocarbon-based economy and provide almost *all* of the energy used on industrial farms. Fertilizers, pesticides, farm equipment, traction fuel, and irrigation, which constitute the very core of all industrialized farming systems, are derived almost entirely from fossil fuels. Can Liebig's paradigm for agriculture still be maintained once cheap fossil fuels are no longer available?

Even without any other challenges, our new energy future may force industrial agriculture, as well as most of the rest of our economy, to change rather swiftly and significantly. As Paul Roberts (2004) puts it, "the real question, for anyone truly concerned about our future, is not *whether* change is going to come, but whether the shift will be peaceful and orderly or chaotic and violent because we waited too long to begin planning for it."

In addition to the energy transition, there are numerous other challenges that will force agriculture to change. Among them are ecological degradation (much of it caused by industrial agricultural practices), climate change, water depletion, and the loss of both genetic diversity and biodiversity. The deteriorating condition of the ecosystem services on which agriculture is heavily dependent was described succinctly in the United Nations Millennium Ecosystem Assessment Synthesis Report (2005).

Produced by 1360 leading scientists from 95 countries, the report detailed some disturbing conclusions about the state of our global ecological resources. The report found that over the last half-century, humans have polluted or overexploited two-thirds of Earth's ecological systems on which life depends, dramatically increasing the potential for unprecedented and abrupt ecological collapses. The report determined that most of these ecosystem damages were the direct or indirect result of changes made to meet rising demands for ecosystem services—in particular the growing demands for food, water, timber, fiber, and fuel. In other words, the industrialization of agriculture which enabled us to dramatically increase production during the past half-century also is a principal cause of the ecological degradation that now threatens our ability to *maintain* productivity.

Climate change likely will be a third driver forcing agriculture to restructure in the decades ahead. Climate change is, of course, caused in part by greenhouse gas emissions, some of which are generated by our industrial agriculture system. But even without human-induced changes, Earth's climate has varied dramatically during its long history. As Stephen Schneider (1976) noted several decades ago, while favorable, stable climate played at least as big a role as technology in producing consistently high crop yields in the past few decades, such favorable climate conditions are not the norm. A sustainable production system probably has to anticipate climate change in any event, but the climate variability we are likely to experience in the immediate future may have significant impacts on how agriculture functions.

Of course, as Cynthia Rosenzweig and Daniel Hillel (1995) point out, there are significant uncertainties involved in making exact climate change projections and, therefore, the impact climate change will have on agriculture and the food supply cannot be predicted precisely. The uncertainties are related to the degree of temperature change, concomitant changes that will likely take place in precipitation patterns, the response of crops to enriched carbon dioxide in the atmosphere, and our inability to anticipate the evolution of complex natural systems like agriculture. But as Rosenzweig et al. (2001) argue, we need to understand what is at stake and "prepare for change wisely."

Despite the uncertainties, some of the short- and long-term consequences of climate change can reasonably be anticipated. In the short term (2020 to 2080), we can expect greater climate fluctuations, especially greater "extremes of precipitation, both droughts and floods" (Rosenzweig et al. 2001). And given that industrial agriculture relies on highly specialized production systems, it requires climate conditions that consistently remain hospitable to monocultures. When 92 percent of Iowa's cultivated land is planted to just two crops—corn and soybeans—climate conditions that are consistently favorable to corn and soybean production will be crucial for maintaining productivity. As climate becomes more unstable, such specialized systems likely will become increasingly vulnerable to climate fluctuations. Additionally, the genetic uniformity that is so indicative of modern industrial agriculture renders it especially vulnerable to climate change (Rosenzweig et al. 2001).

The eventual consequences of climate change could be grim. A few policy makers and agriculturalists still dismiss these dire long-term projections as alarmist. But, the overwhelming majority of climatologists seem to have reached a consensus. Jim Hansen's review of four prominent climate studies in the July 13, 2006, *New York Review of Books* provides some context for understanding the gravity of the situation. Hansen suggests, "If human beings follow a business-as-usual course … the eventual effects on climate and life may be comparable to those at the time of mass extinctions. Life will survive, but it will do so on a transformed planet."

In his 2005 book, *The Weather Makers*, Tim Flannery similarly points out that social and economic chaos likely will result from the effects of climate change. Flannery, a mammalogist, concludes that the speed at which animals and plants need to migrate to remain in suitable thermoclimes (and the speed at which they have to migrate is accelerating) makes it impossible for them to move fast enough to stay ahead of the changing climate to survive. Given the interdependence of species, such species losses likely will cause severe devastation to the biodiversity of the planet—a biodiversity on which agriculture ultimately depends.

How does agriculture operate in a world when significant biodiversity has been eviscerated? Will we continue to have pollinators? We just have no good idea how this unraveling will play out. Can we reasonably continue to assume that we can sustain our agricultural productivity under these changing conditions simply by inventing a few more new technologies?

Additionally, cataclysmic transformations initiated by climate change also may produce general chaos and mayhem, which in turn may have an impact on agriculture. In his "Sustainable Developments" column in *Scientific American,* Jeffrey Sachs provides compelling evidence that the social and political chaos in Darfur likely is attributable to global warming.

Another resource vital to agricultural productivity is water. The effect that irrigation-dependent industrial agriculture has had on our planetary water supply is another indicator that the industrial agriculture system is not sustainable. Lester Brown (2006) points out that while we each drink 4 liters of water daily, modern industrial production systems use 2000 liters of water to produce each of our daily food requirements. In all, 70 percent of all fresh water use today is attributable to agricultural irrigation. We are now using twice as much water for agricultural irrigation as we did in the 1960s, and have been drawing down our freshwater resources at an unsustainable rate.

Water depletion is particularly worrisome in China where 80 percent of grain production is dependent on irrigation, and in India, where 60 percent of grain production relies on irrigation. And, according to Lester Brown (2006), aquifers in some parts of China are dropping at a rate of 10 feet per year and in India at 20 feet per year. In China, some farmers already are pumping from a depth of 1000 feet, and in India from a depth of 3000 feet. China and India are, of course, the two most populous nations in the world.

The Ogallala Aquifer in the central United States, where one-fifth of grain production is dependent on irrigation, has dropped by as much as 100 feet in some places. On the outer edge of the aquifer, some farmers already have had to abandon irrigation due to water depletion.

Many of these environmental challenges are interconnected. For example, in an effort to respond to the need for alternative energy, corn-based ethanol plants are springing up all over the Midwest. Ethanol plants require considerable amounts of water which will further exacerbate water depletion in aquifers like the Ogallala. Longer and more frequent drought periods, spurred by climate change, will encourage more farmers to install center-pivot irrigation systems, especially as land values increase, stimulated by higher corn prices due to rising demands for corn to supply the ethanol plants. Farmers cannot afford crop failures due to drought when they are paying $300 an acre to rent the land. Installing irrigation systems as a hedge against water shortages in drought years will be a rational response.

The 2005 U.N. Millennium Ecosystem Assessment Synthesis Report also highlights a fifth development that will present an additional challenge for industrial agriculture. The loss of both species and genetic diversity has severely damaged ecosystem *resilience*—the level of disturbance that an ecosystem can undergo without crossing a threshold to a different kind of structure or functioning—making it extremely difficult to *restore* ecological health. So, not only have we degraded the productive capacity of the planet, we also have undermined the planet's capacity for self-renewal and self-regulation. At the same time that we are experiencing ecological degradation, we also have diminished nature's capacity to restore ecological resilience.

The combination and interconnected aspects of these challenges likely will force us to abandon the highly specialized, monoculture, industrialized agriculture that has dominated the landscape for the past half-century. The question is, what can take its place?

5.3   A COMPREHENSIVE SYSTEMS APPROACH

As we attempt to imagine a new postindustrial paradigm for agriculture, it will be important to design an alternative using a comprehensive systems approach. Failure to use a systems approach may end up fostering as many unintended consequences as the industrial paradigm. Employing a comprehensive systems approach is a matter of great urgency since we are already aggressively embracing "solutions" to the challenges ahead without paying much attention to comprehensive systems analysis.

The current euphoria associated with the mandate to switch to biofuels to "wean ourselves from Mideast oil" and create "unprecedented opportunities for agriculture" serves as a prime example. A realistic appraisal of energy efficiency ratios (how many kilocalories of energy it takes to make a kilocalorie of energy available in the form of an alternative fuel) has been noticeably absent from the public debate. The potential ripple effects associated with devoting a significant percentage of our land resources to producing energy instead of food, feed, and fiber have not been fully explored. The numerous potentially damaging ecological impacts that may result from higher corn prices, stemming from increased demand for corn to fuel the many new ethanol plants developed to produce alternative energy, have hardly been explored. Farmers likely will switch to continuous corn production, planting fence row to fence row, using maximum nitrogen inputs to ensure the highest possible yields. And who could blame them, given that the higher corn prices also dramatically increase land values and land rental rates?

As Joseph Fiksel (2006) reminds us, "industrial, social and ecological systems are closely linked" and given the dynamic character of all systems, "there is an urgent need for a better understanding of the dynamic, adaptive behavior of complex systems and their resilience in the face of disruptions."

Such comprehensive systems approaches must go beyond "industrial ecology" which is "directed largely at reducing environmental 'burdens' measured in terms of resource consumption and waste emissions." We must focus, instead, on "strengthening sustainability's systemic underpinnings." In other words, we must "design systems with inherent resilience by taking advantage of fundamental properties such as *diversity, efficiency, adaptability,* and *cohesion*" (Fiksel 2006).

But how do we design new agricultural systems based on these more comprehensive systems analyses? The answer may lie in our ability to design farming systems that are less energy intensive, more resilient in the face of unstable climate conditions, and have the potential to out-produce monocultures by virtue of multispecies output. If such farming systems can be created, simple economic advantage may encourage the development of more complex farming systems that substitute "biological synergies"—based on biodiversity—for energy inputs.

5.4 LOOKING TO THE FUTURE: BIODIVERSITY-BASED BIOLOGICAL SYNERGIES

5.4.1 How Shall We Proceed?

First, ecologists and social scientists point out that "adaptive management," especially when emergent properties are involved as they always are in nature and therefore in agriculture, is far more reliable than "control management." Control management, which is part and parcel of industrial agriculture, operates under the assumption that constancy is the rule. But, as C. S. Holling (1995) reminds us, "Assumptions that such constancy is the rule might give a comfortable sense of certainty, but it is spurious. Such assumptions produce policies and science that contribute to a pathology of rigid and unseeing institutions, increasingly vulnerable natural systems and public dependencies."

Second, it perhaps will be necessary to recognize that *the industrial paradigm no longer* works. Fortunately, this awareness already has developed in many sectors of our society, including the science of agriculture. Given the problems of resistance, environmental and human health consequences, and the high cost of industrial pest control, pest management specialists already have begun to see the need for a paradigm shift.

In a landmark essay, Joe Lewis (1997) and his colleagues cogently outline the need for a "paradigm shift" in pest management, while alluding to the fact that the same paradigm shift also is needed in other sectors of our society.

> The basic principle for managing undesired variables in agricultural systems is similar to that for other systems, including the human body and social systems. On the surface, it would seem that an optimal corrective action for an undesired entity is to apply a direct counter force against it. However, there is a long history of experiences in medicine and social science where such interventionist actions never produce sustainable desired effects. Rather, the attempted solution becomes the problem.

> Application of external corrective actions into a system can be effective only for short-term relief. Long-term sustainable solutions must be achieved through "restructuring the system...." The foundation for pest management in agricultural systems should be an understanding and shoring up of the full composite of inherent plant defenses, plant mixtures, soil, natural enemies, and other components of the system.... The use of pesticides and other "treat-the-symptoms" approaches are unsustainable and should be the last rather than the first line of defense (Lewis et al. 1997).

The business world also seems to be recognizing that a paradigm shift is inevitable. John Thackara, business design specialist and director of Doors of Perception, a design futures network based in Amsterdam, suggests that the industrial economy is essentially over and that most business leaders recognize it. The industrial economy, he says, is simply too exploitive and too heavy on the planet to serve the long-term interests of the human community. The new paradigm will be based less on "stuff" and more on "people" as he puts it, and he envisions a world in which food and agriculture systems will be decentralized and based more on biological synergies and less on the industrial economy. "A host of ecological problems in the area of agriculture derive from the fact that the rhythms of nature are displaced by the demands of a higher-speed economy" (Thackara 2006).

5.4.2 New Paradigm Models

To the industrial mind-set, all this may seem like so much theory without any practical application. But a few creative farmers already have designed new complex farming systems based on biological synergies and adaptive management and are demonstrating incredible efficiencies and economic performance. Takao Furuno's duck/fish/rice/fruit farm in Japan serves as a prime example of such productivity and efficiency. He now produces duck meat, duck eggs, fish meat, fruit, and rice—all without any exogenous inputs—in a highly synergistic system of production located on the same acreage where he previously grew only rice. And, in this new production system, his rice yields have increased up to 50 percent over the yields from his former high-input, industrial, monocrop rice farm. His new farm, he writes, is based on the concept of producing "a variety of products within a limited space to achieve maximum overall productivity" by introducing multiple species into the same environment in ways that allow "all components to influence each other positively in a relationship of symbiotic production" (Furuno 2001). Such complex, synergistic systems are proving to be much more productive than monocropping systems, while using far fewer energy-intensive and potentially environmentally damaging inputs.

Many other examples can be cited. Joel Salatin, designer and operator of Polyface Farms near Swoope, Virginia, has developed a rotational grazing production system featuring pastures that contain at least 40 varieties of plants and support numerous animal species. Both plants and animals are linked in a symbiotic set of relationships which allow each species to make a contribution to the vitality and resilience of the system. Consequently, Salatin uses very little fossil fuel on his farm. Yet his 140-acre farm is very productive. It annually produces 30,000 dozen eggs, 10,000 to 12,000 broilers, 100 beef animals, 250 hogs, 800 turkeys, and 600 rabbits (Purdom 2005).

George Boody (2005) and his colleagues have calculated, on a watershed basis, that diverse, synergistic farms can be profitable and simultaneously benefit the environment. Their study demonstrates the transformative value when farms are converted from corn/soybean monocultures to more diverse operations consisting of five crops and including rotational grazing, riparian buffers, and so forth. Net farm income can increase by as much as 108 percent (despite capital costs to purchase animals and install fencing to implement the new biodiverse system), in addition to offering significant environmental and social benefits.

Additional peer-reviewed research seems to corroborate these findings. Research conducted on rice farms in Indonesia confirms Takao Furuno's on-farm experience in Japan. While the research investigated simple rice/fish cultures, as compared with the more complex rice/fish/duck/fruit system on Furuno's farm, the experiment indicated comparable results. According to the study total gross revenues in the rice/fish culture increased by 42 percent compared with rice monocultures and net revenue increased by 47 to 66 percent. The study also concluded that such biodiverse systems "could pave the way to an ecology-sound rice farming due to the reduced or zero use of pesticides" (Dwiyana and Mendoza 2006).

Another study has confirmed the benefits of biodiverse ally cropping in Africa. Ally cropping is a system of farming in which fast-growing leguminous trees are planted in rows with crops planted between the tree rows. Researchers at the International Institute of Tropical Agriculture in Nigeria found that crop yields increased significantly with ally cropping while maintaining soil fertility. Additionally, when farmers fed the prunings from the trees to cattle, milk production and weight gain increased significantly (Ogunlana et al. 2006).

Although these examples may not represent a universally applicable way to transition main-stream agriculture to a new paradigm based on biological synergies, they do represent "working models" that can be used to guide the *research necessary* to scale up new farming systems based on "biodiversity" and "resilience."

As we enter the twenty-first century, mainstream agriculture faces many challenges which may propel agriculture in these new directions. As fossil fuels are depleted, the ratio of energy produced to energy required to produce it continues to diminish, making that source of energy increasingly costly. Agriculture will have to find *alternative energy sources* to sustain its productivity. Agroecologists increasingly are convinced that the most viable alternative technologies will spring from the *"biological synergies" inherent in multispecies systems,* and that additional research might make such systems the next new information technology.

Masae Shiomi and Hiroshi Koizumi (2001) make a strong case for exactly such a transformation in postmodern agriculture. I believe they raise one of the most important questions facing agriculture today: "Is it possible to replace current technologies based on fossil energy with proper interactions operating between crops/livestock and other organisms to enhance agricultural production? If the answer is yes, then modern agriculture, which uses only the simplest biotic responses, can be transformed into an alternative system of agriculture, in which the use of complex biotic interactions becomes the key technology." Farmers like Takao Furuno and Joel Salatin already have answered that question in the affirmative.

It would appear that these new farms of the future will operate on the basis of at least eight principles which are almost diametrically opposed to the assumptions industrial agriculture has taken for granted. Postmodern farms will likely:

1. Be energy conserving
2. Feature both biological and genetic diversity
3. Be largely self-regulating and self-renewing
4. Be knowledge intensive
5. Operate on biological synergies
6. Employ adaptive management
7. Feature ecological restoration rather than choosing between extraction and preservation
8. Achieve optimum productivity by featuring nutrient density and multiproduct, synergistic production on limited acreage

5.4.3 What Else Is Needed?

Naturally, some modification in farm policy could help to move us toward the new paradigm. Most of today's farm policies are designed to *subsidize monoculture production.* A modest shift in farm policy that would encourage transitioning to these new synergistic systems might generate data that would make the transition more attractive. Designating a modest percentage of current research funding to further explore the potential of biological synergies in agriculture in various *watersheds* would, I believe, reveal many additional models of farming in "nature's image." Farming systems based on such "biomimicry" (Benyus 1997) could dramatically reduce dependence on fossil fuel inputs, restore ecological capital, put more diverse, resilient production systems on the landscape, and perhaps make a modest contribution to the reduction of greenhouse gases into the atmosphere.

In addition to policy shifts, we may need to develop a new ethic for agriculture that motivates us to move in these new, ecologically sound directions. Industrialization not only shaped our economy, it also molded our culture and a *cultural transformation* also may be required. In today's industrial culture humans tend to see themselves as the conquerors of nature. Our role is to "bend nature to our will" as Francis Bacon put it at the dawn of the industrial era. A new paradigm cannot emerge in the wake of that ethic. Fortunately an alternative ethic has been proposed. According to Stan Rowe (2002), a Canadian ecologist, we need to fill in two missing values to complete the kind of new ecological paradigm that our future survival depends on.

The missing *concept* is the ecological one of *landscapes-as-ecosystems,* literally "home systems," within which organisms, including people exist. We have been taught that we are separate living *things*, but not so. The realities of the world are ecological systems of which organisms are components and without which no creatures of any kind could exist.

The missing *attitude* is sympathy with and care for the land and water ecosystems that support life. It will come when we make the concept of a planetary home part of our daily thought, part of our hearts and imaginations.

REFERENCES

Benyus, J. M. 1997. *Biomimicry*. William Morrow, New York.

Boody, G. et al. 2005. Multifunctional agriculture in the United States. *BioScience*, 55(1), 33–34.

Brown, L. R. 2006. *Plan B 2.0*. W.W. Norton, New York, 42–44.

Dwiyana, E. and Mendoza, T. C. 2006. Comparative productivity, profitability, and efficiency of rice monoculture and rice-fish culture systems. *Journal of Sustainable Agriculture*, 29(1), 145–166.

Fiksel, J. 2006. Sustainability and resilience: Toward a systems approach. *Sustainability: Science, Practice & Policy*, e-Journal for Sustainability, 2(2), (Fall), 1, 3. http://ejournal.nbii.org/progress/2006fall/0608-028.fiksel.html (Accessed June 5 2008).

Flannery, T. 2005. *The Weather Makers: How Man Is Changing the Climate and What It Means for Life on Earth*. Atlantic Monthly Press, New York.

Furuno, T. 2001. *The Power of Duck*. Tagari Publications, Sisters Creek, Tasmania, Australia, 73.

Hansen, J. 2006. The threat to the planet. *The New York Review of Books*, July 13, 12–16.

Heinberg, R. 2003. *The Party's Over: Oil, War, and the Fate of Industrial Societies*. New Society Publishers, Gabriola Island, BC.

Holling, C. S. 1995. What barriers? What bridges? In *Barriers and Bridges to the Renewal of Ecosystems and Institutions,* Gunderson, L. H., Holling, C. S., and Light, S. S., Eds. Columbia University Press, New York, 34.

Keller, D. R. and Brummer, E. C. 2002. Putting food production in context: toward a post mechanistic agricultural ethic. *BioScience*, 52, 264–271.

Lewis, W. J. et al. 1997. A total system approach to sustainable pest management. In *Proceedings of National Academy of Sciences*, 94, November, 12243–12248.

Liebig, J. 1843. *Chemistry in Its Applications to Agriculture and Physiology*. Taylor and Walton, London.

Ogunlana, E. A., Vilas, S., and Tanghild, L. 2006. Alley farming: a sustainable technology for crops and livestock production. *Journal of Sustainable Agriculture*, 29(1), 131–144.

Purdum, T. S. 2005. High priest of the pasture. *New York Times Style Magazine*, May 1, 76–79.

Roberts, P. 2004. *The End of Oil*. Houghton Mifflin, Boston, 14.

Rosenzweig, C. and Hillel, D. 1995. Potential impacts of climate change on agriculture and food supply. *Consequences,* 1(2), (Summer). http://www.gcrio.org/CONSEQUENCES/summer95/agriculture.html (accessed June 5, 2008).

Rosenzweig, C. et al. 2001. Climate change and extreme weather events: Implications for food production, plant diseases, and pests. *Global Change and Human Health*, 2(2), 90, 100–101.

Rowe, S. 2002. *Home Place: Essays on Ecology*. NeWest Press, Edmonton, 23–24.

Sachs, J. D. 2006. Ecology and political upheaval. *Scientific American*. July, 37. http://www.sciam.com/article.cfm?id=ecology-and-political-uph (accessed June 5, 2008).

Schneider, S. H. 1976. *The Genesis Strategy: Climate and Global Survival*. Plenum Press, New York, 103–112.

Shiyomi, M. and Koizumi, H., Eds. 2001. *Structure and Function in Agroecosystem Design and Management*. CRC Press, Boca Raton, FL.

Thackara, J. 2006. *In the Bubble: Designing in a Complex World*. MIT Press, Cambridge, MA, 32.

United Nations. 2005. *Millennium ecosystem assessment synthesis report* (March). http://www.millenniumassessment.org/en/index.aspx (accessed on June 5, 2008).

6 The Necessity and Possibility of an Agriculture Where Nature Is the Measure[*]

Wes Jackson, Stan Cox, Lee DeHaan, Jerry Glover, David Van Tassel, and Cindy Cox

CONTENTS

6.1 INTRODUCTION

To avoid being convicted of the charge of "grandiose talk" when a major effort or agenda is promoted, one should be required to state both the necessity of the grand scheme and the possibility of achieving it.

6.2 THE NECESSITY

What follows are two of the numerous testaments from knowledgeable and serious students of the impacts of agriculture on the ecosystems of our ecosphere submitted here as evidence of the "necessity" for a new agriculture. The first is from the Millennium Ecosystem Report.

Over the past 50 years, humans have changed ecosystems more rapidly and extensively than in any comparable time in human history, largely to meet rapidly growing demands for food, fresh water, timber, fiber, and fuel.

The changes that have been made to ecosystems have contributed to substantial net gains in human well-being and economic development, but these gains have been achieved at growing costs in the form of degradation of many ecosystem services, increased risks of nonlinear changes, and the exacerbation of poverty for some groups of people. These problems, unless addressed, will substantially diminish the benefits that future generations obtain from ecosystems.

The degradation of ecosystem services could grow significantly worse during the first half of this century and is a barrier to achieving the Millennium Development Goals (MEA 2005).

---

[*] Copyright 2009, Land Institute, Salina, Kansas.

A second testament which speaks to the *necessity* of an agriculture which would feature nature as the measure was published in 2002 (Wackernagel et al. 2002). This time the authors were bent on tracking the ecological overshoot of the human economy. Mathis Wackernagel and ten other authors delivered this summary:

> *Sustainability requires living within the regenerative capacity of the biosphere.* In an attempt to measure the extent to which humanity satisfies this requirement, we use existing data to translate human demand on the environment into the area required for the production of food and other goods, together with the absorption of wastes. *Our accounts indicate that human demand may well have exceeded the biosphere's regenerative capacity since the 1980's.* According to this preliminary and exploratory assessment, humanity's load corresponded to 70% of the global biosphere in *1961*, and grew to *120% in 1999* (Wackernagel et al. 2002).

These authors explicitly state, "Agriculture is the single largest threat to biodiversity and ecosystem function of any single human activity." Let's not argue whether it is rapid climate change or agriculture for even if agriculture is number two, it has already been shown to be a serious threat to the planet.

In writing the Millennium Ecosystem Assessment report, the editor, Stanford ecologist Harold Mooney and others warned, "We are confronting a human dilemma: food production for a rising population will come at the expense of conservation or we will have conservation at the expense of production." It is an either/or assumption and, given the nature of agriculture, it is a fair one. Agriculture is responsible for 70 percent of U.S. water contamination, making 40 percent of our waters unfit for swimming and fishing. There are essentially no pesticide-free zones. Pesticides are present in nearly every water and fish sample for agricultural areas.

- Soil degradation on a global basis has been increasing, little wonder, considering that the increase in the use of nitrogen, phosphorus, and water has been rising exponentially since 1960 (Tilman 1999).
- As a consequence of nitrogen fertilizer, dead zones are on the increase around the world leading to oxygen depletion.

And it is the annual monocultures largely responsible.

A further consequence of this "business as usual" approach will be global agricultural expansion with the following realities necessary to meet that expansion.

- 18 percent increase in cropland, primarily into less resilient soils
- 300 percent increase in fertilizer
- 75 percent increase in pesticide production (Tilman 1999)

It seems safe to assume that the global implications of the "business as usual" approach will lead to the following consequences:

- 2.4- to 2.7-fold increase in eutrophication
- Increased greenhouse gas emissions
- Further soil degradation
- Loss of biodiversity
- Loss of critical ecosystem services: water and nutrient cycling, biocontrol, pollination (Tilman, 1999)

The necessity for a change in course should be apparent. The first order of business should be to have a sense of the size of the planet's agricultural land and the crops it supports worldwide and the food demands of people who eat. To that end, Figure 6.1 shows the total world cropland per person to be 0.62 acres. The total U.S. cropland per person amounts to 1.09 acres. Given that the concerns described by the Millennium Ecosystem Assessment group and the study by Wackernagel et al. are

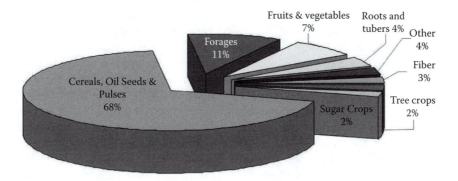

FIGURE 6.1 Global cropland distribution in 2000. (Data from Monfreda, C. et al., *Global Biogeochemical Cycles*, 22, GB1022, 2008.)

TABLE 6.1
Soil Carbon Sequestration in Annual and Perennial Cropping Systems

	Soil Carbon Sequestration (kg ha^{-1} yr^{-1})	Global Warming Potential (kg CO_2 equivalents per year)	Climate Change Impact on Yield (Mg ha^{-1})
Annual crops	<0 to 450	140 to 1140	−1.5 to −0.5
Perennial crops	320 to 1100	−1050 to −200	+5

Source: Summarized from Cox, T. S. et al. *BioScience*, 56, 650–659, 2006.

real, it is our view that the "problem of agriculture" centers on the nearly 70 percent of total acreage of the planet which features monocultures, realizing the consequences of more mouths to feed now and the need for future people even with a reduced population.

When one looks to the soils as a carbon sink, it is clear that annual crops actually contribute to global warming while perennial crops would have the potential to mitigate it (Table 6.1).

6.3 THE ANCIENT CHALLENGES FOR AGRICULTURAL SOCIETIES

The two oldest problems for farmers have been (1) how to obtain an adequate, if not bountiful harvest, every year, and (2) how to assure that future adequate or bountiful harvests are not jeopardized by what we do each year. Simple questions are often generic questions or have corollaries. The first question, therefore, requires farmers to explore ways to maximize sun-sponsored fertility through ample supplies of moisture and protect the crop from insects, pathogens, and weeds. The second question causes us to ask how to minimize soil erosion and avoid chemical contamination of our land and water.

Anyone who died by 1930 never lived through a doubling of the human population. Nor is any person born in 2050 or later likely to live through a doubling of the human population (Cohen, 2005). But since our numbers are still growing and more food is being produced now than ever before, it seems to appear to some that we will continue to solve the food production problem. "But for how long?" the sensible person asks and then adds, "and then what?" Especially since agriculture unintentionally not only worsens the global ecological crisis, but it is heavily reliant on nonrenewable resources. The thoughtful may appreciate this reality, but what is seldom appreciated is that these two faces of agriculture—productivity and destructiveness—did not arise recently, nor did they arise from the conscious decisions of companies that sell fuel, fertilizer, or pesticide, or of farmers, government officials, or grocery shoppers. Productivity and some form of ecological destructiveness are inherent in the way humans have practiced agriculture from the early beginnings. Chemicals and other nonrenewable resources are recent additions to an ancient problem. For that reason we emphasize the "problem of agriculture," rather than "problems in agriculture."

The last sentence above is especially true when considering annual monocultures for grain. Agriculture's foundation of annual plants—grains and legumes—supplies more than two-thirds of human food demands that are grown *from* seed every year and harvested *for* their seed for food and for replanting. This ancient reality requires compromising the soil resource, either by the ancient practice of tilling or by chemical treatment in our industrial time. Tillage can be done without causing great harm when it is on a very small scale, when the "eyes to acres ratio" is favorable. That is not the usual case for most acreage. Nearly everywhere, civilizations that have practiced tillage beyond the level of the kitchen garden have suffered, often catastrophically, from soil erosion.

Measures to prevent soil loss through soil husbandry have been described by F. H. King's *Farmers of Forty Centuries* in the Orient (King 1911). Over the course of those 4000 years, the great works Oriental societies employed to ensure fertility have been staggering. Such practices are no longer the rule, especially on a global scale. We now have satellite images of our planet showing vast swaths of entire continents scoured of their deep-rooted, year-round perennial vegetation, leaving the soil uncovered for months at a time, susceptible to erosion from wind and water. Even during the growing season when the landscape is green, shallow-rooted annual crops fail to manage water and nutrients as did their perennial predecessors. The destruction of deep, massive perennial root systems through tillage has wrecked vast underground ecosystems, subtracting from the soil much of what gives it structure, the below surface ecosystem with diverse living masses of microbes and invertebrates interacting with the nonorganic materials.

6.4   THE POSSIBILITY

Today, we have the scientific knowledge, data, and techniques—fruits of civilization made possible by agriculture—to largely correct the wrong turn our species took. What we have in mind and are at work on here at the Land Institute will not be perfect. Even if we wanted to, we could not return to the crossroads where our ancestors took that wrong turn, or to a Golden Age of folk agriculture that never existed. We can do something far better than what we are now doing through a wholly new way of farming. We can accomplish something never before done in the history of our species. We can make conservation a consequence of, instead of an alternative to, food production. To do so will require a conceptual revolution. Such a conceptual change will require that some scientists begin to envision an agriculture in which the ecological processes embodied within wild biodiversity are brought to the farm, rather than forcing agriculture to relentlessly nick away wild ecosystems.

We are all aware of the historical movements that have most universally been called conceptual revolutions in science. They all contribute to an increased understanding "how the world works" or, if one prefers, "how the world is."

Let us go down the list of conceptual revolutions: the sun-centered Copernican system replacing Ptolemy's Earth-centered theory; Newton's discovery uniting celestial and Earth-bound physics; the oxygen theory of Lavoisier replacing the phlogiston theory; Darwin's theory of evolution replacing the divine creation of species; Einstein's theory of relativity and quantum theory; and most recently the theory of plate tectonics explaining continental drift. The cosmologists, with the aid of the Hubble telescope and other instruments, are providing insights into some of the dynamics in the heavens about which we had no clue less than a decade ago.

The conceptual change we are urging is, we believe, more important than all the others, because it involves our future ability to produce food. It would amount to a marriage of Darwinian ecology/evolutionary biology with agriculture using the ecosystem as the conceptual tool.

To embrace the ecosystem as the conceptual tool requires that we feature perennial mixtures of plants on the landscape. The Land Institute has been working on this idea with the expectation that humans can make conservation a consequence of production—in any region on the planet suitable

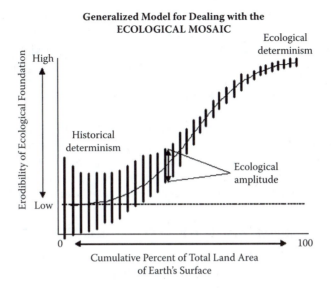

FIGURE 6.2 Ecological mosaic graph.

for agriculture—if we use as our standard the ecosystems that existed in that region before it was managed by humans for food and fiber. We start with the question: "What was here?" We then ask: "What will be required of us here?" Followed by: "What will nature help us to do here?" Some landscapes are more forgiving than others. What we do will amount to a conversation with local nature. Figure 6.2 illustrates the nature of the mosaic's distribution.

Chris Field, member of the National Academy of Sciences at the Carnegie Institute of Washington and Stanford University, has shown that natural ecosystems (and on land, that almost always means mixtures of perennial species) do better than human-managed landscapes in converting sunlight into living tissue (Field 2001). The plants that anchor nature's ecosystems have extensive, long-lived root systems with diverse architectures; they have a longer growing season; and their species diversity protects against epidemics and the vagaries of weather. As a result, such ecosystems can produce, year in and year out, more biomass per acre than nearly all agricultural systems and they do so without requiring a subsidy of fossil fuels and other inputs and without degrading soil and water. And we might add that they are water cleansing.

The Land Institute's mission stops neither at the farm gate nor at the prairie boundaries of the Appalachians, the Rio Grande, or the Rockies. We believe that food worldwide can, indeed must, come to be produced by agroecosystems as close to the efficiency and resilience of those natural ecosystems that were replaced by farms, forest plantations, and fisheries as our nonrenewable energy use.

What's on the line is the necessity to end the spending of Earth's ecological capital. Because we have to eat, we are required to manage landscapes, which makes agriculture the best positioned to take the lead toward this reconciliation with nature. In spite of its current industrial emphasis, agriculture still has the twin disciplines of ecology and evolutionary biology standing in the background, more silently to some than others. The industrial sector has no time-honored discipline to draw on, making these twin disciplines central precisely because ecological processes as properties of ecosystems have long track records of success in building and conserving soil, holding and filtering water, and supporting wildlife diversity. As future agricultural researchers and farmers acknowledge this reality, they can begin to encourage these processes through management. A leap in our collective imagination is necessary about possibilities informed by the rules of a living ecosphere, rather than by reductive machine-like thinking. Nature's economies are real economies. They feature material recycling all the while running on contemporary sunlight and represent baseline reality. As we build agricultural systems designed to square with that baseline reality, we are more likely to miss the mark than hit it. But in so doing we have a chance to mark the distance between nature's standard and our approximation of it. Agriculture is the only cultural product available for measuring our success. It is this pairing of nature and agriculture, regarded as a unit, which is available as an analog or as a reference for those who seek to bring the materials or industrial sector generally away from exploitation of the ecosphere's capital. Capitalism is an economic system designed to dismember and use Earth's capital stock, whether it is after minerals, including fuels, or ecosystems such as prairie or rainforests.

The split between humans and nature began with agriculture. It is fitting that the healing of that split begins with it. By relying on the sciences of ecology and evolutionary biology to help us produce food in properly functioning ecosystems, we are meeting, arguably, the most important of our ancient three basic needs. All visions of a sustainable society rely on renewable resources. Agriculture, broadly defined, is the only artifact of civilization where that potential resides.

6.5   THE ANNUAL REALITY AND THE PERENNIAL OPPORTUNITY

It is in the Great Plains and the Midwest of the United States where we find the largest expanse of the best soils in the world, where we find the top land grant institutions and the most agricultural scientists. Paradoxically, it is also here that we have an agriculture with widespread soil erosion, widespread use of toxic chemicals rendering water unfit to drink, and with nitrogen runoff so serious there is a New Jersey size "dead" zone, hundreds of miles downstream in the Gulf of Mexico.

Mountains of evidence show that reestablishing perennial vegetation across the region would solve most of the ecological problems. Knowing that we humans obtain 70% of our total calories from grains and oilseed crops, knowing that none of them is perennial, and knowing that existing perennial species can produce only a small fraction of the total calories required for direct consumption by a growing human population, it is our mission at the Land Institute to do something about that.

Researchers and farmers are using the only perennial plants available to them. They add perennials devoted to hay and pasture, plant trees and grass along rivers and streams to soak up the contaminants that hemorrhage from cropland; and take more land out of grain production altogether, under the Conservation Reserve Program. None of this puts bread on the table, but annual grains do. Consequently, from an ecological point of view, grain cropping in the interest of human life is a dangerous activity. With no perennial grains on the roster of food plants, farmers have no choice.

6.6   PERENNIAL GRAINS RESEARCH

We at the Land Institute are devoted to the ideal that farms one day need not be ecological sacrifice zones, but, rather, farms that provide our basic calories while protecting soils, water, and biodiversity. The missing link is the perennial grains. That is only half the vision. As these perennials are being developed, plant breeders, agroecologists, and farmers will be working out strategies for growing them in mixtures, to recapture the ecological soundness of preagricultural landscapes.

The genetic raw material is around, ready to be put to use. Plants now in field plots and on greenhouse benches at the Land Institute form the foundation of breeding programs that will, given decades of work, turn out perennial grain crops. It is happening elsewhere as well. Here at the Land Institute most of the current genetic and breeding effort is going into the following species and species hybrids:

> **Wheat** can be hybridized with several different perennial species to produce viable, fertile offspring. We have produced thousands of such plants. Many rounds of crossing, testing, and selection will produce perennial wheat varieties for use on the farm.
> **Intermediate wheatgrass** (*Thinopyrum intermedium*) is one of those perennial relatives of wheat. It is also a potential grain crop on its own. Using parental strains from the Rodale

Institute, U.S. Department of Agriculture (USDA), and other sources, we have established genetically diverse populations that are now in a second cycle of selection for high grain yield and croplike traits. Even at this early stage, large increases in seed size and productivity have been realized.

Grain sorghum is a drought-hardy feed grain in North America and a staple human food in Asia and Africa, where it provides reliable harvests in places where hunger is always a threat. It can be hybridized with perennial species *Sorghum halepense*. We have produced large plant populations from hundreds of such hybrids and have selected perennial strains with seed size and grain yields up to 50 percent those of annual grain sorghum.

Illinois bundleflower (*Desmanthus illinoiensis*) is a native prairie legume that fixes atmospheric nitrogen and produces abundant protein-rich seed. It is one of our strongest candidates for domestication as a crop. We have assembled a large collection of seed from a wide geographical area and have a breeding program.

Sunflower is another annual crop we have hybridized with perennial species in its genus, including *Helianthus maximiliani*, *H. rigidus*, and *H. tuberosus* (commonly known as Jerusalem artichoke). Breeding work has turned out strongly perennial plants. Genetic stabilization will improve their seed production.

There is potential for many more perennial grain species, including **rosinseed maize, eastern gamagrass, rice, chickpea, millets, flax,** and a range of **native plants**. We are studying these and other species and will initiate breeding programs as the scientific staff expands.

6.7   ECOLOGICAL RESEARCH

We need not wait until perennial grain crops are fully developed to begin studying the ecological context in which they will grow. We have established long-term ecological plots of close analogs in which to compare methods of perennial crop management. These perennial-grain prototypes, including intermediate wheatgrass and bundleflower, are allowing us to initiate long-term ecological/production research in these plots. Eventually, true perennial grain crops will succeed them. Additionally, ongoing studies of natural ecosystems, such as tallgrass prairie, provide insight into the functioning of natural plant communities.

6.8   THE ROAD AHEAD

At the Land Institute we have laid out a route to follow in breeding perennial grains and developing the agroecosystems in which they will grow. To foster research on perennial grains across the nation and planet, we will develop and freely distribute germplasm—seed of perennials and hybrids that other plant breeders can use as parents in establishing or enhancing their own perennial grain programs, or for basic research to answer fundamental questions. It will be essential to build a body of knowledge about perennial grain systems through publication in the scientific literature.

6.9   FREQUENTLY ASKED QUESTIONS ABOUT PERENNIAL POLYCULTURE

Over the past three decades, numerous people have asked countless good questions. Some of the most frequently asked follow. Our best answer follows each one.

1. It is expected to take at least 25 years to achieve profitable, productive perennial grain crops. Isn't that too late to address the problems facing the world today?
It is likely that global agricultural acreage will expand over the next two to three decades especially if the human population increases to 8 to 10 billion people. Recent projections predict an 18 percent

or more increase in agricultural land by 2020. The best soils on the best landscapes are already being used for agriculture. Much of the future expansion of agriculture will be onto marginal lands (Class IV, V, and VI) where risk of irreversible degradation under annual grain production is high. As these areas become degraded, expensive chemical, energy, and equipment inputs will become less effective and much less affordable. New strategies are needed that emphasize efficient nutrient use in order to lower production costs and minimize negative environmental impacts. The sooner that successful alternatives are available, the more land we can save from degradation.

Today, 38 percent of global agricultural lands are currently designated as degraded, and the area is increasing. To minimize encroachment onto nonagricultural lands in the future, currently degraded lands will need to be kept in production *and* restored to higher productive potential. In regions of the world where high inputs of fertilizers, chemicals, and fuels are not an option, agricultural systems that are highly efficient, productive, and conservative of natural resources are needed—and will be needed even more 25 years from now.

2. Can we expect perennial grain crops to be as productive as annual grain crops and, if not, won't they actually worsen environmental problems by requiring more land for agricultural production?

Some considerations:

1. Grain yields, typically expressed on a mass or volume per unit area basis (kilograms/hectare or bushels/acre), seldom take into account space and time considerations beyond the immediate farm field and immediate time period. Imagine, for the purpose of illustration, an extreme situation: a new substance allows for a single record-breaking yield of 10,000 kg on a 1-ha farm field but through its impact renders an additional 999 ha nonproductive for 10 years. That single yield would typically be reported as 10,000 kg/ha rather than as 1 kg/ha.
2. Which annual yields are we using as a standard? For example, the world record wheat yield was harvested in the Palouse region of eastern Washington State where wheat yields can top 100 bushels/acre. Annual wheat production in that region, though, has resulted in extensive erosion. All of the topsoil has been lost from more than 10 percent of the region's landscapes. On eroded sites Palouse wheat yields may be less than 25 to 30 bushels/acre. Crop yields that come at such a high cost to the soil resource should not be used as a standard for comparison.
3. If the reference annual yield is attainable only with high inputs of nonrenewable resources and there are not reasonable substitutes for those inputs, then that yield level is not a suitable reference for a long-term comparison. Soil is nonrenewable and nonsubstitutable.

There is sufficient evidence that reasonable reference yields of annual crops can be matched on high-quality lands and exceeded on poor-quality lands by diverse perennial systems with fewer negative impacts.

3. But won't the seed yield of perennials always be limited by the need to save some energy for overwintering that could have been used to produce seed?

The short answer is no. The theoretical limitations to seed yield in perennials are no more serious than in annuals. In annuals, yield is limited by shorter growing seasons, water shortage due to short roots, and poor seedling establishment. In perennials, yield can be constrained by the need to overwinter, but rapid spring growth of perennials, combined with season-long access to water deep in the soil profile, means that perennials such as alfalfa are overall more productive than related annuals like soybeans. Much of the journeywork of plant breeders has been to shift the allocation of resources from leaves, stems, crowns, and roots toward seed in the development of perennial grain crops.

4. With advances in no-till production of annual grain crops, do we need perennial grain production systems to mitigate the environmental problems associated with agriculture?

Unfortunately, yes. Although no-till technology has reduced erosion in many areas, some problems remain due to the biological limitations of annual plants. Chief among the problems associated with no-till is water quality. Annual crops, even in no-till situations, are relatively inefficient in capturing nutrients and water. In the Midwest, as much as 45 percent of precipitation may be lost through the soil profile under annual cropping. Rates of water loss through profiles may be five times greater under annuals than under perennials and losses can be as great or greater under no-till as compared with conventional tillage systems.

Annual crop plants are often not present or are not developed enough to use water at a sufficient rate during times of precipitation. Water flowing through the soil profile transports downward soil nutrients and agrichemicals associated with poor water quality. This problem can be confounded under no-till production which often requires greater inputs of agrichemicals and fertilizers. A 2002 Environment Protection Agency (EPA) survey of the nation's water quality indicates a downward trend from the late 1990s—the problem is getting worse, not better, despite widespread adoption of no-till and minimum-till systems.

Furthermore, adoption of no-till annual cropping systems is also limited by the requirement of seeds for warm, well-drained seedbeds in order to properly germinate. Tillage remains an attractive practice in northern regions because it hastens warming and drying of the seedbed. Although advances in plant breeding may eventually allow for optimal germination in cooler, wetter conditions, the simple biological fact that plants cannot use water and nutrients when they have not yet germinated will be unavoidable.

5. If our farming systems "mimic" natural ecosystems, what level and kind of plant diversity are needed and how will they be deployed?

The answer to both parts of the question is, "It depends." It depends on the resilience and fertility of the soil resource, climate, disease pressures, and types of crops used. Because nearly all land-based ecosystems feature perennial plants grown in diverse mixtures, natural ecosystems, in general, use and manage water and nutrients most efficiently and build and maintain soils. The level and deployment of diversity varies and therefore resilience will depend on the particular crop species being considered and the characteristics of the region in which they are to be grown.

Diversity includes multiple species and genetic diversity within species. Current grain production practices commonly involve planting a single genotype (near-zero genetic diversity) across a field often larger than 100 acres. Furthermore, that single genotype and other genetically similar plants are being grown on millions of acres in a region. Increases in genetic diversity at the species, field, and landscape levels are needed. The final arrangement of that diversity will be determined by what is useful and can be practically achieved by farmers.

6. Several serious attempts have been made in the past to perennialize grain crops and we have none to date. What has changed that offers promise of success now?

A strict focus on achieving the highest possible bushels per acre yield made early perennial wheat varieties, yielding only 70 percent of their annual domesticated relatives, undesirable. In recent years, the costs to the environment incurred through annual cropping are increasingly weighed against bushels per acre yields, making some reductions in yield acceptable. Recent advances in plant breeding and computational ability and the fact that perennials have advantages over annuals (e.g., increased water and nutrient use efficiency, longer growing season) may mean yield reductions are not necessary.

In the case of wheat, most involvement with perennials had to do with bringing desirable genes from a wild perennial relative into the annual crop. The perennialization effort was carried on, more or less as a hobby, by an interested researcher but with no sustained program to guarantee continuity.

7. Since mechanical tillage and annual rotations are eliminated in perennial systems, don't the perennial plants become "sitting ducks" for pests and disease?

Here proof is in the pudding. Perennials dominate most native landscapes and constitute roughly 80 percent of North America's native flora. In other words, perennials have thrived evolutionarily despite the pressures of pests and disease.

In some fields or some regions, some perennial crops will prove to be more problematic than others and breeding for complex traits like yield and perenniality can unintentionally purge genes involved in resistance responses. There will undoubtedly be pest and disease problems. Yet pest and disease problems afflict our most productive annual crops. And there are many examples of herbaceous perennial plants—alfalfa, switchgrass, brome—being highly productive for many years despite exposure to pests or disease. Diversity (whether at the field or landscape scale or over time), field burning, and selecting for resistance in a plant breeding program are seen as essential elements of our work.

8. How do alternative methods of production such as permaculture, biointensive, or organic fit in with perennial grain crops? What about vegetables and fruits? How do CSAs (Community Supported Agriculture) fit in?

We focus on those crops occupying 80 percent of our global cropland: annual grain crops grown primarily in monocultures. Any number of approaches, alternative or conventional, could be used in managing perennial crops and distributing the harvest. Many of the environmental problems associated with large-scale production of vegetables and fruits, currently occupying a small percent of global cropland, could be greatly reduced by local production and distribution mechanisms such as CSAs, which typically focus on small-scale production of fruits and vegetables. Grain production, however, is best done on much larger scales and typically in regions with low population densities. CSAs could distribute some grain but, under practically any scenario, it would only be a very small fraction of total production—much smaller than the currently small percent of fruits and vegetables distributed through CSAs.

This is not to say, however, that efforts aimed at reducing the scale of industrial agriculture and increasing local food security are misguided. They are important and necessary to transform our food system over the long term. While promoting local, small-scale, organic agriculture we must also assess how and where the bulk of our calories can best be produced. If all or even a large portion of the calories consumed by New Yorkers came from New York State there would be very few trees left and the state's thin, poor soils would be quickly degraded. The bulk of the calories consumed by New Yorkers come directly or indirectly from grain crops which grow well in the Midwest and Great Plains states.

9. Will the public eat perennial grains?

Very likely. We see little need for people to significantly change their diets. Greatest short-term success in developing suitable perennial crops will likely come with perennializing current grain crops with which the public are already familiar. Indeed, one of the strongest arguments for perennial grains as an approach to meeting human food needs in the coming centuries is that it does not require large dietary shifts compared to some other approaches that have been suggested.

10. Supposing perennial grain crops are highly productive and become widely planted, won't they require even fewer people on the landscape and thereby worsen rural community life?

Maintaining high seed yields in these systems will require greater skills based on an understanding of the climate, soils, and productive capacity of a particular place. Some type of rotational management may be required where, for example, in one year cool-season grasses of a field are simply grazed by livestock and the warm-season plants harvested later in the season for their seeds. The following year two seed harvests might be possible. The year after that perhaps the entire field is grazed the whole season. A whole array of weed, disease, and fertility challenges will require well-informed, skilled management. Knowing when to graze, burn, harvest, or fallow and knowing when to monitor for pests, disease, and weeds and what to do about them will require more attention than current agricultural production. A sufficiency of people will substitute for a sufficiency of capital in a world of resource scarcity. But it need not be thistles and thorns and the usual "sweat of the brow" described by the Genesis myth-makers. Their psychology will be more like that of a nineteenth-century British naturalist than the modern day grain producing farmer.

Economically, these systems will provide a greater percentage of returns to the farmer and, ideally, to the landscape rather than to the suppliers of inputs. Roughly 90 percent of current annual farm revenue goes off-farm, requiring increasingly larger farms to generate enough income for a family. With a greater percentage of returns going to the farmer, fewer acres will be necessary to comfortably support a family.

11. Finally, how are you going to harvest a perennial grain polyculture?

This question arose so frequently over the years that we finally decided to plant a four species mix consisting of two warm season grasses, corn and sorghum, one legume, soybean, and a member of the sunflower family, the annual sunflower itself. No cool season grasses, the fourth major group of the prairie, were not represented, partly because on the prairie in Kansas it is the smallest represented as a functional group, but mostly because it would not set seed at the same time as the other three species. (Other arrangements involving wheat, the number two crop of the planet in acreage, will have to be made, perhaps involving grazing and burning.)

The seeds were planted with an air drill in a mixture. At harvest time, the air was cut on the combine and the concave opened up. Progress through the field was slow as this instant granola was harvested. Seeds were later separated with a seed cleaner.

We reasoned that given this success with equipment already on the inventory of mechanical equipment that little fine-tuning would be necessary once polycultures were at the stage of being farmer ready. Mechanical harvesting is likely in most places around the world and harvest strategies will depend on planting arrangements and specific crop characteristics. The larger problems are agronomic, not engineering.

REFERENCES

Cohen, J. E. 2005. Human population grows up. *Scientific American,* 293, 48–55.

Cox, T. S. et al. 2006. Prospects for developing perennial grain crops. *BioScience,* 56, 650–659.

Field, C. B. 2001. Sharing the garden. *Science,* 294, 2490–2491.

King, F. H. 1911. *Farmers of Forty Centuries.* Rodale, Emmaus, PA.

MEA (Millennium Ecosystem Assessment). 2005. Millennium Ecosystem Assessment Synthesis Report.

Monfreda, C., Ramankutty, N., and Foley, J. A. 2008. Farming the planet: 2. Geographic distribution of crop areas, yields, physiological types, and net primary production in the year 2000. *Global Biogeochemical Cycles,* 22, GB1022.

Tilman, D. 1999. Global environmental impacts of agricultural expansion: The need for sustainable and efficient practices. *Proceedings of the National Academy of Sciences,* 96, 5995–6000.

Wackernagel, M. et al. 2002. Tracking the ecological overshoot of the human economy. *Proceedings of the National Academy of Sciences,* 99, 9266–9271.

7 Energy and Human Population Growth
The Role of Agriculture

David Pimentel

CONTENTS

7.1 INTRODUCTION

Humans use energy from many sources to grow food, provide shelter, and maintain health. The energy source—be it the sun, animal power, or fossil fuels—influences human activities and agriculture. Currently, ample fossil energy has enabled humans to provide food and services for an ever-increasing global human population. However, the present world population of 6.5 billion is projected to double in about 50 years. The globe faces a major challenge to feed the growing world population with 3.7 billion humans already reported by the World Health Organization to be malnourished.

The consequence of feeding increasing numbers of people and encouraging population growth is a per capita decline in cropland, irrigation, and fertilizer resources. According to the Food and Agricultural Organization of the United Nations, per capita grain production worldwide has been declining continuously for more than 20 years, despite biotechnology; these grains provide more than 80 percent of the world's food.

Humans, energy, and agriculture have always been interdependent. Just as other animals depend on plants, early humans depended on plants for their foods as well as on the animals they hunted (Pimentel and Pimentel 2008). Humans use energy today from many sources to grow food, provide shelter, maintain health, and improve their quality of life. The energy sources—be it the sun, animal power, or fossil fuels—and their relative abundance influence human activities today as in the past (Pimentel and Pimentel 2008).

As societal groupings have changed, so has energy use and sources. Early humans who hunted and gathered their food in the wild depended primarily on their own energies. Although much of the world's population today relies on fossil fuels, many people in developing countries continue to use the energy provided by animal power, human power, firewood, and crop residues for fuel.

During the twentieth and early twenty-first century, ample fossil energy supplies have enabled humans to provide food and services for the ever-increasing global population, with a quarter of a million more people added each day. The present population of 6.7 billion is projected to double in just 58 years (PRB 2005). With this forecast, pressure on food and fuel supplies intensifies.

Agricultural production depends on fertile land, fresh water, and fossil energy. Currently more than 99 percent of the world food comes from the land, while less than 1 percent comes from the oceans and other aquatic ecosystems (FAOSTAT 2006). As the world population continues to increase, limited cropland and freshwater resources have to be divided among more people. This is in part related to the serious malnutrition problem that exists today—according to the World Health Organization, there are more than 3.7 billion humans who are malnourished (WHO 2004). This is the largest number in history.

7.2   ENERGY AND EARLY SOCIETIES

Hunter–gathering societies were relatively small, rarely having more than 500 individuals, and were relatively simple. Because securing food and shelter consumed so much time and energy, other activities scarcely existed. With the development of agriculture about 10,000 years ago, more dependable food supplies were available and some surplus energy was available (Pimentel and Pimentel 2008). As the stability of the food supply increased, societies that had once been seminomadic and followed their food supply, now gained in security and permanence.

Humans gained control of fire about a half million years ago. Fire enabled hunter–gatherers to ward off large animal predators and helped them clear vegetation for the planting of crops (Pimentel and Pimentel 2008). This simple procedure also helped eliminate weeds that competed with their crops. In addition, fires made it possible to cook foods, often making them better tasting, easier to eat, and easier to digest. Perhaps more importantly, cooking reduced the danger of illness from parasites and disease microbes that often contaminated raw foods (Pimentel and Pimentel 1996).

7.3   CURRENT ENERGY USE

The U.S. economy, agriculture, and quality of life are highly dependent on the availability of immense amounts of energy (102 quads per year; USBC 2004–2005). Each year individual Americans use nearly 11,000 liters (2,910 gallons or 101 million kcal) of oil equivalents to maintain their residence, to power their personal transportation, and to indirectly support the nation's food system and other industries. Although Americans make up about 4 percent of the world's population, the United States consumes 25 percent of the world's fossil energy (Dunn 2001). With the U.S. population growing annually by 3.3 million people and expected to double within the next 70 years, meeting the nation's energy demands will become increasingly difficult (USBC 2004–2005).

Already the United States has consumed nearly 90 percent of its proven oil reserves (API 1999) and currently imports 65 percent of its oil (USBC 2007). Petroleum geologists warn that, at the current consumption rate, the world has about 40 to 50 years of world oil supply remaining, approximately the same quantity of natural gas, and anywhere from 50 to 100 years of coal (Youngquist 1997; Youngquist and Duncan 2003; Kerr 2005). We must conclude from these facts that the country's excessive rate of energy consumption simply cannot be continued long into the future.

Forests with ample moisture are the most productive natural ecosystems. In general, natural forests have produced an average of 3 t/ha of biomass per year (Pimentel et al. 2002). This is about maximum, with nitrogen nutrients as the prime limiting factor for natural forests. The annual net primary productivity of U.S. agricultural systems is about 5 t/ha per year with fertilizer nutrients. This figure is higher than natural forests and other natural biomass systems because crops are grown in the best soils with ample water and provided with ample fertilizer nutrients. For example, corn grown under favorable conditions will produce about 9 t/ha of grain and the grain plus the stover total 18 t/ha per year (USDA 2004). Converted into heat energy, this totals about 66×10^6 kcal per hectare. This represents about 0.5 percent of the solar energy reaching 1 ha during the year, a relatively high rate of conversion for crops and natural vegetation. Most crops and natural vegetation under favorable conditions collect about 0.1 percent of solar energy per year (Pimentel and Pimentel 1996).

The U.S. food system requires a massive expenditure of fossil energy; 10 kcal of fossil energy is required for every kilocalorie of food consumed (Pimentel and Pimentel 2008). The food supply (production and imports) provides each American with about 975 kg (2146 lbs) of food annually, or about 3800 kcal per person per day (USBC, 2004–2005; USDA, 2004). Producing, processing, packaging, and distributing food consumes about 1900 liters of oil equivalents per person per year, or a total of approximately 19 quads per year.

The use of fossil energy in U.S. and world agriculture has revolutionized crop and livestock production. Where farmers in the 1940s relied heavily on horses and oxen, today they depend on tractors. Instead of requiring 2 ha to produce 1 ha of corn, we raise corn on both hectares today. In early agricultural practice, 1 ha had to be planted in a legume, like vetch, to collect solar energy and convert the solar energy into nitrogen for the corn crop planted the following year. Also, with commercial nitrogen, we can now double the nitrogen available for the corn crop, plus add ample amounts of phosphorus, potassium, and calcium fertilizers.

The other major change category in crop production has been genetic alteration. For example, hybrid corn helped increase yields more than 25 percent. Of course, hybrid corn production requires energy to produce the hybrid corn seed.

In early agriculture, only about 6 inputs were required to produce the crop, whereas today at least 14 different inputs are essential (Table 7.1). An average investment of about 8.1 million kcal is required to produce a hectare of corn that will provide a yield of about 9 t/ha. In 1940, the yield was only 1.9 t/ha (USDA 1940).

For the past 25 years, some economists and others have been suggesting that the production of ethanol from corn can provide the nation with its supply of liquid fuels and make the United States oil independent (Pimentel and Patzek 2005). By omitting many of the energy inputs and giving excessive credit for the dried distillers grain, a by-product of ethanol production, some investigators claim a net positive energy return (Farrell et al. 2006). However, if all the energy inputs are assessed, then there is a net energy loss of 30 to 40 percent (Pimentel and Patzek 2005; Pimentel et al. 2006) (Table 7.2).

Currently, the United States is producing 6.0 billion gallons of ethanol per year (Kansas Ethanol 2006). This represents about 1 percent of total U.S. petroleum use per year and is using 18 percent of U.S. corn production. If 100 percent of U.S. corn were used, it would provide only 6 percent of current U.S. petroleum use. Statements made by the pro-ethanol lobby suggest that ethanol from corn is going to make the United States oil independent. Clearly, this is an erroneous statement! Concerning the 6.0 billion gallons of ethanol produced, this was assumed to be net positive yield. It is not. More than 5.5 billion gallons of oil equivalents were required to produce the corn and process the corn into ethanol. Thus, from 1.3 to 1.4 gallons of oil equivalents are required to produce 1 gallon of ethanol (see Tables 7.1 and 7.2).

The environmental impacts include increased global warming, soil erosion, freshwater use, large quantities of pesticide, and nitrogen fertilizer applied. The increased global warming is related to two aspects: (1) More fossil fuel is consumed in producing ethanol than is produced as ethanol, and this contributes to global warming, and (2) the "bugs" or yeast organisms when consuming the corn grain and producing ethanol also produce enormous amounts of carbon dioxide; in fact, about one-third of the corn goes off as carbon dioxide during the fermentation process.

Corn production causes more soil erosion in the United States than any crop grown (NAS 2003). Already topsoil in the United States is being lost 10 times faster than is sustainable (Pimentel 2006). More than 1700 gallons of water are required to produce 1 gallon of ethanol. This includes the water required to produce the corn grain, plus the water used in the fermentation and distillation processes (Pimentel et al. 2006). Corn production uses more nitrogen fertilizer than any crop grown, and about 25 percent of the nitrogen applied leaches into ground and surface waters (NAS 2003). The nitrogen applied to corn is the prime cause of the dead zone in the Gulf of Mexico (NAS 2003).

TABLE 7.1
Energy Inputs and Costs of Corn Production per Hectare in the United States

Inputs	Quantity	kcal × 1000	Costs $
Labor	11.4 hr	462[a]	148.20[b]
Machinery	18 kg[c]	333	68.00
Diesel	88 L	1003	34.76
Gasoline	40 L[d]	405	20.80
Nitrogen	155 kg	2480	85.25
Phosphorus	79 kg	328	48.98
Potassium	84 kg	274	26.04
Lime	1,120 kg	315	19.80
Seeds	21 kg	520	74.81
Irrigation	8.1 cm	320	123.00
Herbicides	6.2 kg	620	124.00
Insecticides	2.8 kg	280[e]	56.00
Electricity	13.2 kWh	34[f]	0.92
Transport	204 kg[g]	169	61.20
Total	7,543		$891.76
Corn yield 8,781 kg/ha[h]		31,612	kcal input:output 1:4.19

Source: Pimentel, D. et al. 2007. *Reviews of Environmental Contamination and Toxicology,* 189, 25–41, 2007. With permission.

[a] It is assumed that a person works 2000 hours/year and utilizes an average of 8000 liters of oil equivalents per year.

[b] It is assumed that labor is paid $13 an hour.

[c] Energy costs for farm machinery that was obtained from agricultural engineers—tractors, harvesters, plows, and other equipment that last about 10 years and are used on 160 ha/year. These data were prorated per year per hectare (Pimentel and Patzek 2005).

[d] Estimated.

[e] Input 100,000 kcal/kg of herbicide and insecticide.

[f] Input 860 kcal/kWh and requires 3 kWh thermal energy to produce 1 kWh electricity.

[g] Goods transported include machinery, fuels, and seeds that were shipped an estimated 1000 km.

[h] Corn yield averaged over 3 years.

Corn production requires the application of more insecticide and more herbicide than any other crop grown in the United States (Pimentel et al. 2006). These pesticides, plus fungicides used on corn and other crops, are reported to cause more than 300,000 human nonfatal pesticide poisonings each year (USGAO 1992). These poisonings plus other environmental impacts are reported to cause more than $12 billion in damages each year (Pimentel and Pimentel 1996).

7.4   WORLD POPULATION GROWTH

The current world population of more than 6.7 billion doubled during the last 45 years. Based on its present growth rate of 1.2 percent each year, the world population is projected to double again within a mere 58 years (PRB 2005).

TABLE 7.2
Inputs per 1000 Liters of 99.5% Ethanol Produced from Corn[a]

Inputs	Quantity	kcal × 1000	$
Corn grain	2,690 kg[b]	2,314[b]	273.62
Corn transport	2,690 kg[b]	322[c]	21.40
Water	15,000 L[d]	90	21.16
Stainless steel	3 kg[e]	165	10.60
Steel	4 kg[e]	92	10.60
Cement	8 kg[e]	384	10.60
Steam	2,546,000 kcal	2,546	21.16[f]
Electricity	392 kWh	1,011	27.44
95% ethanol to 99.5%	9 kcal/L[g]	9	0.60
Sewage effluent	20 kg BOD	69	6.00
Distribution	331 kcal/L	331	20.00
Total		7,333	$423.18

Source: Pimentel, D. et al. 2007. *Reviews of Environmental Contamination and Toxicology,* 189, 25–41, 2007. With permission.

[a] Output: 1 liter of ethanol = 5,130 kcal.
[b] Data from Table 7.1.
[c] Calculated for 144 km roundtrip.
[d] 15 L of water mixed with each kilogram of grain.
[e] Estimated.
[f] Calculated based on the price of natural gas.
[g] 95% ethanol converted to 99.5% ethanol for addition to gasoline (T. Patzek, personal communication, University of California, Berkeley, 2004).

Many countries and world regions have populations that are rapidly expanding. For example, China's present population of 1.4 billion, despite the governmental policy of permitting only one child per couple, is still growing at an annual rate of 0.6 percent (PRB 2005). But China, recognizing its serious overpopulation problem, has recently passed legislation that strengthens its one child per couple policy (China 2002). However, because of its young age structure, the Chinese population will continue to increase for another 50 years. India, with nearly 1.1 billion people, living on approximately one-third the land of either of the United States or China, has a current population growth rate of 1.7 percent. This translates to a doubling time of 41 years (PRB 2005). Taken together, the populations China and India constitute more than one-third of the total world population. Given the steady decline in per capita resources, it is unlikely that India, China, or the total world population will double again.

Also, despite the AIDS outbreak, the populations of most of African countries also are expanding. For example, Chad and Ethiopia populations have high rates of increase and are projected to double in 21 and 23 years, respectively (PRB 2005).

The U.S. population also is growing rapidly; it currently stands at about 300 million and has doubled during the past 60 years. Based on its current growth rate of about 1.1 percent, it is projected to double to 600 million in less than 70 years (USBC 2004–2005). It is interesting to note that the U.S. population is growing at a per capita rate nearly double that of China (PRB 2005).

A major obstacle in limiting human population growth is the very young age structure of the current world populations and the population momentum fostered by that pattern. With the population age range of 15 to 40 most prevalent, reproductive rates are high (PRB 2005). Even if all the people in the world adopted a policy of only two children per couple, it would take approximately 70 years before the world population would finally stabilize at approximately 13 billion, which is twice the current level (Weeks 1986; Population Action International 1993). As the world and U.S. populations continue to expand, all vital natural resources will have to be divided among increasing numbers of people and per capita availability will decline to low levels. When this occurs, the maintenance of prosperity, a quality life, and personal freedoms will be imperiled.

7.5   MALNOURISHMENT IN THE WORLD

The present world hunger and shortages of nutrients for many humans alert us to the present serious problem concerning the world food supply and its impact on human health. The report of the Food and Agricultural Organization (FAO) of the United Nations confirms that food per capita has been declining since 1984, based on available cereal grains (FAOSTAT 1961–2004). This is alarming news because cereal grains make up about 80 percent of the world's food supply. Although grain yields per hectare in both developed and developing countries are still increasing, the rate of increase is slowing, while the world population and its food needs escalate (FAOSTAT 1961–2004; PRB 2005). Specifically, from 1950 to 1980 U.S. grain yields increased at about 3 percent per year, but since 1980 the annual rate of increase for corn and other major other grains is only about 1 percent (USDA 1980–2004).

According to the World Health Organization, more than 3.7 billion people are malnourished (WHO 2004). This is the largest number and proportion of malnourished people ever reported! The World Health Organization, in assessing malnutrition, includes deficiencies of calories, protein, iron, iodine, and vitamins A, B, C, and D in its evaluation (Sommer and West 1996; Tomashek et al. 2001).

7.6   AGRICULTURAL PRODUCTION

Most food, estimated to be more than 99 percent, comes from agriculture, while less than 1 percent comes from the oceans and other aquatic ecosystems (FAOSTAT 2006). Cropland is most important now and will become more important in the future, as the world population increases, and fish production declines due to overfishing and pollution.

7.6.1   LAND RESOURCES

Worldwide, food and fiber crops are grown on 11 percent of Earth's total land area of 13 billion ha. Globally, the annual loss of land to urbanization and highways ranges from 10 to 35 million ha (approximately 1 percent) per year, with half of this lost land coming from cropland (Doeoes 1994). Most of the remaining land area (23 percent) is unsuitable for crops, pasture, and forests because the soil is too infertile or shallow to support plant growth, or the climate and land are too cold, dry, steep, stony, or wet (Buringh 1989).

In 1960, when the world population numbered about 3 billion, approximately 0.5 ha of cropland was available per capita worldwide. This half hectare of cropland per capita is needed to provide a diverse, healthy, nutritious diet of plant and animal products—similar to the typical diet in the United States and Europe (Lal 1989; Giampietro and Pimentel 1994). The average per capita world cropland now is only 0.23 ha, or less than half the amount needed according to industrial nation standards (see Table 7.1). This shortage of productive cropland is one underlying cause of the current food shortages and poverty in the world (Leach 1995; Pimentel and Pimentel 2008). For example, in China, the amount of available cropland is only 0.08 ha per capita, and rapidly declining due to continued population growth and extreme land degradation (Leach 1995). This minute amount of arable land forces the Chinese people to consume primarily a vegetarian diet (see Table 7.2).

Currently, a total of 1481 kg/year per capita of agricultural products is produced to feed Americans, while the Chinese food supply averages 785 kg/year per capita (see Table 7.2). By all measurements, the Chinese have reached or exceeded the limits of their agricultural system (Brown 1997). Their reliance on large inputs of fossil fuel–based fertilizers—as well as other limited inputs—to compensate for shortages of arable land and severely eroded soils indicates severe problems for the future (Pimentel and Wen 2004). The Chinese already import large amounts of grain from the United States and other nations, and are planning to increase these imports in the future (Alexandratos 1995).

Escalating land degradation threatens most cropland and pastureland throughout the world (Lal and Pierce 1991; Pimentel 2006). The major types of degradation include water and wind erosion, and the salinization and waterlogging of irrigated soils (Kendall and Pimentel 1994). Worldwide, more than 10 million ha of productive arable land are severely degraded and abandoned each year (Houghton 1994; Pimentel 2006). In addition, approximately 10 million ha of cropland are abandoned each year because of salinization (Thomas and Middleton 1993). Moreover, an additional 5 million ha of new land must be put into production each year to feed the nearly 90 million humans annually added to the world population. Most of the 25 million ha needed yearly to replace lost cropland are coming from the world's forests (Houghton 1994; WRI 1996). The urgent need for more agricultural land accounts for more than 60 percent of the deforestation now occurring worldwide (Myers 1990).

Agricultural erosion by wind and water is the most serious cause of soil loss and degradation. Current erosion rates are greater than ever previously recorded (Pimentel and Hall 1989; Pimentel 2006). Soil erosion on cropland ranges from about 10 t/ha per year in the United States to 40 t/ha per year in China (Wen 1993; McLaughlin 1993; USDA 1994). Worldwide, soil erosion averages approximately 30 t/ha per year, or about 30 times faster than the replacement rate (Pimentel 1993). During the past 30 years, the rate of soil loss in Africa has increased 20-fold (Tolba 1989). Wind erosion is so serious in China that Chinese soil can be detected in the Hawaiian atmosphere during the spring planting period (Parrington et al. 1983). Similarly, soil eroded by wind in Africa can be detected in Florida and Brazil (Simons 1992).

Erosion adversely affects crop productivity by reducing the water-holding capacity of the soil, water availability, nutrient levels and organic matter in the soil, and soil depth (Pimentel 2006). Estimates are that agricultural land degradation alone can be expected to depress world food production between 15 and 30 percent by the year 2020. These estimates emphasize the need to implement known soil conservation techniques, including biomass mulches, no-till, ridge-till, terracing, grass strips, crop rotations, and combinations of these. All these techniques essentially require keeping the land protected from wind and rainfall effects with some form of vegetative cover (Pimentel et al. 1995; Pimentel 2006).

The current high erosion rate throughout the world is of great concern because of the slow rate of topsoil renewal; it takes more than 500 years for 2.5 cm (1 inch) of topsoil to form under agricultural conditions (OTA 1982; Elwell 1985; Troeh et al. 1991; Pimentel et al. 1995). Approximately 3000 years are needed for the natural reformation of topsoil to the 150 mm depth needed for satisfactory crop production.

The fertility of nutrient-poor soil can be improved by large inputs of fossil-based fertilizers or many tons of livestock manure (Pimentel et al. 2005). This practice, however, increases dependency on the limited fossil fuel stores necessary to produce these fertilizers and manures. Even with fertilizer and manure use, soil erosion remains a critical problem in current agricultural production (Pimentel 2006). Crops can be grown under artificial conditions using hydroponic techniques, but the cost in terms of energy and dollars is approximately 10 times that of conventional agriculture (Schwarz 1995).

The arable land currently used for crop production already includes a considerable amount of marginal land, land that is highly susceptible to erosion. When soil degradation occurs, the requirement for fossil energy inputs in the form of fertilizers, pesticides, and irrigation is increased to offset the nutrient and soil quality losses, thus creating nonsustainable agricultural systems (OTA 1982; Follett and Stewart 1985; Pimentel 1993, 2006).

7.6.2 WATER RESOURCES

The present and future availability of adequate supplies of fresh water for human and agricultural needs is already critical in many regions, like the Middle East (Postel 1997; Pimentel et al. 2004). Rapid population growth and increased total water consumption are rapidly depleting the availability of water. Between 1960 and 1997, the per capita availability of fresh water worldwide declined by about 60 percent (Hinrichsen 1998). Another 50 percent decrease in per capita water supply is projected by the year 2025 (Hinrichsen 1998).

All vegetation requires and transpires massive amounts of water during the growing season. Agriculture commands more water than any other activity on the planet. Currently, 70 percent of the water removed from all sources worldwide is used solely for irrigation (Pimentel et al. 2004). Of this amount, about two-thirds is consumed by plant life (nonrecoverable) (Postel 1997). For example, a corn crop that produces about 9000 kg/ha of grain uses more than 6 million liters/ha of water during the growing season (Leyton 1983). To supply this much water to the crop, approximately 1000 mm of rainfall per hectare—or 10 million liters of irrigation—is required during the growing season (Pimentel et al. 2004).

The approximate minimum amount of water required per capita for food is 400,000 liters per year (Postel 1996). In the United States, the average amount of water consumed annually in food production is 1.7 million liters per capita per year (USDA 1996), more than four times the minimum requirement. The minimum basic water requirement for human health, including drinking water, is 50 liters per capita per day (Gleick 1996). The U.S. average for domestic usage, however, is eight times higher than that figure, at 400 liters per capita per day (Postel 1996).

Water resources and population densities are unevenly distributed worldwide. Even though the *total* amount of water made available by the hydrologic cycle is enough to provide the world's current population with adequate fresh water—according to the *minimum* requirements cited above—most of this total water is concentrated in specific regions, leaving other areas water deficient. Water demands already far exceed supplies in nearly 80 nations of the world (Gleick 1993). In China, more than 300 cities suffer from inadequate water supplies, and the problem is intensifying as the population increases (WRI 1994; Brown 1995). In arid regions, such as the Middle East and parts of North Africa, where yearly rainfall is low and irrigation is expensive, the future of agricultural production is grim and becoming more so as populations continue to grow. Political conflicts over water in some areas, such as the Middle East, have even strained international relations between severely water-starved nations (Pimentel et al. 2004).

The greatest threat to maintaining freshwater supplies is depletion of the surface and groundwater resources that are used to supply the needs of the rapidly growing human population. Surface water is not always managed effectively, resulting in water shortages and pollution that threaten humans, as well as aquatic biota. The Colorado River, for example, is used so heavily by Colorado, California, Arizona, and other states, that by the time the river reaches Mexico, it is usually no more than a trickle running into the Sea of Cortes (Sheridan 1983).

Groundwater resources are also mismanaged and overtapped. Because of their slow recharge rate, usually between 0.1 and 0.3 percent per year (UNEP 1991; Covich 1993), groundwater resources must be carefully managed to prevent depletion. Yet, humans are not effectively conserving groundwater resources. In Tamil Nadu, India, groundwater levels declined 25 to 30 m during the 1970s as a result of excessive pumping for irrigation (Postel 1989; UNFPA 1991). In Beijing, the groundwater level is falling at a rate of about 1 m/year; while in Tianjin, China, it drops 4.4 m/year (Postel 1997). In the United States, aquifer overdraft averages 25 percent higher than replacement rates (USWRC 1979). In an extreme case like the Ogallala Aquifer under Kansas, Nebraska, and Texas, the annual depletion rate is three times above replacement (Beaumont 1985). If these rates continue, this aquifer, so vital to irrigation and countless communities, is expected to become nonproductive by 2030 (Soule and Piper 1992). In some regions of Arizona, water is being removed from aquifers 10 times faster than replacement (Pimentel et al. 2004).

High consumption of surface and groundwater resources, in addition to high implementation costs, is beginning to limit the option of irrigation in arid regions. Furthermore, salinized and waterlogged soils—both soil problems that result from continued irrigation (Postel 1997)—that have become unproductive are reducing the amount of possible irrigation area per capita. Per capita irrigated land has declined 10 percent during the last decade (Pimentel and Wen 2004).

Although no technology can double the flow of the Colorado River, or enhance other surface and groundwater resources, improved environmental management and conservation can increase the efficient use of available fresh water. For example, drip irrigation in agriculture can reduce water use by nearly 50 percent (Tuijl 1993). In developing countries, though, equipment and installation costs, as well as limitations in science and technology, often limit the introduction and use of these more efficient technologies.

Desalinization of ocean water is not a viable source of the fresh water needed by agriculture, because the process is energy intensive and, hence, economically impractical. The amount of desalinized water required by 1 ha of corn would cost $14,000, while all other inputs, like fertilizers, cost only $500 (Pimentel and Patzek 2005). This figure does not even include the additional cost of moving large amounts of water from the ocean to agricultural fields.

Another major threat to maintaining ample freshwater resources is pollution. Considerable water pollution has been documented in the United States (USBC 1996), but this problem is of greatest concern in countries where water regulations are less rigorously enforced or do not exist. Developing countries discharge approximately 95 percent of their untreated urban sewage directly into surface waters (WHO 1993). Of India's 3119 towns and cities, only 209 have partial sewage treatment facilities and a mere 8 have full wastewater treatment facilities (WHO 1992). A total of 114 cities dump untreated sewage and partially cremated bodies directly into the sacred Ganges River (NGS 1995). Downstream, the polluted water is used for drinking, bathing, and washing. This situation is typical of many rivers and lakes in developing countries (WHO 1992).

Overall, approximately 95 percent of the water in developing countries is polluted (WHO 1992). There are, however, serious problems in the United States as well. EPA (1994) reports indicate that 37 percent of U.S. lakes are unfit for swimming due to runoff pollutants and septic discharge.

Pesticides, fertilizers, and soil sediments pollute water resources when they accompany eroded soil into a body of water. In addition, industries all over the world often dump untreated toxic chemicals into rivers and lakes (WRI 1991). Pollution by sewage and disease organisms, as well as some 100,000 different chemicals used globally, makes water unsuitable not only for human drinking but also for application to crops (Nash 1993). Although some new technologies and environmental management practices are improving pollution control and the use of resources, there are economic and biophysical limits to their use and implementation (Gleick 1993).

REFERENCES

Alexandratos, N. 1995. *World Agriculture: Towards 2010.* Food and Agriculture Organization of the United Nations and John Wiley & Sons, Rome.

API. 1999. *Basic Petroleum Data Book.* American Petroleum Institute, Washington, D.C.

Beaumont, P. 1985. Irrigated agriculture and groundwater mining on the high plains of Texas. *Environmental Conservation,* 12, 11.

Brown, L. R. 1995. *Who will feed China? Wake-up call for a small planet.* W.W. Norton, New York.

Brown, L. R. 1997. *The Agricultural Link.* Worldwatch Institute, Washington, D.C.

Buringh, P. 1989. Availability of agricultural land for crop and livestock production. In *Food and Natural Resources,* Pimentel, D. and Hall, C. W., Eds., Academic Press, San Diego, 69–83.

China. 2002. One-child per couple legislation.

Covich, A. P. 1993. Water and ecosystems. In *Water in Crisis*, Gleick, P. H., Ed., Oxford University Press, New York, 40–55.

Doeoes, B. R. 1994. Environmental degradation, global food production, and risk for larger-scale migrations. *Ambio,* 23(2), 124–130.

Dunn, S. 2001. Bush energy plan: An oilman's vision. BBC News, Friday, May 18, 2001. http://news.bbc.co.uk/1/hi/sci/tech/1338569.stm (accessed June 5, 2008).

Elwell, H. A. 1985. An assessment of soil erosion in Zimbabwe. *Zimbabwe Science News*, 19, 27–31.

EPA. 1994. Quality of Our Nation's Water 1994. U.S. Environmental Protection Agency, Washington, D.C.

FAOSTAT. 1961–2004. *Statistical Database.* Food and Agricultural Organization of the United Nations, New York.

FAOSTAT. 2006. *Statistical Database.* Food and Agricultural Organization of the United Nations, New York.

Farrell, A. E. et al. 2006. Ethanol can contribute to energy and environmental goals. *Science*, 311, 506–508.

Follett, R. F. and Stewart, B. A. 1985. *Soil Erosion and Crop Productivity.* American Society of Agronomy, Crop Science Society of America, Madison, WI.

Giampietro, M. and Pimentel, D. 1994. Energy utilization. In *Encyclopedia of Agricultural Science,* Arntzen, C. J. and Ritter, E. M., Eds., Academic Press, San Diego, 73–76.

Gleick, P. H. 1993. *Water in Crisis.* Oxford University Press, New York.

Gleick, P. H. 1996. Basic water requirements for human activities: Meeting basic needs. *Water International*, 21(2), 83–92.

Hinrichsen, D. 1998. Feeding a future world. *People and the Planet*, 7(1), 6–9.

Houghton, R. A. 1994. The worldwide extent of land-use change. *BioScience*, 44(5), 305–313.

Kansas Ethanol. 2006. Kansas ethanol: Clean fuel from Kansas farms. http://www.ksgrains.com/ethanol/useth.html (accessed June 5, 2008).

Kendall, H. W. and Pimentel, D. 1994. Constraints on the expansion of the global food supply. *Ambio,* 23, 198–205.

Kerr, R. A. 2005. Bumpy road ahead for world's oil. *Science*, 310, 1106–1108.

Lal, R. 1989. Land degradation and its impact on food and other resources. In *Food and Natural Resources*, Pimentel, D., Ed., Academic Press, San Diego, 85–140.

Lal, R. and Pierce, F. J. 1991. Soil management for sustainability. Soil and Water Conservation Society in Cooperation with World Association of Soil and Water Conservation and Soil Science Society of America, Ankeny, IO.

Leach, G. 1995. *Global Land and Food in the 21st Century.* International Institute for Environmental Technology and Management, Stockholm.

Leyton, L. 1983. Crop water use: Principles and some considerations for agroforestry. In *Plant Research and Agroforestry*, Huxley, P. A., Ed. International Council for Research in Agroforestry, Nairobi, 379–400.

McLaughlin, L. 1993. A case study in Dingxi County, Gansu Province, China. In *World Soil Erosion and Conservation,* Pimentel, D., Ed. Cambridge University Press, Cambridge, 63–86.

Myers, N. 1990. The nontimber values of tropical forests. Working Paper 10. Forestry for Sustainable Development Program, Department of Forest Resources, University of Minnesota, St. Paul.

NAS. 2003. *New Frontiers in Agriculture.* U.S. National Academy of Sciences Press, Washington, D.C.

Nash, L. 1993. Water quality and health. In *Water in Crisis: A Guide to the World's Fresh Water Resources,* Gleick, P., Ed. Oxford University Press, Oxford, 25–39.

NGS. 1995. *Water: A Story of Hope.* National Geographic Society, Washington, D.C.

OTA. 1982. *Impacts of Technology on U.S. Cropland and Rangeland Productivity.* Office of Technology, U.S. Congress, Washington, D.C.

Parrington, J. R., Zoller, W. H., and Aras, N. K. 1983. Asian dust: Seasonal transport to the Hawaiian Islands. *Science*, 246, 195–197.

Pimentel, D. 1993. *World Soil Erosion and Conservation.* Cambridge University Press, Cambridge.

Pimentel, D. 2005. Environmental and economic costs of the application of pesticides primarily in the United States. *Environment, Development, and Sustainability*, 7, 229–252.

Pimentel, D. 2006. Soil erosion: A food and environmental threat. *Environment, Development, and Sustainability,* 8, 119–137.

Pimentel, D. and Hall, C. W. 1989. *Food and Natural Resources.* Academic Press, San Diego.

Pimentel, D. and Patzek, T. 2005. Ethanol production using corn, switchgrass, and wood: Biodiesel production using soybean and sunflower. *Natural Resources Research*, 14(1), 65–76.

Pimentel, D. and Pimentel, M. 2008. *Food, Energy, and Society,* 3rd ed., CRC Press, Boca Raton, FL, 380 pp.

Pimentel, D. and Wen, D. 2004. China and the world: population, food, and resource scarcity. In *Dare to Dream: Vision of 2050 Agriculture in China,* Tso, T. C. and Hang, H., Eds. Agricultural University Press, Beijing, 103–116.

Pimentel, D. et al. 1995. Environmental and economic costs of soil erosion and conservation benefits. *Science*, 267, 1117–1123.

Pimentel, D. et al. 2002. Renewable energy: Current and potential issues. *BioScience*, 52(12), 1111–1120.

Pimentel, D. et al. 2004. Water resources: Agricultural and environmental issues. *BioScience*, 54(10), 909–918.

Pimentel, D., Patzek, T., and Cecil, G. 2007. Ethanol production: Energy, economic, and environmental losses. *Reviews of Environmental Contamination and Toxicology*, 189, 25–41.

Population Action International. 1993. *Challenging the Planet: Connections between Population and Environment.* Population Action International., Washington, D.C.

Postel, S. 1989. Water for agriculture: Facing the limits. *Worldwatch Paper 93.* Worldwatch Institute, Washington, D.C.

Postel, S. 1996. *Dividing the Waters: Food Security, Ecosystem Health, and the New Politics of Scarcity,* Vol. 132. Worldwatch Institute, Washington, D.C.

Postel, S. 1997. *Last Oasis: Facing Water Scarcity.* W.W. Norton, New York.

PRB. 2005. World population data sheet. Population Reference Bureau, Washington, D.C.

Schwarz, M. 1995. *Soilless Culture Management.* Springer-Verlag, New York.

Sheridan, D. 1983. The Colorado—An engineering wonder without enough water. *Smithsonian*, February, 45–54.

Simons, M. 1992. Winds toss Africa's soil, feeding lands far away. *New York Times,* October 29, A1, A16.

Sommer, A. and West, K. P. 1996. *Vitamin Deficiency: Health Survival and Vision.* Oxford University Press, New York.

Soule, J. D. and Piper, D. 1992. *Farming In Nature's Image: An Ecological Approach to Agriculture.* Island Press, Washington, D.C.

Thomas, D. S. G. and Middleton, N. J. 1993. Salinization: new perspectives on a major desertification issue. *Journal of Arid Environments*, 24, 95–105.

Tolba, M. K. 1989. Our biological heritage under siege. *BioScience,* 39, 725–728.

Tomashek, K. M. et al. 2001. Randomized intervention study comparing several regimens for the treatment of moderate anemia in refugee children in Kigoma region, Tanzania. *American Journal of Tropical Medicine and Hygiene*, 64 (3/4), 164–171.

Troeh, F. R., Hobbs, J. A., and Donahue, R. L. 1991. *Soil and Water Conservation,* 2nd ed. Prentice-Hall, Englewood Cliffs, NJ.

Tuijl, W. 1993. *Improving Water Use in Agriculture: Experience in the Middle East and North Africa.* World Bank, Washington, D.C.

USBC. 1996. *Statistical Abstract of the United States 1996,* 200th ed. U.S. Bureau of the Census, U.S. Government Printing Office, Washington, D.C.

USBC. 2007. *Statistical Abstract of the United States.* U.S. Bureau of the Census, U.S. Government Printing Office, Washington, D.C.

USDA. 1940. *Agricultural Statistics.* U.S. Department of Agriculture, Washington, D.C.

USDA. 1980–2004. *Agricultural Statistics.* U.S. Department of Agriculture, Washington, D.C.

USDA. 1994. *Summary Report 1992 National Resources Inventory.* Soil Conservation Service, U.S. Department of Agriculture, Washington, D.C.

USDA. 1996. *Agricultural Statistics.* U.S. Department of Agriculture, Washington, D.C.

USDA. 2004. *Agricultural Statistics.* U.S. Department of Agriculture, Washington, D.C.

UNEP. 1991. *Freshwater Pollution.* Global Environment Monitoring System. United Nations Environment Programme, Nairobi.

UNFPA. 1991. *Population and the Environment: The Challenges Ahead.* United Nations Fund for Population Activities, United Nations Population Fund, New York.

USGAO. 1992. Hired farmworkers' health and well-being at risk. Report to Congressional Requesters. U.S. General Accounting Office, Washington, D.C.

USWRC. 1979. *The Nation's Water Resources. 1975–2000.* Vol. 1–4. Second National Water Assessment, U.S. Water Resources Council, Washington, D.C.

Weeks, J. R. 1986. *Population: An Introduction to Concepts and Issues,* 3rd ed. Wadsworth, Belmont, CA.

Wen, D. 1993. Soil erosion and conservation in China. In *Soil Erosion and Conservation,* Pimentel, D., Ed. Cambridge University Press, New York, 63–86.

WHO. 1992. *Our Planet, Our Health: Report of the WHO Commission on Health and Environment.* World Health Organization, Geneva.

WHO. 1993. Global health situation. *Weekly Epidemiological Record,* World Health Organization, 68 (12 February), 43–44.

WHO. 2004. World Health Report. http://www.who.int/whr/2004/en/ (accessed June 5, 2008).

WRI. 1991. *World Resources 1991–92.* World Resources Institute, Washington, D.C.

WRI. 1994. *World Resources 1994–95.* World Resources Institute, Washington, D.C.

WRI. 1996. *World Resources 1996–97.* World Resources Institute, Washington, D.C.

Youngquist, W. 1997. *Geodestinies: The Inevitable Control of Earth Resources over Nations and Individuals.* National Book Company, Portland, OR.

Youngquist, W. and Duncan, R. C. 2003. North American natural gas: Data show supply problems. *Natural Resources Research*, 12(4), 229–240.

8 Beyond Systems Thinking in Agroecology
Holons, Intentionality, and Resonant Configurations

William L. Bland and Michael M. Bell

CONTENTS

8.1 INTRODUCTION

Goals of scholarly work in agriculture include developing a richer understanding of this essential human endeavor and analyzing potential for interventions into current practice. Such interventions are motivated by a great diversity of issues, for example, less pollution, greater resource efficiency, more stable production, greater social equity. Conway (1985, 1987) introduced "agroecosystem analysis" as a framework for study of farming endeavors, connecting both the concept of the agro-ecosystem and its analysis, leading toward "systems thinking." The idea of the system, that is, "… a group of interacting components, operating together for a common purpose, capable of reacting as a whole to external stimuli: it is unaffected by its own outputs and has a specified boundary based on the inclusion of all significant feedbacks" (Spedding 1988, p. 18) and systems thinking are now commonplace in a wide range of endeavors, from business management, to health care, to ecology. Indeed, the essence of ecosystem ecology is systems thinking about biota and the relationships among themselves and with their physicochemical environment.

To do systems thinking is to adopt an ontology, that is, to make decisions about the entities and their interactions that constitute reality (at least for the purposes of the task at hand). Systems theorists have long warned us to remain cognizant of the fact that just because an analyst chooses to view some portion of the world as a system does not mean that it is, and behaves as such. Rosen (1991, p. 42) takes very seriously the step of positing the existence of a system: "The notion of system-hood … segregates things that 'belong together' from those that do not, at least from the subjective perspective of a … specific observer. These specific things that belong together, and whatever else depends on them alone, are segregated into a single bag called *system*; whatever lies *outside*, like the complement of a set, constitutes *environment*. The partition of ambience into system and environment, and even more the imputation of that partition to the ambience itself as an inherent property thereof, is a basic and fateful step for science. For once the distinction is made, attention focuses on *system*." Checkland and Scholes (1999, p. 22) illuminate the key distinction: "Choosing to think about the world as if it were a system can be helpful. But this is a very different stance from arguing that the world is a system, a position which pretends to knowledge no human being can have." So while one of the motivations for systems thinking is to be inclusive, actually setting to work on a problem requires identifying and making separate a portion of all that is around us.

There can be no question that systems thinking has contributed greatly to agricultural science. Such thinking stimulates even the most narrowly focused research group to articulate its work with the broader world. Indeed, competitive grant funding in the United States related to agriculture is strongly influenced by the imperative of taking a more holistic approach in which research, teaching, and outreach must all be present and connected. On the other hand, we cannot help but worry that some analysts too easily forget the warnings that agricultural reality may not actually be a nicely connected-up system that, for want of greater understanding and passion for systems thinking, we are not yet managing optimally.

In the spirit of the perspective that systems thinking is an ontological choice—often powerful, but confused with reality at the analyst's peril—we propose here a complementary perspective that we believe holds benefits for all manner of students of agriculture. We have adapted Koestler's (1967) notion of the holon—something that is simultaneously a whole and part (Bland and Bell 2007). For Koestler (1967, p. 210), "Parts and wholes in an absolute sense do not exist in the domain of life. The concept of the holon is intended to reconcile atomistic and holistic approaches." Holon agroecology seeks a middle ground between the reductionism that serves purely scientific enterprises so well and the holism that system thinking implies.

In holon agroecology the farmer—be this a multinational corporation or a mother and child with a patch of rice in Bangladesh—is central. Indeed, we submit that agriculture might most usefully be defined as something like, "humans planning and acting to cultivate livelihoods from plant and animal increase." In this definition the importance of human actors is acknowledged first and foremost. For all our scholarly attempts to "design" and "manage" agroecosystems, whether in developed or subsistence settings, any notion that we can do either should be approached skeptically. Whether because of the political lobbying of a commodity producers' association or the reluctance of poor farmers to change practices learned from their forebears, individual human intentionality seems often to override notions of a thoughtfully designed and managed agricultural "system." This human intentionality is in the vast majority of cases directed at profiting from plant and animal increase, that is, that seeds, soil, and water can lead to an excess of seeds that can be sold, or that baby animals grow and multiply, providing some for exchange. But the intentionality of the individual farmer is likely more than simply profit from biological increase. Included might be sensed obligations to generations before to keep the family farm productive, or a desire to rear farm animals according to some ethical code. Whatever the complex intentionality motivating a farmer, the resulting goals are often realized precisely by pursuing courses of action contrary to what others expect and would wish for. We propose that such complex intentionalities make unlikely the idea that the human agricultural endeavor is, or can become, a well-managed system. Holon agroecology offers ways of simultaneously recognizing farmer intentionality and notions of agriculture as a system.

The holon as simultaneously a whole in some senses and a part of things larger makes its definition an unending challenge. What things are usefully thought of as a holon, and how are its boundaries envisioned? If the holon does indeed "reconcile atomistic and holistic approaches," can it even have substantive boundaries? In this chapter we elaborate our interpretation of the holon and the reasons that it offers a useful alternative to systems as an ontology with which to analyze agricultural endeavors. We argue that the holon offers a fresh perspective on questions of boundaries and sources of change in agroecosystems, and on the great variety of farms and their pathways to persistence. Finally, we believe that holon agroecology offers opportunities for the multidisciplinarity that is essential for sustaining and improving the agricultural endeavor.

8.2 INTRODUCTION TO THE HOLON

The holon is simultaneously a part and a whole. Its wholeness is manifested by its capacity for self-governed action, such as when the authors decide to have lunch together lakeside on our campus. Each of us is a whole in the sense that we can decide to do this on most—but not all—days. There are some times that because we are also parts of larger wholes, such as our university, community, and families, we are constrained by our partness, for example, one of our departments may require our participation in a meeting. To most richly understand our noontime behaviors, an observer would need to understand that we have considerable autonomy as wholes, but are often constrained because we are at the same time parts. Similarly an analyst would see the successful farmer acting and planning with appropriate autonomy, but operating within bounds specified by the *ecology of contexts* in which the farm exists. The farmer has an array of options to exercise, but not an infinite number, sharply limited, for example, by the imperative to choose crop species that are adapted to the local climate, or to grow products for which there is a market. We more fully explore this ecology of contexts below, but the usefulness of the holon is to keep foremost in our vision the idea that in order to most richly comprehend many interesting things in biology and society we must simultaneously be aware of both their wholeness and their partness. We have suggested (Bland and Bell 2007) that we might usefully learn to "flicker" between seeing a holon in its wholeness and as a part of an ecology. Our flickered imagining of farm holons helps us see the self-governed whole, as well as the constraining (and enabling) ecology of contexts that so powerfully shape what happens on the farm.

If we are to speak of the wholeness of the holon, there must exist some surface that bounds it, allowing us to envision it as a whole, while simultaneously a part of its ecology of contexts. The problem of where to draw boundaries within systems depictions is long-standing. In a practical sense, in order to study some facet of agriculture the analyst must design a bounded experimental system (Norgaard and Sikor 1995). Consider corn breeding as an example: test plots are typically planted and maintained at multiple sites across a region by a trained and dedicated crew of graduate students. But the hybrids in the experiment are isolated from potentially significant aspects of the actual farms, for example, the challenge of timely planting when there is so much else to be done, or poorly functioning machinery. On-farm research seeks to expand the boundaries of the system to incorporate more of the contexts in which the genetic technology must operate, but boundaries remain present, separating the experiment from the whole of what actual farming involves.

We propose that *intentionality* provides a useful bounding surface for holons. A farmer's intention is (at least in part) to plan and act so as to permit the farm holon to persist and provide livelihoods. Thus, we imagine the farm holon as including the decision makers and the biological, physical, social, economic, and human resources that these managers manipulate and exploit. Some of the components of the farm holon are themselves holons, such as workers, family members, or cows, while others such as tractors and fence posts are not. Thus, holons may be envisioned as *intentionalities* operating in the world—entities that act so as to further goals they possess. What might be understood as intentional behavior ranges from our decision to enjoy a lunch break through phototropism in plants, but we have not found it necessary or useful to attempt to put too fine an edge on what we mean by intentional. Our concern here is with humans doing agriculture—planning and acting to cultivate livelihoods from plant and animal increase—so the degree to which behaviors are understood to be encoded in genetic information is not a concern.

Thus, holon agroecology sees farms as intentionalities consisting of other intentionalities and inanimate material, embedded in an ecology of contexts. This appeal to intentionality as a bounding surface helps us see the farm's wholeness, and reminds us that any agroecosystem analyst should recognize that understanding why a farmer does what he or she does must acknowledge that the farmer is acting on a set of goals that may be impossible to fully articulate in a shared language. Seeking a reasonable return on investment is surely part of the intentionality of many farmers, but things become far more diverse after that. A farm's intentionality might additionally be shaped by love of a particular place, spiritual notions of stewardship, an interpretation of animal sentience, and visions of a better future for the children of the farm family. Yet it seems unimaginable that such a complex stew of influences can be resolved to a clearly defined intentionality. The default assumption should be of *ununified* intentionalities—farmers, like all of us, are typically conflicted, and a simple model of their intentionality, for example, profit, is often misleading. The lack of unity that a farmer-as-holon experiences is even plainer when we consider the farm as a holon, including the intentionalities of the members of a human farm family, livestock, plants, and their human and nonhuman community relations. Although such ununified intentionality may frustrate efforts at reliable analysis, it is a well-spring of novelty and innovation. Indeed, the authors' hope for a more equitable, resource-efficient, and multifunctional agriculture fundamentally depends on the great diversity of intentionalities held and acted upon by farmers.

8.2.1 The Holon's Ecology of Contexts

While intentionality speaks to wholeness, the partness of holons arises because they are *embedded in and help create* an ecology of contexts. A great many contexts in which farms exist are readily envisioned: climate, soil, market access, labor costs, spiritual beliefs, health of household members, debt, and the cost of energy. The farm holon is constrained by many of these contexts, and often helps constitute them. For example, the climate and soil of a place impose a considerable set of constraints on the species that will flourish there; hence farmers do not grow bananas in Wisconsin. Religious tradition is a context that makes Amish farmers unwilling to use certain agricultural technologies, just as swine are uncommon in predominately Islamic countries.

There are other contexts for which farms are constitutive, like the market for corn. The corn market appears as it does in part because of the farms that, for whatever reasons, grow the crop. There are other components to the corn market, of course, for example, demand, and over a century of research and development about how to grow and put it to use. Farms in many cases coevolve (Norgaard and Sikor 1995) with a context, and thus are at different times constrained by it and constitutive of it. An example is labor on Wisconsin dairy farms—these farms have been able to evolve toward herds in excess of 1000 cows in part because of the availability of immigrant labor. The social context in rural Wisconsin for these workers changed to include welcoming ethnic restaurants and markets, perhaps making the employment opportunity more attractive. Thus, the pool of workers expands as the state's dairying system becomes more dependent on their labor.

The theologian Reinhold Niebuhr wrote a famous prayer asking for the wisdom to discern situations that we can change from those that we cannot. In terms of holon agroecology, contexts differ greatly in the degree to which farmers can affect them in the direction of their own intentions or, in sociological terms, can be said to have *capability* with respect to the context (Bland and Bell 2007). There are clear examples where farmer capability is completely lacking, or nearly so, for example, climate or soil texture, and there are some for which there is considerable potential for impact. Many farmers can influence the organic matter in their soil through reduced tillage and crop rotations that include perennials, and this organic matter in turn changes water and nutrient dynamics in the soil. Thus, these farmers have some capability with respect to facets of their soil context. Similarly, treatment of labor may make a farm a more or less desirable place to work, influencing the pool of workers available to the particular farmer. Innovative farmers can foster markets for their production, for example, farmers' markets and community-supported agriculture.

There is a substantial literature on farmer capability, often using the much-debated language of "agency," recently summarized by Higgins (2006), who argued for two main strains of thought: "agri-food globalization theory," and "the actor-oriented approach." Holon agroecology accommodates the first as market contexts in which the farmer has little capability; for example, transnational food corporations have so powerfully determined what will be purchased that what farmers can profitably produce is quite conscribed. Such contexts in holon agroecology are envisioned as *stabilized externally to the farm*, and thus must be taken by the individual farm largely as given. Some of these contexts are stabilized by entities that are themselves holons, for example, agrifood corporations, planning and acting to persist and grow, and among their strategies is to structure markets so that individual farmers see clear and few choices.

Actor-oriented perspectives of farmer capability place emphasis on social relationships, and the interactions that advance each actor's "project." This too is readily accommodated in holon agroecology, as the farmer recognizes opportunities (or imperatives) for planning and acting in those contexts in which social negotiation is central, such as obtaining credit. The capability a particular farmer experiences may be closely tied to his or her social skills. As well, the very intentions of the farmer largely derive from her or his experience of the ecology of contexts: we intend what seems reasonable to us, and reason is always contextual if it is reason at all. But as the actor perspective emphasizes, there are nonhuman components involved in the success of a farm (the actor's project) as well, from the weather to crop susceptibility to disease. Persistence of the farm holon depends on good fortune in avoiding calamity, as well as the farmer's reasoning and perseverance, perhaps emerging from a relatively unified intentionality.

One place of development beyond the actor-oriented approach, though, is holon agroecology's insistence on the importance of intentionality. Perspectives such as Actor Network Theory (ANT) emphasize the equivalence of human and nonhuman actors as, in the terminology of ANT, *actants*. The ANT goal here is worthy: to encourage the conceptual engagement of the human and nonhuman. But the presence of intentionality in some actants and not others, as well as the variety of character and orientation of these intentionalities, suggests that some additional theoretical tools are needed if we are to understand how holons persist in their ecology of contexts.

8.2.2 THE HOLON AS NARRATIVE

Agriculture is an immensely complex human endeavor, connected to more things in more ways than we can ever know, let alone describe to another—human survival, culture, and livelihoods, as well as the fates of nonhuman species, to start a list at a very coarse level. In order for us to communicate with one another about some particular aspect of agriculture, that is, to develop shared perceptions of selected phenomena, we need to make choices about what of its infinite involvements to include in the discussion. In doing so we are crafting a narrative, and this idea is so helpful and important in holon agroecology that we must explore it a bit at this point.

Ultimately, we make sense of what we do and experience, and communicate this appreciation with others, through stories. In creating these stories, or narratives, we must exclude much—that is, make narrative choices. In one sense this is a practical decision: the present authors have a few thousand words in which to share our ideas about the holon as a useful tool for agroecology. To do this most effectively for the audience we anticipate here we chose not to explore its links to philosophical phenomenology, or relationships to issues of power, or contrast it with systems thinking beginning with von Bertalanffy. We all make such decisions within every scholarly communication we create.

But crafting narratives is more than wise communication strategy. Allen et al. (2005) argue that narratives are the only way that truly complex systems, such as ecosystems, can be meaningfully addressed. For them, "The power of narratives is that they can relate in a coherent way contrasting types of things from different scales. Narratives are the device people use to grasp large ideas. Models can be used to calibrate things, and even improve the quality of narratives, but models cannot work with the scope natural to a narrative." We submit that the scope of the issues of true significance in agriculture requires narrative, that is, results of this or that experiment, have meaning only within a larger story.

Elsewhere we use the term narrative in reference to a particular aspect of the challenges of defining system boundaries (Bland and Bell 2007). The narrative boundary problem arises because we can effectively tell but one of the many stories that might be told about a set of items and events. A

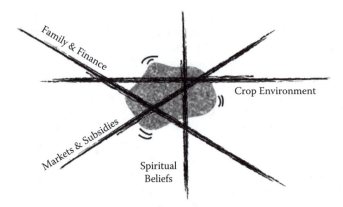

FIGURE 8.1   The farm holon as an entity embedded within and constitutive of an ecology of contexts. The constantly evolving configuration of the farm must be valid in all contexts, although tensions likely exist. The intentionality of the farmer to have the operation persist bounds and animates the holon. (From Bland and Bell, 2007. *Intl. J. Agric. Sustain.* 5(4): 280–294. With permission.)

particular farm holon is part of stories about biogeochemical cycling, local livelihood opportunities, and the quality of life of sentient beings. So narratives are essential to understanding and describing complex systems, and their telling necessarily involves choices about what is relevant to the problem at hand, and, as a result, many parts of the holon remain unacknowledged. The holon idea provides some basic conceptual tools for making and recognizing narrative choices: intentionalities, capabilities, and contexts.

8.2.3   Persistence of the Holon

The ability of the holon to persist (at least long enough for some observer to notice it) depends on its ability to find a configuration—some way of doing business, of choosing and organizing activities—that is simultaneously viable in the many facets of its ecology of contexts (Figure 8.1). Within each of the myriad contexts in which the farm holon exists, such as personal values or soil hydroclimate, there are likely several solutions that are viable, but as additional contexts come into consideration the valid configurations inevitably become fewer.

When the holon fails to persist it may be because the premise of the farm was wrong or that over time one or more contexts change. As much as the community-supported agriculture farmer may cherish living on a pesticide-free farm where his or her children need not be occasionally warned to stay away from the garden, the limits to what folks are willing to pay months in advance for vegetables not yet planted may not be enough to cover the cost of health insurance for a growing family. For farmers in Illinois changing contexts might include the price of soybeans falling as new Brazilian lands come into production, or damaging soil insects that increasingly survive a rotation that formerly kept their populations low. Regardless, the farm holon must change or perish.

While clearly individual farms do find viable solutions in their particular ecology of contexts, none of these solutions is free of internal tensions—viable here means only that particular contexts do not exercise veto power over the project. Tensions within every solution seem likely because a holon's intentionality is rarely unified—any solution is a balancing of at least partially oppositional desires and capabilities with regard to the ecology of contexts. Additionally, an "optimal" solution cannot be calculated because some of the contexts of a farm's ecology are incommensurable; that is, they cannot be directly compared using a single unit of measurement. Much of ecological economics is about addressing the incommensurability of, for example, ready access for urbanites to a stream that supports trout fishing compared to the costs to farmers of mandating particular manure management practices. For the individual holon incommensurability means that many trade-offs can only be understood intuitively, and thus remain always subject to rethinking, for example, the need for labor of farm household children compared to the possibility of injury (Zepeda and Kim 2006). But the search to relax these tensions surely serves as an important source for innovation and change. Here the holon in its unending need to reconfigure as contexts change, and the ever-present impetus to reduce tensions reminds us that farming is not, and can never be, a completely connected-up, finished system. Holon agroecology offers a way to envision both the motivation for and sources of change (Bland and Bell 2007).

8.3 THE VARIETY OF FARMS

The holonic ecology of contexts and imperative for constant reconfiguration offers an explanation for the tremendous variety of farms. Every holon's ecology of contexts is unique, because of location and accidents of history, both personal and environmental. And this unique contextual environment can be seen as only the starting place from which the holonic search for viable configurations leads to ever-increasing diversity of extant solutions. This is the "contextualism" identified by Norgaard and Sikor (1995), in which "phenomena are contingent upon a large number of factors particular to the time and place."

No two farms are identical, yet depending on the problem at hand (i.e., a set of narrative goals) there may be recognizable types, that is, particular configurations that can be seen repeated (although never identically) among the population of farms. Andow and Hidaka (1989) identified "syndromes of production," in which he suggested very diverse strategies for growing rice in Japan could be thought of as integrated packages of practices (some much more intriguing than others to them as ecologists). Giampietro and colleagues (Pastore and Giampietro 1999; Gomiero and Giampietro 2001) developed typologies of small-area farming systems in Vietnam and China based on goods produced, and time and land allocations among a set of activities found to be commonplace in the study regions. Dixon and Gulliver (2001) proposed descriptions of dominant farming systems across much of the world. Eakin (2005) interpreted Mexican farm holon configurations in a "livelihood strategies" framework, identifying four in her study regions. Importantly, though, what characteristics are taken into account in describing a type or strategy is a narrative issue. The analyst makes choices about what factors make farms similar or dissimilar, based on the question at hand, or the data available, and perhaps in ignorance of important issues.

In holon agroecology we interpret these identifiable types as *resonant configurations*. The existence of resonance suggests that for a place and era there may be a limited number of holonic configurations for which there is a contextual "sweet spot" of supportive reverberations. It is not that other configurations of crops, practices, land tenure, and inputs would not work in a region, but that extant farm configurations tend to cluster around a particular set of possibilities. These are partially analogous to "attractors," "domains of attraction," "alternative steady states," or "multiple equilibria" described for ecological systems. The classic example is that of temperate freshwater lakes, typically observed in either an oligotrophic (clear) or eutrophic (turbid) state. Some ecosystems are thought to switch rather abruptly from one such state to another (Scheffer 1999) in the face of continuing stress, for example, increasing P pollution in the case of the oligotrophic lake. The ability of the ecosystem to resist such switches, and thus remain in a particular state in the face of stress, is termed *resilience*, "the capacity of a system to experience shocks while retaining essentially the same function, structure, feedbacks, and therefore identity" (Walker et al. 2006).

A resonant farm holon is relatively resilient to perturbations in its ecology of contexts. Thus, the community-supported agriculture model provides the small-scale vegetable farmer a buffer against failure of a particular crop that the contract grower, committed to delivering 1000 squash on a certain date, does not enjoy. Similarly, government commodity price and weather disaster relief programs for crop and livestock producers soften the impacts of years of bumper crops or droughts. In a biophysical example, deep silt loam soils provide a larger soil water reservoir than do sandy soils, making a crop resilient to rainfall shortages. A large-scale dairy is nearly immune to the health of individual milkers, in marked contrast to a small farm operated by family labor, where injury to a household member can force a considerable reconfiguration of the operation. Antle et al. (2006) demonstrated multiple equilibria in farm productivity arising from the inherent productivity of land, and when, if ever, a farmer decides to make soil conservation investments.

Farms that are quite unlike any notional resonant configuration, that is, appreciably different from others in one or more ways, may be so because of some accident of history, an innovation, or "sunken capital"—infrastructure that generates return with only one set of (potentially obsolete) production practices. In the Wisconsin dairy industry two distinct resonant configurations are readily posited: large-scale confinement facilities in which milking proceeds around the clock by hired labor, and rotational grazing farms. Constituting the vast majority of the 15,000 dairy farms in the state, however, are relatively small (about 100 cows) confinement operations on which the family members supply most of the labor. Lively debate surrounds the future of this vast majority, which are currently declining in percentage terms while the others rise. Are there many viable configurations of dairy farms in Wisconsin, or are those not moving toward one of the two posited resonant configurations in grave danger? Alternatively, might there be under way shifts in the context ecology that few dairy farms or analysts yet appreciate, and that will lead to emergence of new resonant configurations?

But we wish to avoid a top-down sense of system control and determination. The sweetness of a resonance is something each farm holon must judge, listening in its own way, repositioning accordingly, and even discovering and shaping. Thus, the resonance of a farm holon depends on finding a configuration in which multiple contexts are at the least in the same key, as it were.

8.4 HOLON AS A TOOL FOR INTERDISCIPLINARITY

The scholarship of agriculture entails diverse disciplinary perspectives and traditions. Each offers useful and important insights and tools to the agricultural endeavor, but each operates from a unique narrative; for example, soil scientists tell the story of the productivity and health of the soil resource, often abused by human actions, while the sociologist tells the stories of the humans in the diversity of rural settings. The fundamental incommensurability of these narratives is the reason that we have disciplines at all, and each contributes uniquely to a rich understanding of agriculture. Disciplinarity should (and most certainly will) persist in the academy.

Distinctions among these disciplinary narratives are of little concern to stakeholders in the agricultural endeavor, however. The diverse challenges facing agriculture are widely understood to require multidisciplinary approaches, and we propose that holon agroecology offers a powerful tool for this task. The farm holon is a nexus of contexts, and for each of these contexts there is a discipline that is, we trust, cultivating useful and applicable knowledge. Thus, the farm holon offers a meeting place for the disciplines, where each must acknowledge the (at least potential) significance of the others for shaping the what and why of farmer behavior. Choices inexplicable to one discipline (why such big tractors?) are made a bit more understandable by the recognition that the farm holon is shaped by and held together by a complex stew of intentionalities (including garnering respect of peers). We argued earlier that our proposed definition of agriculture as the cultivation of livelihoods from plant and animal increase rightly places the farm holon at the center of the discussion, and it is here that the disciplines can and should articulate with one another.

Holon agroecology begins with the assumption that diverse and even surprising contexts—and thereby disciplinary domains—may be of significance in the farm's unique configuration. We hope that this will make it easier for each discipline to be open to, if not curious about, how its perspective must be applied as a result of other contexts. Again, Norgaard and Sikor (1995) anticipated this necessary "pluralism" in disciplinary approaches. Holon agroecology is complementary—rather than an alternative—to systems thinking as a multidisciplinary approach. The diverse tools of systems portrayals and modeling (e.g., Spedding 1988; Wilson and Morren 1990) are essential for understanding particular contexts; for example, numerical models are arguably the only tractable way of exploring the intersections of soil, climate, and agronomic practice, just as crop calendars help envision the temporal nature of labor requirements. Holon agroecology provides a complementary perspective by placing the context-specific insights provided by systems tools in a broader framework of the agricultural endeavor understood as fundamentally motivated by the play of intentionalities in an ecology of contexts. This seems to us to have potential to help disciplinarians better appreciate the limits of their own perspectives, and to delay normative judgments, as one context helps illuminate the motivating influences for what seems wrongheaded from another perspective.

8.5 SUMMARY

Holon agroecology provides a framework for agroecosystems analysis that is, we propose, complementary to systems thinking. Systems thinking usefully demonstrates important linkages within agriculture and between it and other sectors of society, but the disconnects are important as well. These disconnects are both the source of the endless variety and innovation, as well as the innumerable bad ideas, manifest in the agricultural endeavor. Holon agroecology provides a framework and vocabulary to envision why farmers do what they do, and how farms persist in a world of constant change through the relentless search for viable solutions in an ever-changing ecology of contexts. Holon agroecology provides a meeting place for the diverse disciplinary perspectives that are essential, yet each alone inadequate, to making manifest the possibilities of a multifunctional, equitable, and resource-efficient agriculture.

REFERENCES

Allen, T. F. H., Zellmer, A. J, and Wuennenberg, C. J. 2005. The loss of narrative. In *Ecological Paradigms Lost: Routes of Theory Change*, Cuddington, K. and Beisner, B., Eds. Elsevier, New York, 333–370.

Andow, D. A. and Hidaka, K. 1989. Experimental natural history of sustainable agriculture: Syndromes of production. *Agriculture, Ecosystems, and Environment*, 27, 447–462.

Antle, J. M., Stoorvogel, J. J., and Valdivia, R. O. 2006. Multiple equilibria, soil conservation investments, and the resilience of agricultural systems. *Environment and Development Economics*, 11, 477–492.

Bland, W. L. and Bell, M. M. 2007. A holon approach to agroecology. *International Journal of Agricultural Sustainability*, 5, 280–294.

Checkland, P. and Scholes, J. 1999. *Soft Systems Methodology in Action, Including Soft Systems Methodology: A 30-year Retrospective*. Wiley, New York.

Conway, G. R. 1985. Agroecosystem analysis. *Agricultural Administration*, 20, 31–55.

Conway, G. R. 1987. The properties of agroecosystems. *Agricultural Systems*, 24, 95–117.

Dixon, J. and Gulliver, A. 2001. *Farming Systems and Poverty: Improving Farmers' Livelihoods in a Changing World*. FAO and World Bank, Rome and Washington D.C.

Eakin, H. 2005. Institutional change, climate risk, and rural vulnerability: Cases from Central Mexico. *World Development*, 33, 1923–1938.

Gomiero, T. and Giampietro, M. 2001. Multiple-scale integrated analysis of farming systems: The Thuonng Lo commune (Vietnamese uplands) case study. *Population and Environment*, 22, 315–352.

Higgins, V. 2006. Re-figuring the problem of farmer agency in agri-food studies: A transvaluation approach. *Agriculture and Human Values*, 23, 51–62.

Koestler, A. 1967. *The Ghost in the Machine*. Macmillan, New York.

Norgaard, R. B. and Sikor, T. O. 1995. The methodology and practice of agroecology. In *Agroecology—The Science of Sustainable Agriculture*, 2nd ed., Altieri M. A., Ed. Westview Press, Boulder, CO, 21–39.

Pastore, G. and Giampietro, M. 1999. Conventional and land-time budget analysis of rural villages in Hubei Province, China. *Critical Review in Plant Sciences*, 18, 331–357.

Rosen, R. 1991. *Life Itself: A Comprehensive Inquiry into the Nature, Origin, and Foundation of Life*. Columbia University Press, New York.

Scheffer, M. 1999. Searching explanations of nature in the mirror world of math. *Conservation Ecology*, 3, 11. http://www.consecol.org/vol3/iss2/art11/ (accessed April 18, 2008).

Spedding, C. R. W. 1988. *An Introduction to Agricultural Systems*, 2nd ed. Elsevier Applied Science, New York.

Walker, B. H., Gunderson, L., Kinzig, A., Folke, C., Carpenter, S., and Schultz, L. 2006. A handful of heuristics and some propositions for understanding resilience in social-ecological systems. *Ecology and Society*, 11, 13. http://www.ecologyandsociety.org/vol11/iss1/art13/ (accessed April 18, 2008).

Wilson, K. and Morren, Jr., G. E. B. 1990. *Systems Approaches for Improvement in Agriculture and Resource Management*. Macmillan, New York.

Zepeda, L. and Kim, J. 2006. Farm parent's views of their children's labor on family farms: A focus group study of Wisconsin dairy farms. *Agriculture and Human Values*, 23, 109–121.

Section III

Ecological Foundations of
Agroecosystem Management

9 Ecology-Based Agriculture and the Next Green Revolution

Is Modern Agriculture Exempt from the Laws of Ecology?

P. Larry Phelan

CONTENTS

"We can't solve problems by using the same kind of thinking we used when we created them."

—**Albert Einstein**

9.1   INTRODUCTION

Agriculture around the world is entering a new phase of development. In addition to their traditional roles of providing food and fiber for an ever-growing world population, farmers increasingly will be called on to supply renewable materials for energy and manufacturing and to provide new ecological services, yet will have to do so under unprecedented circumstances of higher energy costs and rapidly changing weather patterns. Therefore, it is an appropriate time to evaluate how we will move forward in designing agricultural systems that will be successful in this new environment.

In keeping with the objective of this volume to honor Ben Stinner, I shall examine the current status and future direction of agricultural research, employing two values for which Ben was well known: (1) a science-based highly integrated systems view of agriculture and its association to the natural world, and (2) a bridge-building approach to developing solutions to long-standing problems by helping rival factions understand opposing points of view, emphasizing common goals rather than assigning blame. I hope the reader will proceed with an open mind, keeping these objectives in view. The extent to which I cause the reader to become defensive should be taken as a failure to live up to standards set by Ben rather than the shortcomings of his philosophy.

In this chapter, I argue for the need to develop a set of principles to guide decision making for improving agriculture. In my view, the current research model, which uses a fragmented approach to agricultural design by addressing components and problems individually (reductionism), is the basis for significant improvements in agricultural productivity, but is also responsible for significant unintended problems and will increasingly prove an inadequate model for future research and agricultural system design. It is insufficient first because it is too narrow to account for many of the true costs of production, some of which are displaced in space and time. Second, it fails to take into account the complexity of interactions that derive from the ecosystem properties of agriculture. Third, by relying heavily on external inputs to control individual system components, access to free ecological services is lost. Finally and most importantly given the changing global climate, our current framework has resulted in a loss of stability inherent in mature natural ecosystems. Thus, as with many significant advances in scientific understanding through history, agricultural research is in need of a new paradigm (Kuhn 1970).

9.2   PHILOSOPHICAL UNDERPINNINGS OF AGRICULTURAL RESEARCH (HOW DID WE GET HERE?)

By most economic measures, the research approach that led to our current agricultural system has been wildly successful, with productivity gains far outpacing most nonfarming sectors of Western economies. Whereas about 20 percent of the U.S. population worked on the farm in 1940, today only about 2 percent supplies the food for a much larger population (Gardner 2002). This steep rise in crop productivity was driven by technological developments in five areas: increased use of chemical fertilizers; high-yield crop varieties with a stronger response to those fertilizers; chemical pesticides for controlling insects, weeds, and diseases that depressed yields; greater use of irrigation; and increased mechanization. These developments were adapted and exported to underdeveloped countries in the form of the Green Revolution, resulting in similar (although not universal) rises in crop productivity. As human populations continue to increase, the leaders of the first Green Revolution now speak of the need for a second Green Revolution, this time based on genetic engineering of crops.

Amid the success in raising productivity, we also need to recognize the significant costs that have been incurred or deferred for future generations. Heavy use of agrichemicals has increased crop nutrient availability and reduced losses to pests, but also has created human health issues and environmental problems. Mechanization of agriculture has reduced labor costs, but in many developing countries, also has displaced populations to overcrowded cities in search of employment. Almost all of the technological developments in agriculture have translated to higher dependence on petrochemicals. As a result, the dominant paradigm underlying agricultural research and development, which focuses on solving individual problems rather than system design, has led to a production system that is highly capital and energy intensive, and not easily transferable to developing countries without a large infusion of governmental support. Moreover, since many of the negative unintended consequences that have attended the celebrated productivity gains are displaced in space and time, they are often overlooked. Some of these costs are external to the production system and are actively ignored, while others are internal, but their cause is not recognized. If we were to adopt a more inclusive view of actual costs and these negative consequences were to be included in an economic analysis, we would certainly see a substantial drop in calculated production efficiencies. There are many well-documented examples of such displaced negative consequences such as soil erosion, loss of wetlands, deforestation, pollution of ground and surface waters caused by herbicides, insecticides, and fungicides, and negative impacts of eutrophication of aquatic and terrestrial food webs caused by off-farm movement of excess nutrients.

An example of temporal displacement is the negative impact on raptor populations that was traced to the disruptive effect of DDT on prostaglandin and calcium metabolism in the eggshell gland (Lundholm 1997). This example illustrates first the interconnections between agriculture and the larger ecosystems, connecting components that can be quite removed. The causal pathway between insect pest control in an agricultural field and eagle population dynamics involves many biotic linkages and is not a connection that anyone could have realistically anticipated with the introduction of DDT. In addition to a spatial separation of system components, it took a number of years for DDT to become concentrated in the upper levels of the food chains, leading to a temporal displacement between cause and effect.

A contemporary example of spatial displacement is the annual formation of a dead zone in the Gulf of Mexico, caused by the leaching of nutrients from chemically fertilized fields of the Midwest United States carried down the Mississippi River where they accumulate in the Gulf (Committee on Environment and Natural Resources 2000). Because this is a cumulative effect of many farms far distant from the problem, it is largely impossible to parse the relative contributions of each farm. Furthermore, the absence of negative feedbacks means there is little economic incentive for the farmers to take on the costs associated with changing their practices to reduce the problem. These are but two examples of significant unintended consequences of agriculture overlooked or intentionally ignored due to the lack of significant economic/ecological feedback loops. In such circumstances, change usually comes only when the problem becomes significant enough to stimulate the imposition of costs externally, such as through government regulation. In ecological terms, such an approach has the effect of shrinking the feedback loop by internalizing the costs.

Going forward, we need to adopt a more inclusive view to assess the true costs and benefits of agricultural practices. In comparing the individual problem-solving approach that has guided the past century of agricultural research with a new approach that emphasizes the systems nature of agriculture, we must consider how we arrived at current practices, examine the prevailing philosophical paradigm of reductionism, and contrast it with an alternative systems perspective.

9.2.1 HISTORICAL PERSPECTIVE OF AGRICULTURAL REDUCTIONISM

Although the term "industrial agriculture" is often used in a pejorative sense, I use it here in reference to the philosophical roots of agriculture as practiced in most of the world today. One of the critical turning points for agriculture can be traced to the early mid-1800s with the push to put farming on a more scientific foundation. Work by Saussure, Sprengel, and others undermined prevailing ideas that manures were essential to crop production by suggesting their value came only in providing essential minerals (Brock 1997). This shift provided the foundation for von Liebig's now-famous "Law of the Minima" that underlies the current mineral theory of plant nutrition. Although many of the ideas Liebig advanced were not original, his prominence was established by his singular contributions to organic chemistry. He wrote and spoke extensively about the application of chemistry to many aspects of life, garnering a broad receptive audience in Europe, England, and the United States (Brock 1997). This time period was also significant because it included one of the philosophical shifts from a vitalistic view to a reductionistic one. Despite their work to refute humus theory and the position that plants required factors other than minerals from manure, Saussure and Sprengel nevertheless held that an organic form of these minerals was best. This position was probably influenced by the residuals of vitalism, but also by the prevailing belief that plant C was derived from the soil. In contrast, von Liebig, guided by his chemistry background and also by advances in understanding of plant photosynthesis, suggested a more mechanistic explanation of plant nutrition that manures could be completely replaced with inorganic sources of minerals. Applying the stoichiometric approaches of Lavoisier and Guy-Lussac, he saw a day when farmers would use a balance-sheet approach to soil fertility, calculating the amounts of nutrients removed during crop harvest and replacing only those minerals to reestablish fertility. Brock (1997) provides a thorough account of von Liebig's application of chemistry and the impact he had on various aspects of contemporary society. Less recognized today are some notable errors of von Liebig, such as his dogged adherence to the belief that plants obtained their N from the atmosphere and that applications of N were a waste of the farmer's money, despite experimental demonstrations to the contrary. However, the greatest irony of his work, given its basis for the chemical fertilizer industry, was the commercial failure of his own "patented manure," which led to a dramatic shift in his philosophy, from what might be characterized as the father of agronomic reductionism to one of warning against the negative consequences of its implementation:

> Unfortunately the true beauty of agriculture with its intellectual and animating principles is almost unrecognized. The art of agriculture will be lost when ignorant, unscientific and short sighted teachers persuade the farmer to put all his hopes in universal remedies, which don't exist in nature. Following their advice, bedazzled by an ephemeral success, the farmer will forget the soil and lose sight of its inherent values and their influence. (von Liebig 1855)

Although von Liebig was fundamentally a reductionist and to him the most satisfying explanation of biological phenomena was a chemical one (Lipman 1967), he still assumed the existence of a vital force. Moreover, he did not see a contradiction here, and viewed vitality as wholly analogous to forces like gravitation, which would never be explained but whose manifestations could be studied. Later in his career, though he remained a reductionist in methodology, he seemed to recognize the limitations of explaining life processes solely by chemistry and despite his earlier writings, he was actively critical of those who tried to do so (Lipman 1967). His later references to natural laws and the inherent value of soil suggests an appreciation of processes that transcended the stoichiometric approach that was his original goal for agriculture. Nevertheless, von Liebig set in motion a train that even he could not slow, and his later more balanced (one might even say more ecological) views on plant nutrition have been largely ignored for his earlier strictly chemical ones.

The reductionistic approach to agricultural research picked up steam in the twentieth century in its goal of improving agriculture by an application of the principles of the Industrial Revolution and the mechanistic view that continues to dominate science. This reductionistic perspective is reflected in the structure of agricultural colleges of U.S. and European universities, which are almost universally divided into disciplines, if not departments, of soil science, agronomy, horticulture, weed science, entomology, and plant pathology. For most of the past century, research has progressed as semiautonomous packets of knowledge, addressing separate aspects of the production system. Although there has been some movement toward a less-fragmented departmental structure in many U.S. universities, with notable exceptions, this trend has been driven more by the realities of declining state and federal budgets than by ideological shifts, and an overarching model with stated principles and presuppositions to structure these efforts is still lacking (Norgaard and Baer 2005).

9.2.2 SCIENTIFIC PHILOSOPHY: REDUCTIONISM AND HOLISM

Understanding the philosophical underpinnings of research and its history is not merely an esoteric exercise. Philosophy is the first layer of the research process (Saunders et al. 2003, p. 83) and one's presuppositions, whether consciously examined or more commonly unconsciously accepted, determine the nature of the questions addressed, the research methodology employed, and the interpretation of the results. For this reason, research cannot be completely objective or value-free, as is commonly claimed. Understanding the philosophic position of others is also essential for intelligent discussions of important issues. As I discuss below, the frustration, name-calling, and caricaturizing that occurs between opponents dealing with agricultural issues is symptomatic of a failure at this more fundamental level.

Much of the current conflict in agricultural research can be attributed to the fundamental differences between the philosophies of reductionism and holism whose history can be traced from the ancient Greek philosophers (Davies 2001) through the long-standing dialectic between mechanists and vitalists, who held that living organic forms were qualitatively distinguished from inorganic forms by the presence of an unexplained "vital or life force." Present-day molecular biology is motivated by the same philosophical underpinnings of the seventeenth- and eighteenth-century mechanists who, in reaction to vitalism, held that organic life could ultimately be explained by the same atomic phenomena governing inorganic chemistry (Davies 2001).

In its purest form, reductionism searches for mechanisms among the constituents of a system and holds that understanding the constituents is sufficient to understanding the system. Reductionism helps us make sense of the world; it is intuitive and generally it works. The beauty of reductionism is its simplicity and the relative ease of experimentally demonstrating cause and effect within system components. By controlling the variables, interpreting experimental results is relatively straightforward. On the other hand, the weakness of reductionism derives from its inability to predict system behavior that arises from interactions among its components.

What has maintained this dichotomy virtually since the advent of scientific philosophy? At its core seems to be an inability to resolve the conflict between those who seek to set biology on equal footing with the "hard" sciences of physics and chemistry by establishing a single set of principles that transcends levels of complexity and those with the sense that somehow such a purely mechanical explanation falls short of fully capturing one's experience with the intricate beauty and amazing interactions that pervade living systems. This philosophical tension may even play out within the individual (Davies 2003). The vitalists resolved the conflict by drawing a line between inorganic and organic, conferring on the latter a life force that could not be dissected into smaller components. Descartes, who advocated a materialist explanation of nature and originated the reductionistic approach, nevertheless, could not accept this model completely, and in a sense only moved the line of demarcation. Although for him all of life could be explained as an elaborate mechanism, in his philosophy of material dualism, the human mind still represented an indivisible and nonmaterialistic construct distinct from the body. Even in modern times, Richard Dawkins, one of the most strident of contemporary reductionists, acknowledges this intuitive difference that sets biological complexity apart: "whatever we choose to call the quality … it is an important quality that needs a special effort of explanation. It is the quality that characterizes biological objects as opposed to the objects of physics" (1987, p. 15).

9.2.3 COMPLEX SYSTEMS

At the two extreme poles of reductionism and holism, the reductionist would assert that knowledge of the parts is both necessary and sufficient, while the holist holds that any deconstruction is likely to result in loss of system properties. Thus, the modern-day holist replaces the intractable "life force" of the vitalist with the concept that special properties emerge from the interaction of system components, properties that cannot be predicted by studying the components in isolation. In its extreme, holism sees no value in isolating components, as all are important; they are so interconnected that elimination of any components or study of these components under artificial conditions would be like pulling a thread from a fabric, causing the whole cloth to unravel. It is this extreme position to which today's reductionists are most likely to react, considering holism as lacking in scientific merit and in some cases akin to New Age mysticism. It is a common logical error to take this view a step further and equate reductionism with the scientific method. However, while most debates try to caricaturize the opposing positions and form qualitative distinctions, these alternative views actually represent points on a continuum. As I argue below, reductionism and holism should be viewed as complementary rather than contradictory. The moderate position recognizes that knowledge of components is not sufficient to understanding system function, but if done properly, reductionist methods can reveal patterns of emergence within systems.

The central aphorism of the holistic view is that "the whole is greater than the sum of its parts." For complex systems, it is more accurately rephrased as: "the whole is *different* than the sum of its parts," in that the result of component interactions may produce a state qualitatively different from its components. This concept of emergent properties is a key trait that characterizes complex systems, which makes them more than just "complicated systems." In natural ecosystems, the number of components or organisms makes the system complicated, but it is the special relations and interactions of particular species in the ecosystem that express its emergent properties, determine its functioning, and thus make it complex, according to this definition. A Swiss mechanical watch is complicated due to the number of components, the precision required for the construction of each of them, and the importance of putting them together in the correct relationship to each other. However, it is not a complex system in the sense that the components interact only to produce two states: a watch that works or a watch that does not work. Of the multitude of possible combinations of parts, none results in a functioning device that is un-watch-like.

In reality, most agricultural researchers do not hold an absolute view of reductionism, but accept the complexity of biological systems and may recognize the possibility of emergent properties, at least on an intellectual level. Dawkins (1987) fervently attempts to dispel the myth of the "nonexistent reductionist—the sort that everybody is against, but who exists only in their imaginations—tries to explain complicated things directly in terms of the smallest part." Nevertheless, we must be mindful that there is a difference between holding a position intellectually and having that position inform our practice of research. In reality, most agricultural research proceeds as if emergence is not all that important, that unintended negative consequences will be rare and will become apparent with the implementation of new technologies. Experiments are designed to isolate the constituents of interest, eliminate all others, and control conditions that might otherwise add "unwanted noise" and reduce statistical sensitivity. Again, this is the power of the reductionist protocol. The problem comes when the conclusions of these controlled experiments are extrapolated to make predictions about the operation of the larger system, where not all the variables can be controlled. Moreover, the nature of the controlled experiment biases against discovery of effects among more-distantly connected components of the system, as illustrated by the example above of DDT causing a loss of raptor birds. In this way, the view that nontarget effects are rare or unimportant becomes self-fulfilling, but a search for transcomponent linkages suggests that unintended consequences are in fact common in agriculture.

Thus, our current paradigm for agricultural research, for all its strengths, falls short on two levels: (1) in extrapolating system function from the study of individual components, we fail to predict the nonlinear effects and special properties that emerge through the interaction of these components, and (2) by constraining the number of parameters that vary and by restricting the number of responses measured, we limit our understanding of how agricultural systems operate and constrain our view of the full consequences of our technologies. When this view informs the production system design, reducing the farm to controllable components of the industrial model, we diminish the benefits that arise from the interactions of a self-organized ecosystem and make it more susceptible to negative unintended consequences. As a result, the fragmented approach of component research becomes primarily reactive, shifting efforts to solving problems as they arise. The danger of the problem-solving perspective is that it sets up a context that focuses more on ameliorating symptoms rather than discovering root causes.

9.3   PRINCIPLES FROM ECOSYSTEM ECOLOGY

The philosophical antecedents of modern agriculture also included the notion that nature needed to be controlled and brought under human dominion (Meine and Knight 1999, p. 88). This approach represents a level of misplaced confidence that with enough work and intelligence, humans can solve all the problems and "triumph" over nature. Peterson (2005) presents a framework to determine the appropriateness of a management approach based on two considerations: controllability of the system and certainty of the conditions. He points out that most cases of ecological management have been developed with the assumptions of a high degree of system control and a low degree of uncertainty. Unfortunately, because ecosystems show complex responses due to the interactions of internal components and are susceptible to the vagaries of abiotic conditions, they are neither highly controllable nor certain. Although contemporary agriculture has its roots in the Industrial Revolution and shares its goals of maximizing productivity, economic return, and efficiency, the agricultural field is not analogous to a factory. Rather it is part of a larger ecosystem that operates according to its own set of rules, so that agriculture is more accurately viewed as a form of ecological management. Going forward, we need to adopt a broader perspective in assessing the success of our farms and bring agricultural research and system design more in line with the realities of the natural world. Systems ecology is a relatively young science that is still on a steep discovery curve; nevertheless there are core principles of this science that can inform an improved design of agricultural systems.

9.3.1   NATURAL ECOSYSTEMS

Natural ecosystems are complex self-organizing systems characterized by nonlinear dynamics with multiple steady states and emergent and chaotic properties (Hearnshaw et al. 2005). The biological complexity of ecosystems derives from the interplay of both deterministic and stochastic forces, which makes predictions about system behavior particularly difficult (Spagnolo et al. 2004). Depending on the relative strength of these forces, a system can show surprising resistance to large perturbations in some circumstances, but can be transformed by seemingly small disturbances in others. How an ecosystem responds to the inherently noisy abiotic environment depends in great measure not only on the number and diversity, but also the structure, of linkages among its biotic components (Kolasa 2006).

Following a major disturbance, natural systems are characterized by an orderly process of ecological succession, in which there is gradual replacement of fast-growing and rapidly reproducing species with slower-growing species that possess greater competitive abilities (Barbour et al. 1999). This competitive progression is accompanied by a narrowing of niche dimensions, an increase in species number and functional diversity (which in turn creates new niches, resources, and microhabitats), a shift from short linear food chains to more complex food webs, greater efficiency in resource utilization, shortening of nutrient cycles, and ultimately greater system stability. During this process, species are under constant selection for a number of traits simultaneously, many of which are in dynamic conflict for resources, resulting in a genotype that optimizes the trade-offs among these traits to maximize fitness. The multidimensionality of selection, which is lacking in our design of agroecosystems, is formalized in the ecological niche concept; however, even this multidimensional perspective of the physicochemical properties of the habitat produces an overly simplistic view of whether a species thrives.

9.3.1.1   Self-Organization in Natural Ecosystems

A concept important to ecosystem function is stability. Community stability is expressed in relation to three parameters (Barbour et al. 1999): (1) resistance—the ability to remain unchanged during a period of stress, (2) resilience—the ability to return to its original state following stress or disturbance, and (3) persistence—the ability to remain relatively unchanged over time, which is largely determined by 1 and 2. Community ecologists have demonstrated that relatively mature ecosystems are not just aggregations of individual organisms that all find a particular abiotic environment favorable, but rather are made up of organisms whose characteristics complement each other and whose functions are highly integrated. The presence of other organisms affects a species' ability to exploit or tolerate conditions, not only through competition or mutualism, but also through their impact on the habitat itself, either opening up new niches or shrinking niche dimensions for other species (Sterelny 2005). The cumulative effect of the biotic component of an ecosystem affects the local environment in a nonlinear fashion.

Although the mechanisms that determine the relationship between biodiversity and stability are still open to debate (McCann 2005), there is general agreement that stability tends to be greater in ecosystems with a larger number of functionally diverse species. The propensity for an ecosystem to persist likely depends less on the number of functional linkages and more on the variation in their strength and the structure of their network of interactions (Moore and Hunt 1988, de Ruiter et al. 1995). Kolasa (2006) argues that the tendency for self-organization and the level of component integration determine the stability of ecosystem communities. The result is that mature ecosystems tend to exist in a state of dynamic equilibrium, maintained in the face of biotic and abiotic variability by the mutualistic and competitive interactions of its component species, until the limits of stability are exceeded and a new equilibrium state is assumed (Solé and Bascompte 2006, p. 27ff). Ecological networks with stabilizing feedbacks also show a greater likelihood to persist than those that do not (Lenton 1998), and simulations by Cropp and Gabric (2002) suggest that within the constraints of the external environment, ecosystems will evolve to maximize resilience, a conclusion confirmed by field observations (Johnson 1990).

Descriptive studies indicate that food webs normally comprise a few strong linkages and many weak ones. With the goal of making systems more understandable, studies usually simplify analysis to focus only on the strong ones. Classic predator-prey models describe the strong out-of-phase oscillations that occur in short linear trophic chains involving a plant, an herbivore, and a specialist predator (Figure 9.1A). However, making these chains more weblike either by introducing weak resource competitors (Figure 9.1B) or particularly by including weak predator–prey associations (Figure 9.1C) significantly dampens population oscillations, which increases system stability by suppressing herbivore outbreaks (McCann 2005). Introducing another weak interaction with the herbivore (e.g., a generalist predator) also stabilizes the system while addition of another strongly interacting (specialist) predator leads to population oscillation of even greater intensity. Thus, pest management based only on strong impacts may contribute to higher-amplitude population fluctuations in the longer term.

9.3.2   Contrasting Natural and Agricultural Ecosystems

The impact of our adherence to reductionist presuppositions and the tenets of the industrial model are made evident by comparison of the properties of today's agricultural systems with those described above for natural systems. Table 9.1 shows how the individual problem-solving approach to management design (from largely independent control of its components) has resulted in an ecosystem different in many ways from the one it replaced.

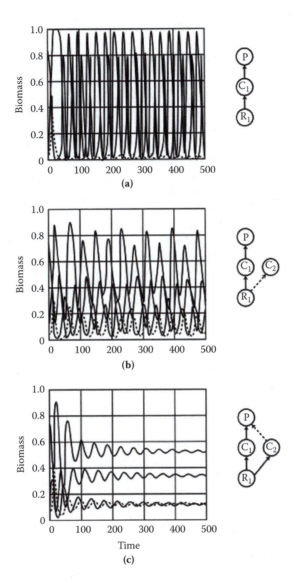

FIGURE 9.1 Model simulation illustrating the impact of trophic-interaction complexity on community stability. (a) Linear food chains involving strong interactions among a plant (R_1), specialist herbivore (C_1), and a specialist predator (P) show chaotic dynamics with wide fluctuations in population densities of all species. (b) With the addition of a generalist herbivore (C_2) that is a weak consumer of the plant, species densities still fluctuate but amplitudes are reduced. (c) The additional broadening of the predator's prey range to include C_2 or inclusion of a generalist predator with a weak impact on herbivore populations stabilizes all trophic interactions and species quickly establish equilibrium. (Based on models and parameters from McCann et al. 1998. Figure reproduced from McCann, K. S. In *Ecological Paradigms Lost: Routes of Theory Change*, Cuddington, K. and Beisner, B., Eds. Elsevier Academic Press, Burlington, MA, 2005, 183–209. With permission.)

Rather than being maintained by biotic interactions and internal feedbacks, the operation of the agricultural system is goal oriented and controlled externally by human decisions and inputs. With the typically singular goal of maximizing yield, agricultural fields are usually managed to prevent ecological succession and maintain an early successional state through tillage. Biocides are then used to replace biological sources of population control and to eliminate competition by other plant species. Reducing the number of biotic components produces broad open niches and creates

TABLE 9.1
Characteristics of Prevailing Agroecosystem Compared to Natural Ecosystems

	Natural Ecosystems	Conventional Agriculture
Control	Internal via subsystem feedback loops	External, goal-oriented
Successional state	Move toward later successional states	Continuously converted back to early successional state
Nutrient cycles	Closed	Open
System composition and function	Determined by the success of biotic components interacting with the environment and each other	Artificially managed to maximize yield
Nature of selection on organisms	Multifactorial	Single traits
Dominant source of nutrients	Soil food web	Chemical salts
System complexity	Complex with many interacting biotic components	Simplified with relatively few biotic components
Microhabitats	High diversity	Low diversity
Niches	Niches mostly occupied	High number of open niches
Food chains	Complex and weblike	Relatively short and linear
Symbiotic relationships	Common	Few
System stability	High (more resistant or resilient to disturbance and invasion)	Low (more susceptible to disturbance or invasion)
Measurement of system efficiency	Energy and resources	Economics

a system with food chains that are relatively short and linear. Accompanying this depopulation with high nutrient inputs creates a habitat with even greater susceptibility to invasion. Conditions meant to enhance crop growth also favor rapidly growing weeds and invasive plant species, reduce ecosystem stability, and increase susceptibility of the crop to outbreaks of phytophagous insects and disease agents.

Another striking feature that distinguishes conventionally managed agroecosystems from natural systems is the openness of their nutrient cycles. The use of highly soluble salts makes nutrients in agricultural soils highly mobile and therefore susceptible to leaching. In contrast, because N and other nutrients in natural ecosystems come through the detrital food web, their levels are highly regulated and a large portion is retained in organic matter fractions of these soils, the turnover of which is driven mainly by microorganisms. Inefficiency in agricultural nutrient utilization also results from the poor synchrony between N applications and crop use. Natural plant communities comprise species with different temporal patterns of nutrient uptake, which tend to reduce seasonal variation in nutrient demands and contribute to local retention. In contrast, because most agricultural fields are monocropped, N uptake follows the temporally more restricted requirements of a single species, resulting in underutilization. Raun and Johnson (1999) calculated N use efficiency (NUE) of cereal crops to be 33 percent worldwide and Cassman et al. (2002) report NUEs of 31 percent for continuous irrigated rice in India and China and 37 percent for maize in the North-Central United States. These inefficiencies have been tolerated due to the relatively low price of fertilizer; however, nutrients leaving the agricultural system lead to eutrophication and significant disruption of natural ecosystems (Matson et al. 1997, Vitousek et al. 1997).

In total, the combined effect of these changes is to maintain agricultural fields in a constant state of primary succession, whereas natural systems move toward greater species diversity, complexity of interactions, and stability. Sustaining this disturbed state not only reduces the resistance and resilience of agroecosystems to abiotic perturbation, but also requires significant inputs of chemicals and energy.

More than 20 years ago, ecologists studying a wide range of degraded ecosystems determined that irrespective of the source of stress and the ecosystem type, communities responded in a remarkably similar fashion (Odum 1985, Rapport et al. 1985, Rapport and Whitford 1999). Rapport et al. (1985) suggested that all dysfunctional ecosystems are characterized by an "ecosystem distress syndrome," showing significant change in the following parameters:

1. Nutrient cycling—generally, unhealthy ecosystems show an increase in leakiness of nutrient cycles.
2. Community diversity—a reduction in species diversity is one of the most widespread correlates of strong ecosystem distress.
3. Successional retrogression—stressed ecosystems show increased invasibility with a shift in species composition toward more r-selected opportunistic species, characteristic of early ecological succession.
4. Primary productivity—stressed systems may show either an increase or decrease in net primary productivity. Higher primary productivity is associated with early successional stages, while stress due to extreme resource limitations may cause a decline in productivity.
5. Organism size distribution—severe stress generally results in a reduction in the average size of dominant species.
6. Increased disease incidence—stress on organisms reduces their defensive capabilities and/ or the absence of biological control of pathogens releases them to thrive.
7. Amplitude of population fluctuation—with a reduction in species diversity and an increase in invasion by new species, there is a breakdown in stable linkages among species, a reduction in the complexity of food webs, and a loss of stability, allowing outbreaks of certain species, which rapidly crash.

Looking at the list of symptoms associated with ecosystem distress, one cannot help but note how many of them characterize our agricultural systems.

9.4 EXAMPLES OF SYSTEM BEHAVIOR IN AGRICULTURE

Deconstruction of the ecosystem, with independent management of its component parts for agriculture, has come at the price of lower system stability, the loss of free beneficial services, and unintended consequences for other system components. Despite our efforts to simplify the system, biological linkages remain. Here we consider how some of the specific management practices create problems in other system components.

9.4.1 BELOWGROUND AND ABOVEGROUND LINKAGES

The most dramatic impacts of agricultural management and its unintentional consequences are expressed in the soil. The soil is where humans have caused the greatest alteration of the environment and where management requires the greatest inputs and energy costs for crop production. The focus on increasing nutrient levels to maximize agricultural productivity has meant that soil science has been dominated by study of the physical and chemical aspects of soil. By contrast, soil management based on principles of systems ecology requires a greater understanding of soil biology, and how management practices affect biological function. It was not long ago that soil ecology was considered an esoteric field (Coleman 2008), but there has been a growing recognition of the importance of the biological component to soil and ecosystem function. The soil is the literal foundation for plant communities and contains the "central organizing centers of terrestrial ecosystems" (Huang et al. 2005; Coleman 2008). In unmanaged temperate ecosystems, cycling of plant nutrients and carbon, creation of soil structure, and maintenance of soil moisture levels are as much biological as chemical processes. The food web of soil fauna and microflora occupies a central position in many aspects of a fully functioning soil: providing both a sink and source of plant nutrients, transforming nutrients for plant use, forming mutualisms with plants, suppressing pathogenic organisms, and contributing to soil formation and structure stabilization (Dalal 1998).

Although additions of inorganic N can elevate soil C initially, there are generally long-term negative impacts on biologically active soil organic matter, microbial biomass, and soil N pools compared to inputs of organic nutrient sources (Fauci and Dick 1994; McCarty and Meisinger 1997). These impacts on the biological component of the soil are responsible for loss of organic matter, leading to soil compaction, reduced resource utilization efficiency, and disruption of internal nutrient cycling with increased leaching of nutrients and production of greenhouse gases (Weil and Magdoff 2004). In contrast, factors that enhance the influx of organic matter to the soil increase numbers and diversity of soil fauna, both in natural and managed ecosystems. In a woodland habitat, a positive relationship between organic matter and soil microarthropods is often found (Poole 1964; Mitchell 1978). Scholte and Lootsma (1998) measured significantly larger spring populations of fungivorous *Collembola* in agricultural soils amended with farmyard manure. Empirical comparisons of conventional and organic soil management generally find higher and more diverse populations of soil fauna associated with C-based nutrient management than with purely inorganic nutrient inputs. In replicated plots under different long-term crop management, El Titi and Ipach (1989) found higher levels of *Collembola* and predatory mites in soils whose fertility input was almost solely manure, compared with conventional plots, which received only mineral fertilizers. Populations of saprophytic and predatory nematodes were also higher, while plant-parasitic nematodes were reduced. Similar results have been obtained from numerous comparisons of organic and conventional farms (Paoletti et al. 1992; Krogh 1994; Moreby et al. 1994; Yeates et al. 1997). In a 21-year study of replicated plots comparing two organic and two conventional management systems, Mäder et al. (2002) found that the organic systems had higher biological activity, with greater earthworm and epigeic predator densities, higher mycorrhizal colonization of plant roots, and greater diversity of weeds and microbial communities, which had a respiratory quotient more characteristic of successionally mature ecosystems. Although yields were on average 20 percent lower in the organic systems, energy expenditures per unit area were 36 to 53 percent lower and nutrient inputs were 34 to 51 percent lower. The greater biodiversity both above- and belowground, the more efficient utilization of energy and nutrients, along with other positive edaphic effects such as greater soil aggregate stability, should contribute to making these systems more resistant to perturbations.

Less obvious effects of disrupting the soil food web can also be seen for plant health and aboveground interactions with pests and disease. As an example of the application of reductionist presuppositions to this area, it is a commonly stated view that nutrients from chemical fertilizers are no different from those derived from organic matter (IFA 2003):

> Plants cannot distinguish between nutrients supplied from organic or inorganic sources. All nutrients to be absorbed by plants have to be available in their inorganic form, irrespective of their source.

Or as one leading agriculturalist puts it: "the plant doesn't give a damn whether nitrogen and phosphorous come from manure or from a bag of commercial fertilizer" (http://www.capmag.com/article.asp?ID=2106).

Aside from the inaccuracy that plants can only take up N in the inorganic form (Näsholm et al. 2000), the primary inference is that since elements are equivalent at the atomic level irrespective of their source, they are interchangeable. The error in this logic comes from its extrapolation to higher levels of the system hierarchy. In reality, the addition of animal and plant manures affects the soil environment as well as the plant quite differently than does application of soluble minerals, and in ways that are not immediately apparent. First, is the increase in soil organic matter, with the physical benefits accrued to soil as described above. In addition, there are differences in the temporal dynamics of mineral availability to plants (Zogg et al. 2000). Both of these effects on the soil will have significant impacts on plant health and metabolism. Also affected are associations with beneficial microbes, such as rhizobacteria and mycorrhizae (Nichols and Wright 2004), which change the plant nutritional state and can trigger aboveground secondary metabolism through defensive-signaling pathways (Audenaert et al. 2002).

One central tenet of most organic agricultural management is the use of C-based fertility. Proponents of this farming system have long claimed that it produces crops with a "healthier" constitution, which makes them more resistant to insects and plant diseases (Howard 1943). Although this claim was long reputed or ignored by the scientific community, renewed interest in organic farming has stimulated a number of on-farm studies and replicated plots or controlled greenhouse experiments that generally support the claim for a range of crops and foliar pests: leaf beetles and planthoppers on rice (Andow and Hidaka 1989), various pests on tomato (Drinkwater et al. 1995), European corn borer on maize (Phelan et al. 1995, 1996), European corn borer and aphids on maize (Bedet 2000), aphids on maize (Morales et al. 2001), and Colorado potato beetle on potato (Alyokhin et al. 2005).

My colleagues and I conducted a series of controlled greenhouse studies to test the claim that organic soil management reduces plant susceptibility to insects, and to eliminate alternative explanations such as differential predatory activity, crop rotation, and planting dates (Phelan et al. 1995, 1996). Amending soils collected from neighboring organic and conventional farms with compost, cow manure, or chemical fertilizer allowed us to separate short-term fertilizer effects from long-term soil management history effects. After planting maize, female *Ostrinia nubilalis* (European corn borer) were released into the greenhouse to determine egg-laying preferences. In each of four experiments, females consistently laid fewer eggs on corn plants in soil from organic farms than on plants in conventional soil (Figure 9.2A). In addition, there were no differences in egg laying on plants receiving different amendments in organic soil, whereas in conventional soil, response to plants receiving different amendments varied in an unpredictable manner. As a result, variance in egg laying across the four experiments was >18 times higher among plants in conventional soil than among those in organic soil (Figure 9.2B).

9.4.2 BIOLOGICAL BUFFERING

The dampening of plant susceptibility to insects and disease led to the concept of biological buffering (Phelan et al. 1995; 1996), which asserts that a more complex soil community supported by the influx of active organic matter tends to moderate fluctuation in the soil environment. Biological buffering highlights the ecological processes operating in soils and posits that ecosystems whose nutrient cycling is predominantly modulated by the soil food web possess greater ecological stability, with resilience and resistance to perturbation. Thus, it extends the capacity of soil organic matter to buffer physical parameters, such as moisture levels and pH, to biological interactions involving soil communities and plants, which in turn influences plant interactions with microbes, herbivores, and other plants. I argue that this stability derives from the greater complexity of biological interactions. The greater soil species diversity and linkages created by an active detrital food web may temper soil population fluctuations (as discussed above), while additional mechanisms that transfer this stability above ground through greater plant resistance may include (1) modulation of plant mineral nutrient availability by the soil food web, and/or (2) an enhanced plant systemic defense induced by beneficial microbes interacting with plant roots.

With regard to the first mechanism, the mineral balance hypothesis (Phelan et al. 1996; Phelan 1997, 2004) conceives an optimal mineral balance for each plant species whose limits are defined by both sufficiency and by proportions among the minerals. When provided its optimal mineral balance, a plant not only grows well but possesses enhanced resistance to herbivory. Occupying a habitat that provides this optimal blend of nutrients is probably the exception for plants, and therefore they have evolved a number of physiological mechanisms or biological associations to better meet their nutrient requirements (Marschner 1995). A key prediction of the hypothesis is that when

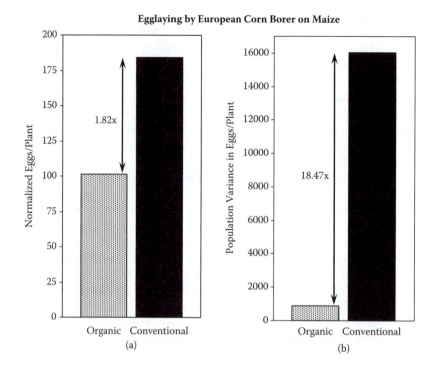

FIGURE 9.2   Meta-analysis of egg laying by *Ostrinia nubilalis*, the European corn borer, on maize planted in the greenhouse in soils collected from neighboring organic or conventional farms. Analysis conducted on results from four replicated factorial experiments with amendments of dairy cow manure, cow manure compost, or chemical fertilizer in each soil type: (a) mean egg laying by soil type across fertilizer treatments, normalized to account for differences in total egg laying among experiments, and (b) variance (sum of squares) in egg laying across fertilizer treatments and experiments.

mineral nutrition is optimal, resistance to herbivory does not require a trade-off with plant growth; the opposite of that predicted by most models of plant defense allocation (Herms and Mattson 1992; Stamp 2003). As herbivores are generally N-limited, a positive association between N level and insect performance is predicted (White 1984) and generally found (Mattson 1980; Waring and Cobb 1992); levels of N are 2 to 4 percent in nonleguminous plants, while those of insect tissues are 7 to 14 percent (White 1984). Therefore, susceptibility to insect pests is generally viewed as a necessary trade-off of providing sufficient N fertilizer to maximize crop yield. However, the mineral balance hypothesis emphasizes that qualitative aspects of plant N (form of N and its proportion to other nutrients) must also be considered (Phelan 2004). When nutrient imbalances occur in the plant, the metabolic machinery is not able to work at peak efficiency, and various mineral imbalances can result in elevated levels of free amino acids (Court et al. 1972; Mengel and Kirkby 1987; Marschner 1995; Amancio et al. 1997; Alam et al. 2001). Elevated levels of amino acids and other simple soluble metabolites can affect insect herbivores in multiple ways (Cockfield 1988): (1) they are metabolically more accessible to insect herbivores than are proteins, nucleic acids, and structural carbohydrates; (2) due to their greater solubility and mobility in the plant, they are more available to sucking insects feeding on phloem or xylem tissue; (3) their presence reduces the effectiveness of some plant defensive compounds, such as proteinase inhibitors (PINs); and (4) they act as feeding and oviposition stimulants for many herbivorous insects. Testing the mineral balance hypothesis, Beanland et al. (2003) found a nearly inverse relationship between the growth of soybeans and two insect herbivores in response to different proportions of boron (B), iron (Fe), and zinc (Zn). Plant biomass was greatest at intermediate proportions (B:Fe:Zn 2:2:1), whereas Mexican bean beetles (*Epilachna varivestis*) were 45 percent smaller and soybean loopers (*Pseudoplusia includens*) were 20 percent smaller after developing on leaves from these plants compared with leaves from B-deficient plants. Also consistent with the predictions of the mineral balance hypothesis, free amino acid levels were around 10 times higher in the nutritionally unbalanced plants (Phelan et al., unpublished data).

Soils with an active food web are hypothesized to better approximate the optimal plant nutrient balance because they are closer to the soil environment in which plants evolved and because of their capacity to buffer mineral availability. In unmanaged and organically farmed soils where sufficient carbon is available, a greater proportion of soil N is contained in the microbial soil fraction (Jackson et al. 1989), which acts as a biological storehouse for N and other nutrients that are made available to plants in a continuous low-level supply through microbial turnover. By contrast, in conventional agricultural soils where N inputs are primarily applied as large pulses of inorganic compounds early in the season and organic matter inputs are low, carbon is limiting for microbes, the potential for biostorage in greatly restricted, and thus N remains mobile in the soil. This high availability stimulates plant growth along with rapid uptake in excess of metabolic needs, likely putting the plant in a state of nutrient imbalance. The ability of organically managed soils to buffer nutrient uptake was demonstrated by a principal-component mapping of leaf minerals in maize plants grown in soil collected from organic versus conventional farms (Phelan 1997). When grown in organic soils, tissue mineral profiles showed little variation among compost-, ammonium nitrate (NH_4NO_3)-, or unamended plants, whereas those in conventional soil showed dramatic shifts, particularly when amended with NH_4NO_3.

Alyokhin and Atlihan (2005) report that potatoes grown in manure-amended soil were poorer hosts for Colorado potato beetle than potatoes receiving only chemical fertilizer. In a no-choice test, females laid fewer eggs, larvae had lower survivorship to 2nd instar, showed slower development to adult, and consumed less foliage when held on manured plants compared to chemically fertilized plants. Similar patterns were seen in the field, where potato plots fertilized primarily with cow manure had lower densities of Colorado potato beetle than plots receiving only chemical fertilizer (Alyokhin et al. 2005). Potato plants receiving manure were similar in size to chemical-only potatoes, but had higher tuber yields. In support of the mineral balance hypothesis, multiple regression models of leaf-mineral profiles showed strong association with beetle populations, accounting for up to 57 percent of the variation in beetle densities.

Bypassing the detrital food web and maintaining high levels of soil nutrients can also contribute to plant stress by increasing susceptibility to moisture deficits. Because important plant nutrients such as N and P typically occur in suboptimal levels and are heterogeneously distributed in natural soils, plants have evolved the plasticity to increase the effectiveness and efficiency of nutrient foraging. For many species, when the root of a nutrient-deficient plant encounters a patch of high N or P, lateral growth is stimulated over vertical growth, causing roots to be concentrated in the nutrient patch. In nutrient-rich soil, when all the roots encounter high nutrients, plants reduce allocation to root growth and resources are preferentially allocated aboveground (Drew and Saker 1978; Fitter et al. 1988; Ericsson 1995; Zhang and Forde 1998). Although this strategy for optimizing resource allocation is adaptive under natural conditions of low nutrients and interspecific plant competition, in chemically fertilized agricultural soils where mineral N levels are high near the soil surface, it causes plants to produce a shallow root system and relatively large shoot biomass. This combination reduces their capacity to cope with suboptimal moisture levels.

9.4.2.1   Soil Communities

Led by the precepts of agronomic reductionism, it is easy to see why soil quality came to be defined almost solely by its physical and chemical characteristics, with little consideration of management effects on the biotic component of soils (Karlen et al. 1997; van Bruggen and Semenov 2000). This dominant view came about partly as an outcome of the shift from organic sources of plant nutrition to chemical ones and partly as a result of our lack of understanding of how soil communities function. The shift to inorganic fertility management has resulted in a loss of the biological complexity hypothesized to underlie higher-level biological buffering and system stability. Studies generally find higher diversity of soil organisms in undisturbed native soils compared to agricultural soils (Curry and Good 1992; Wardle 2002; Cole et al. 2002; Sousa et al. 2006) or in organically managed compared to conventionally farmed soils (Lagerlöf and Andrén 1991; Krogh 1994; Petersen 2000). In addition to the destructive physical impacts of tillage on selective members of the macro- and mesofauna (Wardle 1995; Hülsmann and Wolters 1998; Marasas et al. 2001), differences in soil biodiversity are also traced to fertilizer use. Elevated N causes a reduction in the numbers and diversity of nematodes, collembola, and soil mites (Doelman and Eijsackers 2004, pp. 111–112), as well as saprophytic fungi and mycorrhizae (Berg and Verhoef 1998). The reduction in mesofauna diversity is due to strong dominance by a relatively small number of nitrophilous species (Doelman and Eijsackers 2004). Nitrogen enrichment also causes a shift in the dominance of fungal-based energy channels toward bacteria (Moore and Hunt 1988; Bardgett and McAlister 1999), which has other effects on the decomposition process and detrital food web. Whereas the early stages of organic matter decomposition are accelerated by N enrichment, the later stages are slowed (Fog 1988) and the positive effect of microarthropods on nutrient mineralization from litter is neutralized (Verhoef and Meintser 1991). Ecosystems can evolve toward more steady states at either low- or high-nutrient conditions; however, models predict that ecosystems that are adapted to low-nutrient conditions are severely destabilized by the influx of nutrients (Cropp and Gabric 2002), such as would describe the transition of natural systems to intensive agriculture.

Evidence of the biological buffering concept at the microbial level is provided by the work of van Bruggen and colleagues on fine-scale soil microbial population dynamics (van Bruggen et al. 2006 and references therein). Because soil microbial communities are under constant flux, use of keystone (indicator) species or even species-compositional analysis has not proved to be a reliable predictor of soil health and stability. Following a temporary disturbance in the soil, composition of the microbial community moves through a succession of functional groups. For example, with an infusion of organic matter, the succession is led by a spike in populations of the fast-acting copiotrophic microbes, which are primarily responsible for the fast breakdown of substrates. The rapid crash of these populations is followed by a smaller rise in the functionally heterogeneous hydrolytic bacteria, which is followed in turn by oligotrophic bacteria (Zelenev et al. 2006). The latter group shows a slower response and is suppressed by high resource availability, but has a dampening effect on the population fluctuations of copiotrophs. Several studies have demonstrated a regular wave-like oscillation in microbial population density through both space and time in response to nutrient influx that is amenable to harmonics analysis (Semenov et al. 1999; van Bruggen et al. 2000; Zelenev et al. 2005). The amplitude of the population fluctuations and the time required for populations to return to their predisturbance levels varies with soil parameters, as well as the resilience of the microbial community, leading van Bruggen et al. (2006) to recommend wave parameters as a functional indicator of soil health. Microbial populations also exhibit this wavelike spatial distribution along a growing root, which appears to reflect the temporal cycles of exponential growth and death stimulated by the C-rich exudates from the root tip as it grows through the soil. The amplitude of the wavelike microbial response along a growing wheat root was greater in soil low in organic matter than in high organic matter soil (Semenov et al. 1999) and was also greater in conventional soil than in organically managed soil (Figure 9.3).

9.4.2.2   Aboveground Food Webs

The biological buffering concept and mineral balance hypothesis offer mechanisms to explain why the high-nutrient conditions associated with chemical fertilizer can create outbreak conditions for N-limited insect herbivores. Our research would suggest this bottom-up side effect of nutrient enrichment as the primary cause of destabilized aboveground food webs (Phelan 2004); however, Scheu (2001) and others have suggested a second connection, through which top-down control

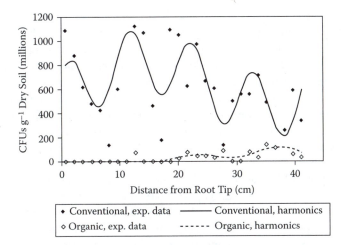

FIGURE 9.3 Wavelike pattern of population density and harmonics analysis of a green fluorescent protein (gfp)-labeled strain of *Pseudomonas fluorescens* along a wheat root growing in conventionally or organically managed soils. Note the strong dampening effect of the organic soil environment on the population density and oscillation amplitude of the experimenter-introduced bacterial strain. (From van Bruggen, A. H. C. et al. 2006. *European Journal of Plant Pathology*, 115, 2006, 105–122. With permission.)

of pests is also enhanced by soil organic matter. As discussed above, making linear food chains more weblike by introduction of less specialized trophic associations dampens the amplitude of predator–prey cycles. Spiders and other generalists are the dominant predators in aboveground food webs. However, their effectiveness in controlling phytophagous insects may be diminished by intraguild predation. Because epigeic predators are food-limited in agricultural systems, stimulation of the detrital food web by addition of organic matter may provide an alternative source of prey to reduce intraguild predation and maintain higher predator densities in the field (Wise et al. 1999). Augmentation of generalist predator populations by this "allochthonous energy subsidy" has been demonstrated by a number of studies, although evidence of enhanced trophic cascades has been more limited. Hines et al. (2006) determined that the addition of thatch detritus in a salt-marsh meadow negatively affected aboveground herbivores by increasing spider populations. Addition of detritus in a cucumber/squash system increased *Collembola* and other detritivores, as well as carabid beetles and spiders, although an increase in fruit yield was not recorded (Halaj and Wise 2002). Rypstra and Marshall (2005) increased the density of spiders in soybean plots by the addition of compost. Spiders in the compost plots had larger abdomens, suggesting a greater availability of prey, and leaf damage was significantly reduced although herbivore numbers were not. They suggest the mechanism may have been behavioral disruption of herbivore feeding rather than outright mortality.

The unintended negative effect of broad-spectrum pesticides on beneficial arthropods is widely recognized and contributes to destabilizing predator–prey cycles. In the absence of stable predator populations, not only do primary pests reach economic thresholds before predators can catch up, populations of secondary pests arise that are normally held in check by natural enemies, requiring additional intervention. An additional consideration working against the use of insecticides is the propensity for the evolution of insecticide resistance. The metabolic mechanisms for resistance are often the same as those used for detoxifying plant-defensive chemicals, such as cytochrome P450 monooxygenases, glutathione *S*-transferases, and carboxylesterases (Després et al. 2007). Because of the evolutionary history of exposure to plant-defensive chemistry, these enzyme systems are more prevalent and more easily induced in herbivores. Because most zoophagous species do not have an evolutionary history of dealing with these compounds, they also have a much less active enzyme system for detoxifying xenobiotics (Ode 2006). A less-recognized unintended consequence of synthetic pesticide use is the phenomenon known as *hormoligosis*, in which beneficial effects on pest performance is recorded (James and Price 2002; Marcic 2003; Abdullah et al. 2006). The mechanism underlying hormoligosis may be the result of a sublethal effect of pest exposure (James and Price 2002; Marcic 2003) or it may be due to changes in plant chemistry caused by pesticides. Abdullah et al. (2006) found treating cotton plants with a number of common insecticides increased free amino acid levels, reduced sugar levels, and altered total phenolics. Although this effect has not been widely reported, it is not clear whether this is because it is rare or simply because it has not been examined. It may be that emergence of secondary pests or resurgence of primary pests that have been assumed to be caused by mortality of natural enemies may in some cases have been due to hormoligosis effects.

The negative impacts of broad-spectrum insecticides on beneficial arthropods have motivated the development of other means of control, particularly using genetic engineering. The advantages of such control strategies that may be very effective when tested against pest species become less evident when higher trophic levels are considered, in part because nontarget effects are difficult to predict. For example, exudates from glandular trichomes of pubescent crop varieties that provide resistance to early instars of pest species may be equally effective in killing small natural enemies (van Emden 2002). Pest resistance based on plant toxins may cause mortality of natural enemies that feed on the insects that ingest these compounds. Higher-trophic group interactions with genetically engineered plant resistance to pests have not yet been well studied in the field. However, according to van Emden's "golden rules" of pest management, when a single method provides adequate control, there is a greater likelihood of the evolution of pest tolerance to that method and less opportunity for compatibility with other methods (van Emden 2003). Generally speaking, the probability of positive synergy with biological control in reducing pest populations declines as host resistance increases from partial to full immunity (van Emden 2002). Partial host-plant resistance resulting from polygenic expression achieved through traditionally breeding methods usually enhances biological control. In contrast, because plant resistance achieved through genetic engineering is usually based on a single toxin, it is more likely to select for resistant/tolerant pests, and because it is more absolute in its efficacy, has a negative impact on biological control. Even when introduced resistance is based on selective toxins such as the δ-endotoxins from *Bacillus thuringiensis*, impacts on higher trophic groups may be profound. *Plutella xylostella* larvae die so quickly that there are either no hosts available for hymenopterous parasitoids of these pests or those that are found do not live long enough for the parasitoid to complete development (Schuler et al. 1999).

Most commercial genetically modified (GM) insect-resistant varieties are the result of a single mechanism, usually a toxin and usually expressed by a single gene. Recognizing the greater potential for the evolution of resistance to these varieties by pests has motivated efforts to improve the durability of crop resistance by stacking multiple genes with different mechanisms (Gatehouse 2002). The need to overcome multiple mechanisms of toxicity simultaneously significantly reduces the probability of pests developing a tolerance to these crop varieties, and this strategy is more easily achieved by genetic engineering than by traditional crop breeding; however, while the effectiveness of resistance is extended, the likelihood of negative impacts on higher trophic levels remains.

9.4.2.3 Plant Pathogens and Parasitic Nematodes

Control of soil-borne diseases by suppressive soils and various forms of organic matter is well established (see reviews of van Bruggen and Termorshuizen 2003; Stone et al. 2004; Noble and Coventry 2005), and the effectiveness against pathogens may be either general or limited to a few pathogen species. General suppression is usually not the result of a particular group of microbes, but due to the total microbial activity stimulated by a material, which reduces the availability of nutrients for pathogenic species. Natural soils (Buckley and Schmidt 2001) or cultivated soils receiving high organic matter are generally characterized by a higher microbial diversity than conventionally managed soils and are less conducive to pathogens (Workneh and van Bruggen 1994; van Bruggen et al. 2004). On the other hand, high levels of soil mineral N are often associated with a higher incidence of root fungal diseases and may amplify the oscillations of various microbial populations as described above and in Figure 9.3. Eutrophication also causes similar undesirable shifts in nematode communities toward higher numbers of plant-parasitic nematodes (PPI) and a decline in the maturity index (MI), a measure of non-PPI diversity and successional state. Bongers et al. (1997) found that the ratio of the indices (PPI/MI) rises with intensity of agricultural practice, with lowest values in natural relatively undisturbed ecosystems to intermediate for pastures and organic farms to highest for intensive conventional farms.

As with aboveground insect pests, the detrital food web appears also to provide biological buffering capacity for foliar diseases, possibly via the same two mechanisms listed for insect herbivores. Although many leaf diseases were thought to be determined primarily by climatic factors, there are numerous studies that, consistent with the mineral balance hypothesis, show a significant positive association between aboveground pathogens and plant N (Walters and Bingham 2007). Pathogen developmental success, like that of herbivores, is determined by the combined effect of nutrient availability and defensive chemistry. Newton and Guy (1998) found that the partial resistance of some barley varieties to powdery mildew was lost by increasing N. Neumann et al. (2004) found a highly significant positive relationship between leaf N and the severity of yellow rust on winter wheat such that leaf chemistry was a far stronger determinant of the carrying capacity for this disease than leaf canopy or microclimate.

With regard to the second mechanism hypothesized for biological buffering, plants respond to herbivore and pathogen attack with a rapid accumulation of secondary metabolites. Inducible plant resistance represents a just-in-time strategy that might provide an effective defense at a lower cost than continuous maintenance of defensive chemical levels. In general terms, biotrophic phytopathogens and many sucking insects stimulate production of salicylic acid (SA), which enhances defense in other parts of the plant via the systemic acquired resistance (SAR) path. Jasmonic acid (JA) mediates the induced systemic resistance (ISR) path in response to feeding by chewing insect with elevated levels of proteinase inhibitors and various oxidative enzymes (Thaler et al. 2001) and/or higher levels of secondary defensive metabolites and volatile production. Numerous studies have unveiled great complexity in these pathways, with evidence that each triggers a cascade of biochemical responses, including downregulation of basic processes such as photosynthesis and indications of extensive cross-talk between the signaling pathways (reviewed by Stout et al. 2006).

In addition to suppression of soil pathogens by organic-matter-induced microbial competition, some root and foliar diseases are suppressed via these signaling pathways. Although initially studied in relation to induction by pathogens, SAR can also be triggered by commensal or mutualistic microbes. Vallad et al. (2003) showed that resistance to the foliar disease, bacterial speck, could be conferred on tomatoes growing in soil amended with composted paper mill residual and that these plants showed increased expression of pathogenesis-related genes before pathogen inoculation. Two plant growth-promoting rhizobacteria (PGPR) strains produced systemic resistance in tomato against *Phytophthora* via a JA- and ethylene-dependent, but SA-independent response (Yan et al. 2002). PGPR-elicited JA-dependent ISR against phytopathogens has also been demonstrated in other plant species, including *Arabidopsis* (Lavicoli et al. 2003) and tobacco (Zhang et al. 2002), while fluorescent *Pseudomonas* induced an SA-mediated resistance to root-knot nematodes in tomatoes (Siddiqui and Shaukat 2004). On the other hand, Shaul et al. (1999) found that leaves of mycorrhizal-colonized tobacco plants showed a higher incidence and severity of necrotic lesions caused by *Botrytis cinerea* or tobacco mosaic virus than those of nonmycorrhizal plants. Higher susceptibility to pathogens may be caused by a suppression of plant-defense response associated with the establishment of the symbiotic infection.

9.4.2.4   Weeds

At present, agricultural weed management in developed countries is energy, chemical, and capital intensive; however, herbicides are also responsible for significant savings in labor costs and

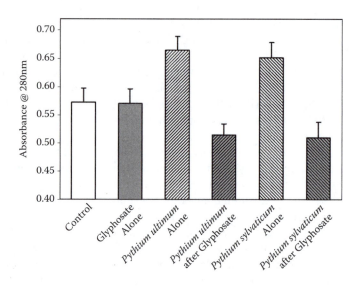

FIGURE 9.4 Glyphosate reduces the induced defensive response of bean roots, which normally increase their production of lignin when exposed to pathogenic *Pythium*. Bean seedlings were treated with glyphosate (shaded bars) or were left untreated prior to inoculation with one of two *Pythium* species (cross-hatched bars). Glyphosate alone had no effect on root lignin levels, while roots exposed to either *Pythium* species showed significantly higher lignin levels than either control-plant roots or roots of plants pretreated with glyphosate. Lignin levels determined by absorbance at 280 nm. (Modified from Liu et al. 1997.)

mechanical tillage. The convenience and commercial success of herbicides have in large measure determined the direction of weed research and to some degree stifled fundamental research on weed ecology. Despite their value for controlling weeds, the use of herbicides is another example of component focus, where problems may be created or exacerbated in other components of agricultural management. Herbicides can have negative effects on numerous nontarget organisms in the agroecosystem, including earthworms (Lydy and Linck 2003; Mosleh et al. 2003; Muthukaruppan et al. 2005; Frampton et al. 2006; Xiao et al. 2006) and beneficial soil microbes (Kremer et al. 2005). In contrast, promotive effects on root pathogens have been reported (Lévesque and Rahe 1992; Meriles et al. 2006), so that their use can enhance outbreaks of root diseases. For example, glyphosate can increase *Pythium* root rot in soybeans, for which Liu et al. (1997) found evidence of two complementary mechanisms. First, germination and growth of *Pythium ultimum* germ tubes were significantly greater in root exudates from bean plants treated with glyphosate than in exudates from nontreated plants, possibly as a response to higher levels of amino acids. Second, glyphosate suppressed the pathogen-induced lignification response of bean roots, making them more susceptible to pathogen penetration (Figure 9.4). Although glyphosate breaks down readily in the soil, this does not mean its pathogen-synergizing effects do not persist. Examining the contribution of crop production factors to the increase in *Fusarium* head blight in spring wheat, Fernandez et al. (2005) found glyphosate use within the past 18 months to be the most consistent predictor. Lynch and Penn (1980) demonstrated that treating quackgrass with glyphosate increased colonization by *Fusarium culmorum*, which then damaged the subsequent barley crop.

Somewhat further removed from the agricultural ecosystem, studies suggest a link between herbicide use (particularly atrazine) and the widespread decline of amphibian populations, due to endocrine disruption (Hayes et al. 2006), immune disruption (Brodkin et al. 2007), and/or increased susceptibility to parasites (Koprivnikar et al. 2006). Exposure to atrazine can also reduce mosquito sensitivity to insecticides (Boyer et al. 2006). Most significantly, numerous studies have linked herbicide exposure with an increased incidence of non-Hodgkin's lymphoma and leukemia in humans

(Hoar et al. 1986; Woods et al. 1987; Zahm and Ward 1998; McDuffie et al. 2001; Chiu et al. 2004; Lee et al. 2004).

Theory as well as history demonstrates that widespread use of pesticides, particularly a single compound, almost invariably leads to the evolution of resistant species. Resistance to glyphosate, the most widely used broad-spectrum herbicide, has been documented from a taxonomically varied list of weeds worldwide (http://www.weedscience.org/Summary/UspeciesMOA.asp?lstMOAID=12) that is increasing both in the acreage infested and in the number of species. Glyphosate-resistant weeds economically important to U.S. agriculture include horseweed (*Conyza canadensis*), hairy fleabane (*Conyza bonariensis*), Palmer pigweed (*Amaranthus palmeri*), common waterhemp (*Amaranthus rudis*), rigid ryegrass (*Lolium rigidum*), Italian ryegrass (*Lolium multiflorum*), common ragweed (*Ambrosia artemisiifolia*), and giant ragweed (*Ambrosia trifida*). A deeper understanding of weed ecology is needed in agricultural systems as part of the systems approach to determine why the weeds are there in the first place, to what degree we are enhancing their success by other aspects of agricultural management, and how current practices of herbicide use and tillage are creating problems for other system components.

There have been numerous attempts to identify plant traits that would allow us to predict invasiveness (Baker 1974; Grime 1977). In general terms, it is agreed that agricultural weeds are likely to possess high growth rates with short-generation times and produce large crops of small seeds, traits that are characteristic of early-successional species. Daehler (1998) found that plant families that were overrepresented among a global survey of agricultural weeds comprised herbaceous, rapidly reproducing, and abiotically dispersed species. Plant species invading unmanaged natural areas possessed different characteristics, and only 25 percent of natural-area invasives were also agricultural weeds. Thus, it is apparent that plant traits responsible for weediness are also dependent on interactions with environmental parameters.

Two main aspects of farm soil management determine the invasibility of fields to weeds and their ability to persist. Regular soil disturbance by tillage creates open niches, with conditions most favorable to early successional species and "*r*"-strategists. When biotic competition is reduced or eliminated by tillage or herbicide, resistance to invasion is minimized and the number of propagules needed for a population to become established is expected to be low (D'Antonio et al. 2001). So the number of propagules available may be less important for establishment of a weed in agricultural fields than in nondisturbed natural areas. Global-scale studies of the habitat correlates of invasibility have revealed robust patterns, most relevant of which here is that temperate agricultural and urban habitats represent the most invasible of all habitats (Lonsdale 1999).

Equally or more important in promoting invasion of these systems is the application of chemical fertilizer. Eutrophication of soils, intended to increase crop productivity, also favors weed production. To parse the relative impact of colonization opportunity and resource availability on the make-up of plant communities, Tilman (1987) established plots with a range of N inputs in abandoned fields of different histories as well as a native plant community that had never been cultivated. These sites varied significantly in their initial species composition, and included both undisturbed and newly disturbed (disked) plots. Dramatic changes in plant communities were observed within 3 years in plots receiving high annual N inputs, with >60 percent of species being displaced. Species response to the N gradient was similar between disturbed and undisturbed plots, and species composition converged on all sites, despite large difference in initial abundance, suggesting that species abundance was more dependent on resource availability than on history and initial species composition. Similarly, Hobbs (1989) found that disturbance alone did not make a plant community susceptible to invasion, but the addition of nutrients did. These observations suggest an effect termed the "paradox of enrichment," first predicted from predator–prey systems by Rosenzweig (1971), who warned of the destabilizing effects of enriching ecosystems. As a further illustration of this effect on plant community structure, Tilman (1987) demonstrated that high-N plots tended to be dominated by a few species. When fertilizer is added to plots, the relationship between primary productivity and species diversity most commonly forms an inverted "U," irrespective of initial productivity (Gurevitch et al. 2002). At very low soil fertility levels, plant species diversity is low as few can cope with the lack of nutrient resources. Species diversity then increases with productivity as resources increase, but only up to a point. At high soil fertility, total plant productivity continues to increase, but plant diversity usually declines (Grime 1973), as biomass becomes concentrated in a few fast-growing species. The significance for agricultural systems is that the concentrated use of chemical fertilizers is likely to have an impact on the total productivity and community structure of weeds by selecting for the most aggressive growing species. Thus, the unintended consequence of managing crop production by high inputs of soluble nutrients is to increase weed competitive ability, which reduces crop yields unless external measures of weed control are employed.

An observation from organic farming systems is that despite heavy weed biomass, the yield losses predicted from conventional systems are often not realized (Davis and Liebman 2001; Ryan and Hepperly 2005). In a 2-year study of three paired organic and conventional farms in Ohio, Bedet (2000) measured mean weed biomass of 129 kg/ha per year on organic farms compared to 7.6 kg/ha on neighboring conventional farms, a 17-fold difference. However, mean corn yields from these fields were almost identical: 8733 kg/ha for the organic farms and 8783 kg/ha for the conventional. Similarly, Ryan and Hepperly (2005) applied university agronomy guidelines developed from conventional systems to predict the impact of weeds in long-term plots at the Rodale Institute. Weed densities in organic plots were projected to result in a 40 percent reduction in corn yield, yet their yield was no different from that of conventionally managed plots. This provides another example of how the form of N fertility does matter and can have a significant affect on system function. In keeping with the biological buffering concept, soils with C-based fertility that maintains N in the soil solution at lower and less variable levels may allow a greater tolerance for weeds as a result of the "paradox of enrichment." Partial support for this hypothesis is provided by Davis and Liebman (2001) who compared the impact of wild mustard on sweet corn yield in plots fertilized with compost + green manure or NH_4NO_3, the latter supplied as either one early season application or two split applications. In the second year of the 2-year study, corn yields were reduced by 20 and 35 percent by mustard competition in the plots receiving single and split application of NH_4NO_3, respectively. In contrast, there was no yield reduction due to the wild mustard in plots receiving organic forms of fertility.

The current view of weeds as being wholly incompatible with crop species because they compete for the same water, light, and nutrient resources may be too simplistic. A more ecological approach would characterize as many niche dimensions as possible for the weeds and crops and create an environment that either favors the tolerance of crops toward weeds or that enhances their competitive advantage. For example, due to their relative compact form, soybeans are more susceptible to fast-growing weeds than corn. Although they share many niche dimensions with weeds, as legumes, soybeans show clear niche differentiation along one dimension, that of soil N utilization. With this in mind, we have conducted a series of experiments to determine the feasibility of niche differentiation to improve soybean competitiveness with weeds (Phelan, Stinner, and Nacci, in prep.). In controlled greenhouse studies, soybeans were planted with three graminoid and three broadleaf weeds in farm-collected soils to which straw and/or sawdust was added to increase soil C:N. As predicted, shoot biomass was inversely related to the level of C addition for weed species, particularly broadleaves, while soybean plant biomass and yield was positively correlated. Soybeans planted with weeds in soil with a high C:N were comparable in growth and yield to soybeans grown in the absence of weeds.

Controlling weeds by spraying an herbicide is far easier in the short run than an integrated approach, and even among some organic farmers, one gets a sense of "herbicide envy"; however, irrespective of whether they are synthetic or "natural" herbicides allowed by organic farming standards, an ecological approach to weed management should not rely solely on herbicides and "curative" tools. A long-term perspective of reducing weed impacts requires the application of principles of plant ecology to guide multiple cultural practices, or as Liebman and Gallandt (1997) describe it, "many little hammers."

9.5 REDESIGNING AGRICULTURAL PRODUCTION

9.5.1 NEW PARADIGM FOR AGRICULTURAL DESIGN (FUNDAMENTAL PRINCIPLES)

The examples discussed in the previous section illustrate the connectedness among agroecosystem components, and how addressing components in isolation commonly results in unintended consequences for other components. Well over a century of agricultural research has significantly improved yields, but has not diminished the number of problems that need to be addressed. Like the multiheaded Hydra of Lerna, as one problem is addressed, new ones continuously appear in their place. Examining the long history of research, there is no reason to believe this pattern will change in the future if we follow the same research paradigm. A new systems-based paradigm is needed guided by the following principles:

1. As a type of ecosystem, we must design future farming systems or practices based on an understanding of how ecosystems are organized and function.
2. As with other ecosystems, the farm operates as a complex system with emergent properties, and therefore management decisions must allow for the special behaviors and properties of complex systems.
3. A philosophy of research and experimental design must be employed that better reveals the relative importance of system constituents and their interactions.
4. We need to adopt a perspective of costs and benefits that is more inclusive in space and time to assess the true value of technologies.

In addition, redesigning agriculture cannot be driven solely by the goal of maximizing yield. Yield is only one of a suite of parameters that should be considered in our research and design (Table 9.2). Attending all of these goals will not necessarily mean significant trade-offs, but will require significantly different approaches. Of paramount importance in anticipating future challenges of crop production is the need to increase the stability of agroecosystems; success on this front will likely lead to progress toward other goals. Population and ecosystem dynamics is a major theme in ecology. Just as the resistance and resilience of natural ecosystems determine their propensity to persist, agroecosystems must be designed to cope with fluctuating environmental conditions, which are outside our control and likely to be more extreme in the future. In addition to the absence of characteristics of ecosystem distress discussed above, Rapport et al. (1998) suggest that ecosystem health should be defined in terms of vigor or primary productivity, diversity, and number of interactions among system components, and the capacity to maintain structure and function in the face of stress.

Any ecologically based approach to designing agricultural systems must start with the soil. Evidence has accumulated from both natural and managed ecosystems that the shift from a C-based system of fertility to an inorganic one did more to reduce the ecological stability of agricultural

---

TABLE 9.2
Goals for Designing an Ideal Agricultural System

Maximize long-term productivity (efficiency of land and energy use)

Maximize resistance to environmental variability

Maximize resilience to perturbations

Minimize inputs—energy, chemicals, costs

Support human systems/communities

Minimize negative environmental impacts

Maximize sensory and nutritional quality of products

---

systems than any other practice. Understanding the operation of detrital food webs and designing agricultural nutrient management that is more consistent with the nutrient cycles of natural systems is the single most important step that can be taken to increase the economic sustainability, environmental compatibility, and biological resilience of agricultural systems. An active detrital food web has the potential to increase the resistance and resilience of agroecosystems via a number of different mechanisms. There is evidence for each of these mechanisms from natural ecosystems and from agroecosystems with C-based fertility, although the relative contribution to system stability has not been determined and probably depends on other system parameters:

1. Better soil structure: In addition to the physicochemical benefits of recalcitrant organic matter, an active biological community significantly improves soil structural stability and pore space, leading to better soil moisture balance and air penetration, which results in less physiological stress on plants (Weil and Magdoff 2004).

2. Bottom-up effects on the aboveground autotrophic food chain: The extended slow mineralization of plant nutrients from the detrital food web provides better synchrony with plant requirements, providing better mineral balance for the plant metabolic machinery, which optimizes growth while suppressing herbivory and disease (Phelan et al. 1996).

3. Paradox of enrichment: Whether addressing the lateral interactions of competition among plant species or the vertical interactions of the food chain, eutrophication has broad destabilizing effects on ecosystems. Compared to chemical fertilizer, C-based nutrient sources are less likely to produce eutrophic conditions, reducing pressures by fast-growing early-succession weeds that are responsive to nutrients (Tilman 1987).

4. Top-down effects on aboveground autotrophic food chain: The presence of an active soil food web enhances epigeic fauna, which in turn make aboveground trophic interactions more weblike through an "allochthonous subsidy" that stabilizes populations of generalist predators when herbivorous prey are limited (Polis and Strong 1996; Wise et al. 1999).

5. Stabilization of soil microbial populations: The regular influx of organic matter increases resilience in microbial populations and suppresses pathogens and plant-parasitic nematodes by supporting functional redundancy in soil microbial communities and increasing the competitive advantage of nonpathogenic microbes (van Bruggen et al. 2006).

6. Induction of plant-defensive pathways: Elicitation of plant signaling systems by beneficial rhizobacteria and fungi increases resistance of aboveground portions of the plant to foliar pathogens (Vallad et al. 2003).

7. Tightening of nutrient cycles: High nutrient levels, above that needed by the crop, lead to losses from the system through soil leaching of NO_3^- (Kramer et al. 2006) and atmospheric release of NO_2 (Cassman et al. 2002; Petersen et al. 2006). With the availability of C energy sources, a greater portion of nutrients are stored in the microbial component of the soil. When soluble nutrient levels are synchronous with plant demands, more nutrients are retained locally. In agricultural systems, this process may be enhanced by the use of fall-planted cover crops to continue nutrient capture after crop demands have declined (Torstensson et al. 2006).

Given the importance of the detrital food web in natural ecosystems, there is good reason to expect a more efficient use of nutrients in agricultural systems that provide a regular influx of organic matter. From the biological side, the fate of soluble N depends on the amount of uptake by plants and microbes. Soil detrital microbes represent an important short-term buffering mechanism for N in the soil solution (Zogg et al. 2000), which outcompete plants in the short term for N. The capacity for microbial uptake depends on the level of biological active C available. With microbial turnover, N can become part of the longer-term storage in soil organic matter, remineralized and assimilated by plants, or recycled by the microbial biomass. Therefore, the amount of N retained locally depends on the activity of the soil food web and by plant demand as determined by species-specific seasonal phenology.

Many studies indeed find lower nutrient leaching from soils managed organically or receiving organic amendments compared to mineral fertilizers (Goss and Goorahoo 1995; Eltun and Fugleberg 1996; Eltun et al. 2002). In a 15-year study of three fertility systems (conventional, legume-only, and legume + cattle manure), Drinkwater et al. (1998) measured 50 percent higher nitrate leaching in the conventional system compared to the C-based fertility systems. The organic fertility also generated significantly greater increases in soil organic matter accumulations, while maize yields were not significantly different among the systems. Kramer et al. (2006) found an approximately fivefold lower nitrate leaching in long-term organic plots than in conventional plots due to more efficient microbial denitrification. Ma et al. (1999) measured reduced N loss potential for maize in soils with a history of application of dairy manures compared with NH_4NO_3 because the former showed better synchrony between N mineralization and plant requirements. However, other researchers have concluded that organic forms of fertility show similar levels of nutrient loss or even lower nutrient efficiencies (Bergström and Kirchmann 2004; Dahlin et al. 2005; Torstensson et al. 2006). These contradictory findings suggest that it is not sufficient to rely on broad descriptors such as organic and conventional in drawing conclusions about nutrient efficiency, but that one must also consider the specifics of the soil management practices and history. For example, in a comparative analysis of management systems for nutrient leaching, Torstensson et al. (2006) concluded that organic management held no advantage for retaining nutrients, whether the nutrients came through animal or green manures since 6-year averages showed similar leaching from these systems compared to a conventional one receiving only chemical fertilizer; however, a closer analysis shows that nutrient losses from the organic plots actually were lower than from the conventional except for one year in which potato blight, left untreated, caused a near-complete crop failure in organic plots, but was controlled with the application of fungicides in conventional plots. On the other hand, they also demonstrated that an active soil food web alone does not guarantee high nutrient retention, and that use of a cover crop even with chemical fertilizer provides strong nutrient retention. Taken together, these studies suggest that multiple parameters must be considered in the goal of tightening nutrient cycles: (1) an active soil food web, driven by the availability of labile C, (2) timing of nutrient availability with the demands of a healthy crop, and (3) a diversity of plant species with complementary nutrient demands to extend the period of plant assimilation. Natural ecosystems exhibit a high level of synchrony and synlocation between N mineralization and plant uptake potentials (Christensen 2004). However, even with an active soil food web, one must consider that the capacity of these systems is not unlimited and excessive additions of organic matter or an imbalance in its C:N will result in nutrient losses.

9.5.2 Opposition to Ecologically Principled Agriculture—Roadblocks to Change

One roadblock to an ecologically principled agriculture is the fact that proponents of this model do not share some of the presuppositions of the prevailing reductionist paradigm. Vocal proponents of conventional agriculture are frustrated by what they see as an unscientific and illogical position of those pushing for ecologically based agricultural practices, particularly as embodied in organic farming. Unfortunately, their response frequently degenerates to a similar level of irrationality. As a result, what could be a fruitful opportunity to improve our understanding of ecological processes instead deteriorates into name calling, questioning of motives, and hardening of positions. Although some critics of conventional agriculture in the lay community may employ loaded terms such as "poisons" in relation to agrichemicals or describe genetically engineered crops as "Frankenfoods," it seems more surprising when those who stake out the scientific high ground for conventional agriculture seem to prefer the same emotionalism over rational analysis when it comes to discussion of alternative farming systems. These attacks take on a moral, almost religious stance, as in this quote from Norman Borlaug (http://www.highyieldconservation.org/):

> We cannot choose between feeding malnourished children and saving endangered wild species. Without higher yields, peasant farmers will destroy the wildlands and species to keep their children from starving…. We aren't going to feed six billion people with organic fertilizer.

Taverne (2005) levels his moral hammer on those scientists he sees as standing on the sideline for not joining the fray:

> Scientists, who know that there is no intellectual case for organic farming and who are aware that its principles are based on myths and untruths, frequently say they have nothing against it…. I believe this position is morally untenable.

Or this quote by C. S. Prakash (cited by Taverne 2005, p. 60):

> Organic farming is sustainable. It sustains poverty and malnutrition.

Arguments against organic or ecological farming seem to take three forms. First, there is the view that it has no scientific foundation. For example, Taverne (2005) states:

> The organic movement has murky origins, its basic principle is founded on a scientific howler, it is governed by rules that have no rhyme or reason, it is steeped in mysticism and pseudo-science, and when it seeks to make a scientific case for itself, the science is shown to be flawed.

The second criticism characterizes ecologically based farming as a return to the past, a position that Taverne (2005) illustrates:

> The poorest farmers in Africa and Asia are already organic farmers: they do not use pesticides or artificial fertilizers because they cannot afford them…. The organic movement seeks to go back to the days before the Green Revolution.

In a similar vein, while acknowledging the problems of other forms of farming, Trewavas (2002) nevertheless saves his greatest disdain for the backwardness of organic farming:

> From the perspective of ten thousand years, however, some look back to the hunter–gatherer lifestyle with wistful nostalgia. They argue for a retreat from modern technology, so that humankind can achieve some kind of balance with nature.

The third criticism of organic farming is related to the second and is probably the most widespread, the issue of comparative crop yield. Dennis Avery, director of the Center for Global Food Issues, asserts that the latest research shows organic food yields are nearly 50 percent lower per acre than modern methods (a claim that is hard to document) so that if Europe were to switch to exclusively organic farming methods, the cropland needed to produce the resulting lower yields would equal "all the forest area in Germany, France, Denmark and the UK…. How many people in Western Europe would vote for organic farming if it was put in terms of clearing all their forests?"

Use of the actual words of these writers is necessary as they illustrate the tenor of the conflict that has arisen around the call for a different way of farming. These views reflect ones that are often encountered not only in the literature but also in conversations with colleagues. Quite frankly, they do not sound like voices of confidence or of reason. They sound more like voices of those whose core beliefs are under attack, in this case, by a group that they themselves describe as representing only 2 percent of the food market. Consideration of a couple of factors may help explain the highly charged tone that this discussion has taken. First, we must recognize how our set of presuppositions frames our view of the world. In large measure, participants in this debate who hold different assumptions see the other camp as being illogical. In the absence of logic, one is left with appeals to emotion. This is reflected in assertions that a scientific basis for ecological farming is lacking, for example, Taverne (2005):

> Rejecting the methods of science as "reductionistic" makes assessment of the effectiveness of organic farming impossible, because only by changing one factor or variable at a time can cause be related to effect. But the organic farming lobby, like supporters of alternative medicine, do not believe in the scientific method. Both practices have virtues, it seems, that can only be detected by intuition; they are both revealed as based on a belief in magic or mysticism, not reason.

Here we can see that reductionism is equated with the scientific method, and therefore any method that does not break the system down one variable at a time is unscientific and is a reversion to mysticism. Even from my perspective as a scientist, my experience with long-time organic farmers whose livelihoods depend on their productivity is that they are far from being antiscience. Although they may not be trained in empirical methods, they actively seek scientific explanations for observations they have made on their own farms. Nor can they afford to operate on nostalgic yearnings or a desire to turn back the clock.

The second cause of the hardened positions is the sense that many believe their work, and in some cases their career, is being attacked. In their vocation, many agricultural researchers have dedicated themselves to helping farmers be more productive and reduce hunger in the world. The call here for a new paradigm for agriculture should not be misinterpreted as an attempt to devalue this body of work. Nor is the point to idealize organic farming. From an ecological perspective, organic farming often also falls short of mimicking natural ecosystems, particularly with regard to its heavy reliance on soil disturbance for weed control. Comparative system-level studies of organic, no-till, and conventional farms have been instructive in understanding agroecosystem function, particularly with regard to the soil processes. Research in organic systems is on the rise; unfortunately, there remains a somewhat uneasy tension between organic and conventional agricultural research. As a result, we see parallel paths of research being conducted with relatively little cross-fertilization, and often studies are cited selectively to bolster one's preconceived notions. For example, as discussed above, lower yields are often recorded for organic production. However, there are also many controlled or empirical on-farm studies that find little or no difference in yields between organic and conventional systems (Stanhill 1990; Drinkwater et al. 1995, 1998; Bedet 2000; Colla et al. 2000; Davis and Liebman 2001; Ryan and Hepperly 2005). Rather than dismissing studies inconsistent with our beliefs, these conflicting results should be viewed as an opportunity for understanding those factors that limit yield. Furthermore, virtually all the research on crop yields has been conducted in conventional systems, but we cannot assume that the conclusions translate to an organic management context. For example, crop varieties selected for high yield under conventional management often perform poorly in organic farming systems (Murphy et al. 2007). Also, given the profound differences in mineral release dynamics between inorganic and C-based sources, different information on nutrient management is needed for organic systems, such as for optimizing the timing, quality, and quantity of organic matter inputs (Seiter and Horwath 2004). Future research based in the context of organic soil management will likely narrow yield differences.

9.6 ADOPTING A SYSTEMS APPROACH FOR AGRICULTURAL RESEARCH METHODOLOGY

If we accept the goals of the ecological paradigm (Table 9.2), how do we actually implement the principles given in Section 9.5.1 into the research program? The reductionist approach has been successful in advancing our understanding in agricultural research, just as it has in most aspects of science. However, the ability to reduce phenomena to causal relations between system components does not imply the ability to predict the behavior of a complex system through reconstruction of individual cause-and-effects relations. Human consciousness is dependent on neuronal mechanisms, but understanding how neurons work does not predict how humans think. So are we to give up on the goal of understanding agroecosystems by component research? Are we left in the intellectually unsatisfying conclusion that everything is important?

In fact, when approached properly, reductionism and a systems perspective are not contradictory but complementary (Bauchau 2006). "Hierarchical reductionism" represents a middle ground that has the potential to provide the best of both perspectives. Dawkins (1987, p. 13) describes the approach as explaining "a complex entity at any particular level in the hierarchy of organization, in terms of entities only one level down the hierarchy; entities which, themselves, are likely to be complex enough to need further reducing to their own component parts; and so on." Therefore, if we view agricultural production as hierarchically nested systems of organization and interactions, applying the approach of hierarchical reductionism provides a means by which to reveal emergent properties. Simplifying the system by elimination of certain components and then comparing the behavior of the resulting system to that of the whole highlights the importance of those components. In the hierarchical structure of systems, understanding cause and effect at one level helps us understand emergence at higher levels; that is, to the extent that fundamental principles deduced from lower levels fail to explain the larger system, emergence is revealed. So, for example, growing maize in the greenhouse using field-collected soils revealed the same pattern of insect susceptibility seen in the field between organic and conventional farm (Phelan et al. 1995, 1996), suggesting a significant role for bottom-up regulation of pest populations and indicating the appropriateness of this context for further stepwise comparisons down the hierarchy. As obvious as this approach might appear, it is rarely implemented. Instead, most agricultural research addresses relations within components and extrapolates these findings to the larger system. This approach is susceptible to what I would term "upward errors," in that it is blind to both unintended negative consequences and the loss of beneficial system properties that arise only through interactions, such as mutualism, competition, and feedback loops. Upward errors are the basis for many of the problems now being addressed in agricultural research.

There are many on-farm studies comparing the dominant farming systems: conventional, organic, no-till, and integrated farming. These studies can provide a starting point for the hierarchical reductionist approach by suggesting testable hypotheses on ecosystem function. First and foremost, the objective of these studies should not be to demonstrate the superiority of one system over another. Rather, we need to approach them with the objective of understanding how whole farming systems function differently. These studies have some unique challenges. Since the initial steps of whole-farm studies are descriptive rather than controlled experiments, the researcher has limited control either in the practices used or their timing when comparing operating farms. When farms are geographically separated, there is a greater potential for confounding factors that have nothing to do with the farming system. On the other, the use of replicated plot designs on research land to study system-level response, as is characteristic of traditional agricultural research has its own limitations of which we must be aware.

1. Expertise. The experiential knowledge of farmers can be invaluable, as they develop or adopt practices that through trial and error have worked for their farms. The common absence of organic management experience in replicated plot research comparing farming systems can lead to erroneous conclusions since the plots do not accurately mimic an organic system.
2. Soil management history. Although enhancing soil organic matter may produce a system with greater stability and higher resilience to perturbations, it may take a number of years for the system to establish this new dynamic equilibrium.
3. Ghosts of research past. High fertility management typical of research land will continue to affect soil biology and crop performance many years after implementation of new management methods. Previous research on plant diseases and weeds may have involved inoc-

ulations of pathogens and planting of weed seeds, populations of which remain during the transition to organic practice.

Also essential to developing a systems approach to agricultural research is a synergistic form of collaboration among researchers of different expertise. There has been a favorable trend emphasizing *interdisciplinary* research, with a growth in the number of interdisciplinary programs and centers at universities. Although this movement has the potential to counter atomistic thinking and many researchers consider themselves interdisciplinary, much of this work in agriculture is better described as *multidisciplinary*. Researchers from different disciplines are brought together to work on a common project; however, division of labor still has researchers assigned to different components, rather than considering how the components interact. As a result, these efforts frequently fall short of a systems-level understanding. Rapport (1997) suggests that a new term more specific to the goals of ecosystem health is needed. *Transdisciplinarity* is unique in its emphasis on seeking out higher-level system processes that transcend component or discipline-limited research. This perspective is necessary if we are to avoid the upward errors that come from extrapolation. Only when the participants are committed to understanding how the system works as a set of interacting components, not just an assemblage of components, and have an active view for emergent behavior, will a synergy of effort result. Transdisciplinarity encourages one to think about how what one is doing might affect other system components and what interactions among components produce unexpected behavior.

9.7 FUTURE OF AGRICULTURE

9.7.1 The Next Green Revolution

One of the largest technological developments in agriculture in recent times is the application of biotechnology to introduce new traits to crops. Unlike traditional breeding programs, genetic engineering permits direct manipulation of a cell's genetic makeup and the transfer of genes across species barriers, and eventually even the possibility of inserting "designer genes" that have been synthesized in the laboratory. Thus, this technology promises almost unlimited possibilities for new crop traits. As the crop productivity gains from the first Green Revolution have slowed, genetic engineering of crops is seen as the basis for a second Green Revolution (Viswanathan et al. 2003; Sakamoto and Matsuoka 2004). With the potential to introduce new crop traits not previously possible through conventional breeding methods, genetic engineering is conceived as a quantum advance in crop improvement. Notwithstanding the controversy surrounding the use of genetically modified crops, genetic engineering is likely to continue to drive the development of new varieties. It remains to be seen whether genetic engineering of crops can live up to the level of expectation created by its proponents; however, two predictions seem relatively certain:

1. Genetic engineering will be increasingly employed in the development of new crop varieties, and
2. Genetic engineering cannot be the basis for a revolution in agriculture since it represents a technology based on the same single-component approach of problem solving of the dominant paradigm. Thus, although this technology may speed the development of new traits, whether by single genes or "stacking" multiple genes, the same philosophical assumptions and goals underlie genetic engineering that have guided conventional breeding and pest management tools: reductionism in addressing individual challenges, usually driven by the sole goal of increasing production. As such, the upward error of introducing a crop with a new trait is likely to have unintended consequences for the system. In fact, the term "engineering" is somewhat misleading in this context, as it connotes a high level of precision and control. As Peterson (2005) points out, when it comes to managing ecosystems, we

have neither certainty nor control. The next Green Revolution needs more than technology; it will require a fundamental change in the design of agricultural systems and a conceptual change in what we expect from them.

I have tried to argue here that a new agricultural context is needed to lower the overall energy state of the system as well as a broader perspective for assessing costs versus benefits of practices and technologies. This change will not come about simply by replacing current pesticides with more environmentally benign ones or by genetically engineering new traits into crops to overcome specific hurdles to production. The major advances in scientific thinking have not come about through gradual modifications in the prevailing paradigm (Kuhn 1970), but through a fundamental challenge to it, accompanied by a shift to a new way of seeing things. These paradigm shifts usually encounter strong resistance, much as we see today to efforts to change agricultural systems.

Although the phrase paradigm shift has become banal by overuse, its appropriateness to understanding the current tension between the ecology-based framework and the prevailing reductionistic paradigm can be illustrated. In his now-classic essay on the nature of scientific revolutions, Kuhn (1970) describes a shared paradigm as the set of assumptions and standards to which the group of scientists is committed. The paradigm creates structure for what would otherwise be a collection of "mere facts." The consensus created by the paradigm is required for "normal science," which describes the period of gradual advance in scientific understanding during which the paradigm guides and informs the process of fact gathering and interpretation of research. During the period when the paradigm is successful in explaining facts, "the profession will have solved problems that its members could scarcely have imagined" (p. 25). On the other hand, the execution of normal science can also be viewed as "an attempt to force nature into the preformed and relatively inflexible box that the paradigm supplies. No part of the aim of normal science is to call forth new sorts of phenomena; indeed those that will not fit the box are often not seen at all" (p. 24). This accurately describes the situation for agricultural design in which the implementation of solutions to individual component problems creates unintended consequences for other system components. Not only does our reductionist paradigm not seek to discover these unintended problems, but when revealed they are largely ignored and the causative practice or technology is retained. As von Liebig (1855) observed after his own paradigm shift: "The human spirit, however, is a strange thing. Whatever doesn't fit into the given circle of thinking, doesn't exist."

The value of the paradigm is that by committing (usually unconsciously) to a common set of assumptions, science advances because the scientist is free to concentrate on the task at hand: "the articulation of those phenomena and theories that the paradigm already supplies" (p. 24). Normal science does not seek out new paradigms and is often intolerant of those who do. Adoption of new paradigms requires a reconstruction of prior assumptions and the reevaluation of prior observations. This is difficult and time-consuming, and since it calls into question one's work, it is strongly resisted. Thus, the strong reactions to organic farming discussed in Section 9.5.2, which seem unduly strident, are consistent with the response seen for other paradigm conflicts in history. Because by definition, those committed to different paradigms do not share the same assumptions about how to conduct science, it is easy to see how charges of being unscientific may emerge. Overlay on this the fact that most scientists cannot articulate the set of assumptions that make up their own paradigm, let alone those of competing paradigms, and it is easy to understand why arguments abandon logic and become emotional.

Given the resistance integral to paradigms, how do paradigm shifts occur? Generally, observations counter to the prevailing paradigm do not bring about shifts, but are considered anomalies. However, eventually either a large problem persists or the model fails to prepare the researcher for a number of unexpected observations, and it ceases to serve its purpose. This creates a period of professional unease; however, once the paradigm shift occurs, "a scientist's world is qualitatively transformed as well as quantitatively enriched by fundamental novelties of either fact or theory" (Kuhn 1970, p. 7).

9.7.2 NEW CHALLENGES AND NEW OPPORTUNITIES

I have provided numerous examples from the literature of agroecosystem behavior that is not predicted by the reductionistic paradigm and to which it was largely blind. We need to move agriculture toward a system more in line with the "laws" of ecology, with the potential to analyze new technology with a more inclusive frame of reference. Reevaluating our efforts and redirecting agricultural design at this point in time is needed for a number of reasons. First, measures of success have been skewed in that they have excluded significant temporally and spatially displaced costs, such as environmental contamination, eutrophication of natural ecosystems, and contributions to greenhouse gases. These costs, which have been ignored for a long time, must be included when designing new agricultural systems, not only on ethical grounds and for scientific honesty, but also because in the not-to-distant future, these environmental loads may represent real costs that must be paid, for example, carbon taxes or credits. In the past, environmental costs of farming have been viewed as a trade-off necessary to the economics of productive farming. Second, the ecological inefficiencies of our present agricultural system have been permitted by the availability of cheap energy. The recently rapid rise in energy costs means that the costs of current external inputs will rise sharply and be directly borne by farmers. Finally, we are entering an unprecedented stage of humankind in terms of climate change, which does not simply mean a rise in temperatures, but greater fluctuations in climatic conditions, with new extremes and environmental stresses for crop production likely.

I am not suggesting that an agricultural design based on natural systems will be free of pests and disease, and without environmental problems. Even natural systems are subject to biotic stress. However, it is clear that the problem-solving approach of the past century of agricultural research has unintentionally created a managed ecosystem that is highly susceptible to invasive plants, insects, and disease, and that lacks the resistance to fluctuating abiotic conditions and resilience in subsequent response that is characteristic of most natural ecosystems. We will not be able to continue in the same manner; within a paradigm of individual problem solving, we will fall farther behind as we enter an era where new and greater challenges will arise, changing weather patterns will create more frequent stress and crop failures, and some of the real costs, previously ignored, will have to be paid. Unless we are willing to provide massive governmental support or accept environmental damage as a necessary trade-off, we must prepare for the future by changing the paradigm from one of problem solving to one of ecosystem management.

In addition to increasing stability and reducing causes of environmental damage, adoption of an ecologically based paradigm will position agriculture to provide additional environmental services that will provide additional income to farmers (Jordan et al. 2007; Lal 2007). Agricultural soils have been recommended as a potentially important C sink (Bruce et al. 1999). Many studies have indicated that farming systems that rely on C-based fertility have a greater potential to increase soil organic matter and carbon sequestration. Even more significant than CO_2 is the emission of N_2O from farms, with a global warming potential 300 times that of CO_2 (IPCC 2001). While farming is estimated to contribute only 1 percent of global CO_2 emissions, it is responsible for 60 percent of N_2O emissions (Chu et al. 2007). Use of C-based fertility typically results in lower N_2O emissions than chemical fertilizer, although the magnitude of the difference depends on composition of the organic biomass and particularly C:N (Sarkodie-Addo et al. 2003). As a result, organic soil management can result in lower N_2O emissions, likely due to the correlation between soil N levels and N_2O emissions, irrespective of farm type (Petersen et al. 2006).

9.8 CONCLUSIONS

The agricultural research community can look with pride to the technological developments that have allowed the great improvements in productivity of the past century and that brought about the first Green Revolution. However, will the same paradigm serve us best as we look forward? What will the next Green Revolution look like? What will be its structure and underlying philosophy? How will the value of practices and new technologies be assessed? Will change continue as it has with research efforts focused on individual problem solving to reduce roadblocks that reduce yield, while others work to ameliorate the unintended negative consequences of these practices and technologies? I would argue that a paradigm shift is required, with a fundamental change in our approach to optimize production quantity and quality. We need to establish the primacy of ecological principles and systems thinking in the design of production systems. In so doing, we must reject the hubris of the human goal to subjugate nature with technology and the simplistic perspective of agricultural research as a never-ending series of problems to be fixed. On the other hand, just as quantum mechanics did not negate Newtonian physics, but rather incorporated it into a more inclusive view of reality, a systems perspective for agricultural research and design will not require a wholesale abandonment of reductionism and component research. The ecosystem perspective will provide the context within which the goals of research will be determined and its success evaluated.

No other human activity has transformed our planet as much as agriculture, with both positive and negative consequences for humanity (Millennium Ecosystem Assessment 2004). Stability is a characteristic important to both the economics and environmental impacts of agroecosystems that has been almost completely ignored, and one whose significance is expected to rise in the future. Agricultural fields should be resistant to stress and resilient to return to a steady state when external perturbation displaces the system. With the expectations of rapid future climate change and increased fluctuations in climatic conditions, building resistance and resilience into our agricultural fields to buffer the negative impacts of weather extremes will become essential to the economic survival of farming. As we look back over 20 years of efforts to advance agroecology (Lowrance et al. 1984), we must acknowledge that an understanding of the function of natural ecosystems still does not inform our prevailing agroecosystems, and we must recognize that Pierce and Lal's (1991) admonition that, "Soil management practices in the twenty-first century must be formulated based on an understanding of the ecosystem concept," remains an aspiration rather than reality.

REFERENCES

Abdullah, N. M. M., Singh, J., and Sohal, B. S. 2006. Behavioral hormoligosis in oviposition preference of *Bemisia tabaci* on cotton. *Pesticide Biochemistry and Physiology*, 84, 10–16.

Alam, S., Kamei, S., and Kawai, S. 2001. Effect of iron deficiency on the chemical composition of the xylem sap of barley. *Soil Science & Plant Nutrition*, 47, 643–649.

Alyokhin, A. and Atlihan, R. 2005. Reduced fitness of the Colorado potato beetle (Coleoptera: Chrysomelidae) on potato plants grown in manure-amended soil. *Environmental Entomology*, 34, 963–968.

Alyokhin, A. et al. 2005. Colorado potato beetle response to soil amendments: A case in support of the mineral balance hypothesis? *Agriculture, Ecosystems, and Environment*, 109, 234–244.

Amancio, S. et al. 1997. Assimilation of nitrate and ammonium by sulfur deficient *Zea mays* cells. *Plant Physiology & Biochemistry*, 35, 41–48.

Andow, D. A. and Hidaka, K. 1989. Experimental natural history of sustainable agriculture: Syndromes of production. *Agriculture, Ecosystems, and Environment*, 27, 447–462.

Audenaert, K. et al. 2002. Induction of systemic resistance to *Botrytis cinerea* in tomato by *Pseudomonas aeruginosa* 7NSK2: Role of salicylic acid, pyochelin, and pyocyanin. *Molecular Plant–Microbe Interactions*, 15, 1147–1156.

Baker, H. G. 1974. The evolution of weeds. *Annual Review Ecology & Systematics*, 5, 1–24.

Barbour, M. G. et al. 1999. *Terrestrial Plant Ecology*, 3rd ed, Addison Wesley Longman, Menlo Park.

Bardgett, R. D. and McAlister, E. 1999. The measurement of soil fungal: bacterial biomass ratios as an indicator of ecosystem self-regulation in temperate grasslands. *Biology and Fertility of Soils*, 29, 282–290.

Bauchau, V. 2006. Emergence and reductionism: From the game of life to science of life, In *Self-organization and Emergence in Life Sciences*, Feltz, B., Crommelinck, M., and Goujon, P., Eds. Springer, Dordrecht, 29–40.

Beanland, L., Phelan, P. L., and Salminen, S. 2003. Micronutrient interactions on soybean growth and the developmental performance of three insect herbivores. *Environmental Entomology*, 32, 641–651.

Bedet, C. 2000. Soil fertility, crop nutrients, weed biomass, and insect populations in organic and conventional field corn (*Zea mays* L.) agroecosystems. Ph.D. dissertation, Ohio State University, Columbus.

Berg, M. P. and Verhoef, H. A. 1998. Ecological characteristics of a nitrogen-saturated coniferous forest in the Netherlands. *Biology and Fertility of Soils*, 26, 258–267.

Bergström, L. and Kirchmann, H. 2004. Leaching and crop uptake of nitrogen from nitrogen-15-labeled green manures and ammonium nitrate. *Journal of Environmental Quality*, 33, 1786–1792.

Bongers, T., van der Meulen, H., and Korthals, G. 1997. Inverse relationship between the nematode maturity index and plant parasite index under enriched nutrient conditions. *Applied Soil Ecology*, 6, 195–199.

Boyer, S. et al. 2006. Do herbicide treatments reduce the sensitivity of mosquito larvae to insecticides? *Chemosphere*, 65, 721–724.

Brock, W. H. 1997. *Justus von Liebig: The Chemical Gatekeeper*. Cambridge University Press, New York.

Brodkin, M. A. et al. 2007. Atrazine is an immune disruptor in adult northern leopard frogs (*Rana pipiens*). *Environmental Toxicology & Chemistry*, 26, 80–84.

Bruce, J. P. et al. 1999. Carbon sequestration in soils. *Journal of Soil & Water Conservation*, 54, 382–389.

Buckley, D. H. and Schmidt, T. M. 2001. The structure of microbial communities in soil and the lasting impact of cultivation. *Microbial Ecology*, 42, 11–21.

Cassman, K. G., Dobermann, A., and Walters, D. T. 2002. Agroecosystems, nitrogen-use efficiency, and nitrogen management. *Ambio*, 31, 132–140.

Chiu, B. C. H. et al. 2004. Agricultural pesticide use, familial cancer, and risk of non-Hodgkin lymphoma. *Cancer Epidemiology Biomarkers & Prevention*, 13, 525–531.

Christensen, B. T. 2004. Tightening the nitrogen cycle. In *Managing Soil Quality: Challenges in Modern Agriculture*, Schjønning, P., Elmholt, S., and Christensen, B. T., Eds. CABI Publishing, Cambridge, 49–67.

Chu, H., Hosen, Y., and Yagi, K. 2007. NO, N_2O, CH_4, and fluxes in winter barley field of Japanese Andisol as affected by N fertilizer management. *Soil Biology & Biochemistry*, 39, 330–339.

Cockfield, S. D. 1988. Relative availability of nitrogen in host plants of invertebrate herbivores: Three possible nutritional and physiological definitions. *Oecologia*, 77, 91–94.

Cole, L. J. et al. 2002. Relationships between agricultural management and ecological groups of ground beetles (Coleoptera: Carabidae) on Scottish farmland. *Agriculture Ecosystems and Environment*, 93, 323–336.

Coleman, D. C. 2008. From peds to paradoxes: Linkages between soil biota and their influences on ecological processes. *Soil Biology & Biochemistry*, 40, 271–289.

Colla, G. et al. 2000. Soil physical properties and tomato yield and quality in alternative cropping systems. *Agronomy Journal*, 92, 924–932.

Committee on Environment and Natural Resources. 2000. An Integrated Assessment: Hypoxia in the Northern Gulf of Mexico. National Science and Technology Council, Washington, D.C.

Court, R. D., Williams, W. T., and Megarty, M. P. 1972. The effect of mineral nutrient deficiency on the content of free amino acids in *Setaria sphacelata*. *Australian Journal of Biological Science*, 25, 77–87.

Cropp, R. and Gabric, A. 2002. Ecosystem adaptation: Do ecosystems maximize resilience? *Ecology*, 83, 2019–2026.

Curry, J. P. and Good, J. A. 1992. Soil fauna degradation and restoration. *Advances in Soil Science*, 17, 171–215.

Daehler, C. C. 1998. The taxonomic distribution of invasive angiosperm plants: Ecological insights and comparison to agricultural weeds. *Biological Conservation*, 84, 167–80.

Dahlin, S. et al. 2005. Possibilities for improving nitrogen use from organic materials in agricultural cropping systems. *Ambio*, 34, 288–295.

Dalal, R. C. 1998. Soil microbial biomass: What do the numbers really mean? *Australian Journal of Experimental Agriculture*, 38, 649–665.

D'Antonio, C. M., Levine, J. and Thomsen, M. 2001. Ecosystem resistance to invasion and the role of propagule supply: A California perspective. *Journal of Mediterranean Ecology*, 2, 233–245.

Davies, K. G. 2001. What makes genetically modified organisms so distasteful? *Trends in Biotechnology*, 19, 424–427.

Davies, K. G. 2003. Zones of inhibition: Interactions between art and science. *Endeavour*, 27, 131–133.

Davis, A. S. and Liebman, M. 2001. Nitrogen source influences wild mustard growth and competitive effect on sweet corn. *Weed Science*, 49, 558–566.

Dawkins, R. 1987. *The Blind Watchmaker*. W. W. Norton, New York.

de Ruiter, P. C., Neutel, A.-M., and Moore, J. C. 1995. Energetics, patterns of interaction strengths, and stability in real ecosystems. *Science*, 269, 1257–1260.

Després, L., David, J.-P., and Gallet, C. 2007. The evolutionary ecology of insect resistance to plant chemicals. *Trends in Ecology and Evolution*, 22, 298–307.

Doelman, P. and Eijsackers, H. J. P. 2004. *Vital Soil: Function, Value, and Properties*. Elsevier, Boston.

Drew, M. C. and Saker, L. R. 1978. Nutrient supply and the growth of the seminal root system in barley. III. Compensatory increase in growth of lateral roots, and in rates of phosphate uptake, in response to a localized supply of phosphate. *Journal of Experimental Botany*, 29, 435–451.

Drinkwater, L. E. et al. 1995. Fundamental differences between conventional and organic tomato agroecosystems in California. *Ecological Applications*, 5, 1098–1112.

Drinkwater, L. E., Wagoner, P., and Sarrantonio, M. 1998. Legume-based cropping systems have reduced carbon and nitrogen losses. *Nature*, 396, 262–265.

El Titi, A. and Ipach, U. 1989. Soil fauna in sustainable agriculture: Results of an integrated farming system at Lautenbach, F.R.G. *Agriculture Ecosystems & Environment*, 27, 561–572.

Eltun, R. and Fugleberg, O. 1996. The Apelsvoll cropping system experiment: VI. Runoff and nitrogen losses. *Journal of Agricultural Science* (Norway), 10, 229–248.

Eltun, R., Korsæth, A., and Nordheim, O. 2002. A comparison of environmental, soil fertility, yield, and economical effects in six cropping systems based on an 8-year experiment in Norway. *Agriculture, Ecosystems, and Environment*, 90, 155–168.

Ericsson, T. 1995. Growth and shoot:root ratio of seedlings in relation to nutrient availability. *Plant & Soil*, 168–169, 205–214.

Fauci, M. F. and Dick, R. P. 1994. Soil microbial dynamics: Short and long-term effects of inorganic and organic nitrogen. *Soil Science Society of America Journal*, 58, 801–806.

Fernandez, M. R. et al. 2005. Crop production factors associated with fusarium head blight in spring wheat in eastern Saskatchewan. *Crop Science*, 45, 1908–1916.

Fitter, A. H., Nichols, R., and Harvey, M. L. 1988. Root system architecture in relation to life history and nutrient supply. *Functional Ecology*, 2, 345–351.

Fog, K. 1988. The effect of added nitrogen on the rate of decomposition of organic matter. *Biological Reviews*, 63, 433–462.

Frampton, G. K. et al. 2006. Effects of pesticides on soil invertebrates in laboratory studies: A review and analysis using species sensitivity distributions. *Environmental Toxicology & Chemistry*, 25, 2480–2489.

Gardner, B. L. 2002. *American Agriculture in the Twentieth Century: How It Flourished and What It Cost*. Harvard University Press, Cambridge. MA.

Gatehouse, A. M. R. 2002. Durable resistance in crops to pests. In *Encyclopedia of Pest Management*, Pimentel, D., Ed. CRC Press, Boca Raton, FL, 207–209.

Goss, M. J. and Goorahoo, D. 1995. Nitrate contamination of groundwater: Measurement and prediction. *Fertilizer Research*, 42, 331–338.

Grime, J. P. 1973. Competitive exclusion in herbaceous vegetation. *Nature*, 242, 344–347.

Grime, J. P. 1977. Evidence for the existence of three primary strategies in plants and its relevance to ecological and evolutionary theory. *American Naturalist*, 11, 1169–1194.

Gurevitch, J., Scheiner, S. M., and Fox, G. A. 2002. *The Ecology of Plants*. Sinauer Press, Sunderland, MA.

Halaj, J. and Wise, D. H. 2002. Impact of a detrital subsidy on trophic cascades in a terrestrial grazing food web. *Ecology*, 83, 3141–3151.

Hayes, T. B. et al. 2006. Pesticide mixtures, endocrine disruption, and amphibian declines: Are we underestimating the impact? *Environmental Health Perspectives*, 114(Suppl. 1), 40–50.

Hearnshaw, E. J. S., Cullen, R., and Hughey, K. F. D. 2005. Ecosystem health demystified: An ecological concept determined by economic means. Paper presented at the 2005 EEN National workshop, Australian National University, Canberra. http://een.anu.edu.au/e05prpap/hearnshaw.doc (accessed December 10, 2007).

Herms, D. A. and Mattson, W. J. 1992. The dilemma of plants: To grow or defend. *Quarterly Review of Biology*, 67, 283–335.

Hines, J., Megonigal, J. P., and Denno, R. F. 2006. Nutrient subsidies to belowground microbes impact aboveground food web interactions. *Ecology*, 87, 1542–1555.

Hoar, S. K. et al. 1986. Agricultural herbicide use and risk of lymphoma and soft-tissue sarcoma. *Journal of the American Medical Association*, 256, 1141–1147.

Hobbs, R. J. 1989. The nature and effects of disturbance relative to invasions. In *Biological Invasions: A Global Perspective*, Drake, J. A. et al, Eds. John Wiley & Sons, Chichester, UK, 389–403.

Howard, A. 1943. *An Agricultural Testament*. Oxford University Press, Oxford.

Huang, P.-M., Wang, M.-K., and Chiu, C.-Y. 2005. Soil mineral–organic matter–microbe interactions: Impacts on biogeochemical processes and biodiversity in soils. *Pedobiologia*, 49, 609–635.

Hülsmann, A. and Wolters, V. 1998. The effects of different tillage practices on soil mites, with particular reference to *Oribatida*. *Applied Soil Ecology*, 9, 327–332.

IFA. 2003. Use of fertilizers in organic farming. http://www.fertilizer.org/ifa/Form/pub_position_papers_9.asp (accessed December 10, 2007).

IPCC, 2001. Atmospheric chemistry and greenhouse gases. In *Climate Change 2001. The Scientific Basis.* Houghton, J. T. et al., Eds. Cambridge University Press, Cambridge, 239–287.

Jackson, L. E., Schimel, J. P., and Firestone, M. K. 1989. Short-term partitioning of ammonium and nitrate between plants and microbes in an annual grassland. *Soil Biology & Biochemistry,* 21, 409–415.

James, D. G. and Price, T. S. 2002. Fecundity in two-spotted spider mite (Acari: Tetranychidae) is increased by direct and systemic exposure to imidacloprid. *Journal of Economic Entomology,* 95, 729–732.

Johnson, L. 1990. The thermodynamics of ecosystems. In *The Handbook of Environmental Chemistry: The Natural Environment and the Biogeochemical Cycles,* Hutzinger, O., Ed. Springer-Verlag, Berlin, 1–47.

Jordan, N. et al. 2007. Sustainable development of the agricultural bio-economy. *Science,* 316, 1570–1571.

Karlen, D. L. et al. 1997. Soil quality: A concept, definition, and framework for evaluation. *Soil Science Society of America Journal,* 61, 4–10.

Kolasa, J. 2006. A community ecology perspective on variability in complex systems: The effects of hierarchy and integration. *Ecological Complexity,* 3, 71–79.

Koprivnikar, J., Baker, R. L., and Forbes, M. R. 2006. Environmental factors influencing trematode prevalence in grey tree frog (*Hyla versicolor*) tadpoles in southern Ontario. *Journal of Parasitology,* 92, 997–1001.

Kramer, S. B. et al. 2006. Reduced nitrate leaching and enhanced denitrifier activity and efficiency in organically fertilized soils. *Proceedings of the National Academy of Sciences,* 103, 4522–4527.

Kremer, R. J., Means, N. E., and Kim, S. 2005. Glyphosate affects soybean root exudation and rhizosphere micro-organisms. *International Journal of Environmental Analytical Chemistry,* 85, 1165–1174.

Krogh, P. H. 1994. Microarthropods as bioindicators. A study of disturbed populations. In *Terrestrial Ecology.* Natural Environmental Research Institute, Silkeborg.

Kuhn, T. S. 1970. *The Structure of Scientific Revolutions,* 2nd ed. University of Chicago Press, Chicago.

Lagerlöf, J. and Andrén, O. 1991. Abundance and activity of *Collembola, Protura,* and *Diplura* (Insecta, Apterygota) in four cropping systems. *Pedobiologia,* 35, 337–350.

Lal, R. 2007. *Soil science and the carbon civilization. Soil Science Society of America Journal,* 71, 1425–1437.

Lavicoli, A. et al. 2003. Induced systemic resistance in *Arabidopsis thaliana* in response to root inoculation with *Pseudomonas fluorescens* CHA0. *Molecular Plant–Microbe Interactions,* 16, 851–858.

Lee, W. J. et al. 2004. Cancer incidence among pesticide applicators exposed to alachlor in the Agricultural Health Study. *American Journal of Epidemiology,* 159, 373–380.

Lenton, T. M. 1998. Gaia and natural selection. *Nature,* 394, 439–447.

Lévesque, C. A. and Rahe, J. E. 1992. Herbicide interactions with fungal root pathogens, with special reference to glyphosate. *Annual Review of Phytopathology,* 30, 579–602.

Liebman, M. and Gallandt, E. R. 1997. Many little hammers: Ecological management of crop–weed interactions. In *Ecology in Agriculture,* Jackson, L. E., Ed. Academic Press, San Diego, 291–243.

Lipman, T. O. 1967. Vitalism and reductionism in Liebig's physiological thought. *Isis,* 58, 167–185.

Liu, L., Punja, Z. K., and Rahe, J. E. 1997. Altered root exudation and suppression of induced lignification as mechanisms of predisposition by glyphosate of bean roots (*Phaseolus vulgaris* L.) to colonization by *Pythium* spp. *Physiological and Molecular Plant Pathology,* 51, 111–127.

Lonsdale, W. M. 1999. Global patterns of plant invasions and the concept of invasibility. *Ecology,* 80, 1522–1536.

Lowrance, R., Stinner, B. R., and House, G. J., Eds. 1984. *Agricultural Ecosystems: Unifying Concepts.* John Wiley, New York.

Lundholm, C. E. 1997. DDE-induced eggshell thinning in birds: Effects of *p,p´*-DDE on the calcium and prostaglandin metabolism of the eggshell gland. *Comparative Biochemistry and Physiology C, Pharmacology, Toxicology, and Endocrinology,* 118, 113–128.

Lydy, M. J. and Linck, S. L. 2003. Assessing the impact of triazine herbicides on organophosphate insecticide toxicity to the earthworm *Eisenia fetida. Archives of Environmental Contamination and Toxicology,* 45, 343–349.

Lynch, J. M. and Penn, D. J. 1980. Damage to cereals caused by decaying weed residues. *Journal of the Science of Food and Agriculture,* 31, 321–324.

Ma, B. L., Dywer, L. M., and Gregorich, E. G. 1999. Soil nitrogen amendment effects on seasonal nitrogen mineralization and nitrogen cycling in maize production. *Agronomy Journal,* 91, 1003–1009.

Mäder, P. et al. 2002. Soil fertility and biodiversity in organic farming. *Science,* 296, 1694–1697.

Marasas, M. E., Sarandon, S. J., and Cicchino, A. C. 2001. Changes in soil arthropod functional group in a wheat crop under conventional and no tillage systems in Argentina. *Applied Soil Ecology,* 18, 61–68.

Marcic, D. 2003. The effects of clofentezine on life-table parameters in two-spotted spider mite *Tetranychus urticae*. *Experimental and Applied Acarology*, 30, 249–263.

Marschner, H. 1995. *Mineral Nutrition of Higher Plants*, 2nd ed. Academic Press, San Diego.

Matson, P. A. et al. 1997. Agricultural intensification and ecosystem properties. *Science*, 277, 504–509.

Mattson, W. J. 1980. Herbivory in relation to plant nitrogen content. *Annual Review of Ecology & Systematics*, 11, 119–162.

McCann, K. S. 2005. Perspectives on diversity, structure, and stability. In *Ecological Paradigms Lost: Routes of Theory Change*, Cuddington, K. and Beisner, B., Eds. Elsevier Academic Press, Burlington, MA, 183–209.

McCarty, G. W. and Meisinger, J. J. 1997. Effects of N fertilizer treatment on biologically active N pools in soils under plow and no tillage. *Soil Biology & Biochemistry*, 24, 406–412.

McDuffie, H. H. et al. 2001. Non-Hodgkin's lymphoma and specific pesticide exposure in men: Cross-Canada study of pesticides and health. *Cancer Epidemiology Biomarkers and Prevention*, 10, 1155–1163.

Meine, C. and Knight, R. L. 1999. *The Essential Aldo Leopold: Quotations and Commentaries*. The University of Wisconsin Press, Madison.

Mengel, K. and Kirkby, E. A. 1987. *Principles of Plant Nutrition*, 4th ed. International Potash Institute, Bern.

Meriles, J. M. et al. 2006. Glyphosate and previous crop residue effect on deleterious and beneficial soil-borne fungi from a peanut-corn-soybean rotations. *Journal of Phytopathology*, 154, 309–316.

Millennium Ecosystem Assessment. 2004. *Ecosystems and Human Well-Being: Our Human Planet*. Island Press, Washington, DC.

Mitchell, M. 1978. Vertical and horizontal distributions of *Oribatid* mites (Acari: Cryptostigmata) in an aspen woodland soil. *Ecology*, 59, 516–525.

Moore, J. C. and Hunt, H. W. 1988. Resource compartmentation and the stability of real ecosystems. *Nature*, 333, 261–263.

Morales, H., Perfecto, I., and Ferguson, B. 2001. Traditional fertilization and its effect on corn insect populations in the Guatemalan highlands. *Agriculture, Ecosystems, and Environment*, 84, 145–155.

Moreby, S. J. et al. 1994. A comparison of the flora and arthropod fauna of organic and conventionally grown winter wheat in southern England. *Annals of Applied Biology*, 125, 13–27.

Mosleh, Y. Y. et al. 2003. Comparative toxicity and biochemical responses of certain pesticides to the mature earthworm *Aporrectodea caliginosa* under laboratory conditions. *Environmental Toxicology*, 18, 338–346.

Murphy, K. M. et al. 2007. Evidence of varietal adaptation to organic farming systems. *Field Crops Research*, 102, 172–177.

Muthukaruppan, G., Janardhanan, S., and Vijayalakshmi, G. S. 2005. Sublethal toxicity of the herbicide butachlor on the earthworm *Perionyx sansibaricus* and its histological changes. *Journal of Soils and Sediments*, 5, 82–86.

Näsholm, T., Huss-Danell, K., and Hogberg, P. 2000. Uptake of organic nitrogen in the field by four agriculturally important plant species. *Ecology*, 81, 1155–1161.

Neumann, S. et al. 2004. Nitrogen per unit area affects the upper asymptote of *Puccinia striiformis* f.sp. *tritici* epidemics in winter wheat. *Plant Pathology*, 53, 725–732.

Newton, A. C. and Guy, D. C. 1998. Exploration and exploitation strategies of powdery mildew on barley cultivars with different levels of nutrients. *European Journal of Plant Pathology*, 104, 829–833.

Nichols, K. A. and Wright, S. F. 2004. Contributions of fungi to soil organic matter in agroecosystems. In *Soil Organic Matter Management in Sustainable Agriculture*, Magdoff, F. and Weiler, R. R., Eds. CRC Press, Boca Raton, FL, 179–198.

Noble, R. and Coventry, E. 2005. Suppression of soil-borne plant diseases with composts: A review. *Biocontrol Science and Technology*, 15, 3–20.

Norgaard, R. B. and Baer, P. 2005. Collectively seeing complex systems: The nature of the problem. *Bioscience*, 55, 953–960.

Ode, P. J. 2006. Plant chemistry and natural enemy fitness: Effects on herbivore and natural enemy interactions. *Annual Review of Entomology*, 51, 163–185.

Odum, E. P. 1985. Trends expected in stressed ecosystems. *BioScience*, 35, 419–422.

Paoletti, M. G. et al. 1992. Biodiversita in pescheti forlivesi. In *Biodiversita Negli Agroecosystemi*, Paoletti, M. G. et al., Eds. Wafra Litografica, Cesena, Italy, 30–80.

Petersen, H. 2000. *Collembola* populations in an organic crop rotation: Population dynamics and metabolism after conversion from clover-grass ley to spring barley. *Pedobiologia*, 44, 502–515.

Petersen, S. O. et al. 2006. Nitrous oxide emissions from organic and conventional crop rotations in five European countries. *Agriculture, Ecosystems, and Environment*, 112, 200–206.

Peterson, G. 2005. Ecological management: Control, uncertainty, and understanding, In *Ecological Paradigms Lost: Routes of Theory Change*, Cuddington, K. and Beisner, B., Eds. Elsevier Academic Press, Burlington, MA, 371–395.

Phelan, P. L. 1997. Soil-management history and the role of plant mineral balance as a determinant of maize susceptibility to the European corn borer. *Biological Agriculture & Horticulture*, 15, 25–34.

Phelan, P. L. 2004. Connecting belowground and aboveground food webs: The role of organic matter in biological buffering. In *Soil Organic Matter Management in Sustainable Agriculture*, Magdoff, F. and Weiler, R. R., Eds. CRC Press, Boca Raton, FL, 199–225.

Phelan, P. L., Mason, J. R., and Stinner, B. R. 1995. Soil-fertility management and host preference by European corn borer, *Ostrinia nubilalis* (Hübner), on *Zea mays* L.: A comparison of organic and conventional chemical farming. *Agriculture, Ecosystems, and Environment*, 56, 1–8.

Phelan, P. L., Norris, K., and Mason, J. R. 1996. Soil-management history and host preference by *Ostrinia nubilalis* (Hübner): Evidence for plant mineral balance as a mechanism mediating insect/plant interactions. *Environmental Entomology*, 25, 1329–1336.

Pierce, F. J. and Lal, R. 1991. Soil management in the 21st century. In *Soil Management for Sustainability*, Lal, R. and Pierce, F. J., Eds. Soil and Water Conservation Society, Ankeny, IA, 175–179.

Polis, G. A. and Strong, D. R. 1996. Food web complexity and community dynamics. *American Naturalist*, 147, 813–846.

Poole, T. B. 1964. A study of the distribution of soil *Collembola* in three small areas in a coniferous woodland. *Pedobiology*, 4, 35–42.

Rapport, D. J. 1997. Transdisciplinarity: transcending the disciplines. *Trends in Evolution and Ecology*, 12, 289.

Rapport, D. J. and Whitford, W. G. 1999. How ecosystems respond to stress. *Bioscience*, 49, 193–203.

Rapport, D. J., Regier, H. A., and Hutchinson, T. C. 1985. Ecosystem behavior under stress. *American Naturalist*, 125, 617–640.

Rapport, D. J., Costanza, R., and McMichael, A. J. 1998. Assessing ecosystem health. *Trends in Evolution and Ecology*, 13, 397–402.

Raun, W. R. and Johnson, G. V. 1999. Improving nitrogen use efficiency for cereal production. *Agronomy Journal*, 91, 357–363.

Rosenzweig, M. L. 1971. Paradox of enrichment: Destabilization of exploitation ecosystems in ecological time. *Science*, 171, 385–387.

Ryan, M. and Hepperly, P. 2005. Can organic crops tolerate more weeds? http://www.newfarm.org/depts/NFfield_trials/0705/weeds.shtml (accessed December 10, 2007).

Rypstra, A. L. and Marshall, S. D. 2005. Augmentation of soil detritus affects the spider community and herbivory in a soybean agroecosystem. *Entomologia Experimentalis et Applicata*, 116, 149–157.

Sakamoto, T. and Matsuoka, M. 2004. Generating high-yielding varieties by genetic manipulation of plant architecture. *Current Opinion in Biotechnology*, 15, 144–147.

Sarkodie-Addo, J., Lee, H. C., and Baggs, E. M. 2003. Nitrous oxide emissions after application of inorganic fertilizer and incorporation of green manure residues. *Soil Use & Management*, 19, 331–339.

Saunders, M., Lewis, P., and Thornhill, A. 2003. *Research Methods for Business Students*, 3rd ed. Prentice-Hall, Englewood Cliffs, NJ.

Scheu, S. 2001. Plants and generalist predators as links between the below-ground and above-ground system. *Basic and Applied Ecology*, 2, 3–13.

Scholte, K. and Lootsma, M. 1998. Effect of farmyard manure and green manure crops on populations of mycophagous soil fauna and *Rhizoctonia* stem canker of potato. *Pedobiology* 42, 223–231.

Schuler, T. H. et al. 1999. Parasitoid behaviour and Bt plants. *Nature*, 400, 825–829.

Seiter, S. and Horwath, W. R. 2004. Strategies for managing soil organic matter to supply plant nutrients. In *Soil Organic Matter Management in Sustainable Agriculture*, Magdoff, F. and Weiler, R. R., Eds. CRC Press, Boca Raton, FL, 269–293.

Semenov, A. M., van Bruggen, A. H. C., and Zelenev, V. V. 1999. Moving waves of bacterial populations and total organic carbon along roots of wheat. *Microbial Ecology*, 37, 116–128.

Shaul, O. et al. 1999. *Mycorrhiza*-induced changes in disease severity and PR protein expression in tobacco leaves. *Molecular Plant-Microbe Interactions*, 12, 1000–1007.

Siddiqui, I. A. and Shaukat, S. S. 2004. Systemic resistance in tomato induced by biocontrol bacteria against the root-knot nematode, *Meloidogyne javanica,* is independent of salicylic acid production. *Journal of Phytopathology*, 152, 48–54.

Solé, R. V. and Bascompte, J. 2006. *Self-organization in Complex Ecosystems*. Princeton University Press, Princeton, NJ.

Sousa, J. P. et al. 2006. Changes in *Collembola* richness and diversity along a gradient of land-use intensity: A pan European study. *Pedobiologia*, 50, 147–156.

Spagnolo, B., Valenti, D., and Fiasconaro, A. 2004. Noise in ecosystems: A short review. *Mathematical Biosciences and Engineering*, 1, 185–211.

Stamp, N. 2003. Out of the quagmire of plant defense hypotheses. *Quarterly Review of Biology*, 78, 23–55.

Stanhill, G. 1990. The comparative productivity of organic agriculture. *Agriculture, Ecosystems, and Environment*, 30, 1–26.

Sterelny, K. 2005. Made by each other: organisms and their environment. *Biology and Philosophy*, 20, 21–36.

Stone, A. G., Scheuerell, S. J., and Darby, H. M. 2004. Suppression of soilborne diseases in field agricultural systems: Organic matter management, cover cropping, and other cultural practices. In *Soil Organic Matter Management in Sustainable Agriculture*, Magdoff, F. and Weiler, R. R., Eds. CRC Press, Boca Raton, FL, 131–178.

Stout, M. J., Thaler, J. S., and Thomma, B. P. H. J. 2006. Plant-mediated interactions between pathogenic microorganisms and herbivorous arthropods. *Annual Review of Entomology*, 51, 663–89.

Taverne, D. 2005. *The March of Unreason: Science, Democracy, and the New Fundamentalism.* Oxford University Press, New York.

Thaler, J. S. et al. 2001. Jasmonate-mediated induced plant resistance affects a community of herbivores. *Ecological Entomology*, 26, 312–324.

Tilman, D. 1987. Secondary succession and the pattern of plant dominance along experimental nitrogen gradients. *Ecological Monographs*, 57, 189–204.

Torstensson, G., Aronsson, H., and Bergström, L. 2006. Nutrient use efficiencies and leaching of organic and conventional cropping systems in Sweden. *Agronomy Journal*, 98, 603–615.

Trewavas, A. J. 2002. Malthus foiled again and again. *Nature*, 418, 668–670.

Vallad, G. E., Cooperband, L., and Goodman, R. M. 2003. Plant foliar disease suppression mediated by composted forms of paper mill residuals exhibits molecular features of induced resistance. *Physiological & Molecular Plant Pathology*, 63, 65–77.

van Bruggen, A. H. C. and Semenov, A. M. 2000. In search of biological indicators for soil health and disease suppression. *Applied Soil Ecology*, 15, 13–24.

van Bruggen, A. H. C. and Termorshuizen, A. J. 2003. Integrated approaches to root disease management in organic farming systems. *Australasian Plant Pathology*, 32, 141–156.

van Bruggen, A. H. C., Semenov, A. M., and Zelenev, V. V. 2000. Wavelike distributions of microbial populations along an artificial root moving through soil. *Microbial Ecology*, 40, 250–259.

van Bruggen, A. H. C. et al. 2004. Suppression of take-all disease in soils from organic versus conventional farms in relation to native and introduced *Pseudomonas fluorescens*. *Phytopathology*, 94, S105.

van Bruggen, A. H. C. et al. 2006. Relation between soil health, wave-like fluctuations in microbial populations, and soil-borne plant disease management. *European Journal of Plant Pathology*, 115, 105–122.

van Emden, H. F. 2002. Interaction of host-plant resistance and biological control. In *Encyclopedia of Pest Management*, Pimentel, D., Ed. Marcel Dekker, New York, 420–422.

van Emden, H. F. 2003. GM crops: A potential for pest mismanagement. *Acta Agriculturae Scandinavica, Section B, Soil and Plant Science*, Suppl. 1, 26–33.

Verhoef, H. A. and Meintser, S. 1991. The role of soil arthropods in nutrient flow and the impact of atmospheric deposition. In *Advances in Management and Conservation of Soil Fauna*, Veeresh, G. K., Ed. Vedam Books, New Delhi, 497–506.

Viswanathan, V., Jaffery, F. N., and Viswanathan, P. N. 2003. Agrobiotechnology for a more sustainable and environment-friendly second Green Revolution in India. *Biological Memoirs*, 29, 1–20.

Vitousek, P. M. et al. 1997. Human alteration of the global nitrogen cycle: Sources and consequences. *Ecological Applications*, 7, 737–750.

von Liebig, J. 1855. *Principles of Agricultural Chemistry, with Special Reference to the Late Researches Made in England.* Walton & Maberly, London.

Walters, D. R. and Bingham, I. J. 2007. Influence of nutrition on disease development caused by fungal pathogens: Implications for plant disease control. *Annals of Applied Biology*, 151, 307–324.

Wardle, D. A. 1995. Impacts of disturbance on detritus food webs in agro-ecosystems of contrasting tillage and weed management practices. *Advances in Ecological Research*, 26, 105–185.

Wardle, D. A. 2002. *Communities and Ecosystems: Linking the Aboveground and Belowground Components.* Princeton University Press, Princeton, NJ.

Waring, G. L. and Cobb, N. S. 1992. The impact of plant stress on herbivore population dynamics In *Insect–plant Interactions*, vol. IV, Bernays, E., Ed. CRC Press, Boca Raton, FL.

Weil, R. R. and Magdoff, F. 2004. Significance of soil organic matter to soil quality and health. In *Soil Organic Matter Management in Sustainable Agriculture*, Magdoff, F. and Weiler, R. R., Eds. CRC Press, Boca Raton, FL, 1–43.

White, T. C. R. 1984. The availability of invertebrate herbivores in relation to the availability of nitrogen in stressed food plants. *Oecologia*, 63, 90–105.

Wise, D. H. et al. 1999. Spiders in decomposition food webs of agroecosystems: Theory and evidence. *Journal of Arachnology*, 27, 363–370.

Woods, J. S. et al. 1987. Soft tissue sarcoma and non-Hodgkin's lymphoma in relation to phenoxyherbicide and chlorinated phenol exposure in western Washington. *Journal of the National Cancer Institute*, 78, 899–910.

Workneh, F. and van Bruggen, A. H. C. 1994. Microbial density, composition, and diversity in organically and conventionally managed rhizosphere soil in relation to suppression of corky root of tomatoes. *Applied Soil Ecology*, 1, 219–230.

Xiao, N. et al. 2006. The fate of herbicide acetochlor and its toxicity to *Eisenia fetida* under laboratory conditions. *Chemosphere*, 62, 1366–1373.

Yan, Z. et al. 2002. Induced systemic protection against tomato late blight elicited by plant growth-promoting rhizobacteria. *Phytopathology*, 92, 1329–1333.

Yeates, G. W. et al. 1997. Faunal and microbial diversity in three Welsh grassland soils under conventional and organic management regimes. *Journal of Applied Ecology*, 34, 453–470.

Zahm, S. H. and Ward, M. H. 1998. Pesticides and childhood cancer. *Environmental Health Perspectives*, 106, 893–908.

Zelenev, V. V., van Bruggen, A. H. C., and Semenov, A. M. 2005. Short-term wave-like dynamics of bacterial populations in response to nutrient input from fresh plant residues. *Microbial Ecology*, 49, 83–93.

Zelenev, V. V. et al. 2006. Oscillating dynamics of bacterial populations and their predators in response to fresh organic matter added to soil: The simulation model 'BACWAVE-WEB.' *Soil Biology & Biochemistry*, 38, 1690–1711.

Zhang, H. and Forde, B. 1998. An Arabidopsis MADS box gene that controls nutrient-induced changes in root architecture. *Science*, 279, 407

Zhang, S. et al. 2002. The role of salicylic acid in induced systemic resistance elicited by plant growth-promoting rhizobacteria against blue mold of tobacco. *Biological Control*, 25, 288–296.

Zogg, G. P. et al. 2000. Microbial immobilization and the retention of anthropogenic nitrate in a northern hardwood forest. *Ecology*, 81, 1858–1866.

10 Agroecosystem Integrity and the Internal Cycling of Nutrients

Michelle Wander

CONTENTS

10.1 INTRODUCTION

This chapter evaluates the origins of our soil stewardship instinct and asks whether or not the soil health paradigm embraced by agroecology is principally derived from current science or, instead, from social movements developed in reaction to modern agriculture and reductionist approaches to research. Humans associated soil organic matter with fertility and sustainability long before scientific evidence gave support to this notion. Early beliefs in the lithic origin of life that resulted in the use of soil as a metaphor for human well-being explain why soils are revered. The attack on the humus theory that took place in the nineteenth century was a rejection of this spiritual or metaphysical view of soils. Historical recountings emphasize that the humus theory wrongly argued that plants obtained their mass by consuming soil but leave out the fact that the theory considered questions about the origins of life and decay that trace back to Aristotle.

The battle over the humus theory was as much about our assumptions about science and the way we pursue it as it was about how matter was cycled between plants, soils, and the atmosphere. Aristotelian approaches to science that considered the influence of human actions on the greater whole were excluded from agricultural research when the humus theory and its romantic views of nutrient cycling were deposed by the mineralist theory and the subsequent rise of modern agriculture.

Liebig's attack of the humus theory marks a turning point in the history of science wherein physical mechanisms and quantitative methods were embraced as the principal tools for modern science and so segregated chemistry and physics from spiritual and biological themes (see Chapter 5 for further discussion of Liebig). As a result of this transformation, science was separated from philosophy. By replacing natural philosophy, which addressed all knowledge and sought to understand the broader workings of nature, with Newtonian determinism, we redefined how we conceive of, and manage, soil. During the late nineteenth century, new scientific and statistical methods and specialization in the sciences narrowed the focus of agriculture and diminished the importance of theoretical and applied empirical work.

Concerns resulting from reductionism and land degradation caused by agricultural intensification spawned alternative agriculture movements that reclaimed a romantic view of soils and stewardship. The biodynamic, organic, permaculture, and natural systems agriculture movements all placed soil organic matter and nutrient recycling at the center of sustainable management. During the twentieth century, agronomists and ecologists studying agricultural systems have pursued different lines of evidence, each emphasizing different aspects of performance. Agroecologists seeking to evaluate the tenets of sustainable agriculture that were outlined by alternative agriculture movements also looked to nature as a model. Maximization principles, plant strategy, and facilitation theories inspired by Eugene Odum's "Strategy for Ecosystem Development" may provide a conceptual foundation for agroecology's soil quality paradigm. Collectively these theories hold that soil biology and the resulting feedbacks on systems determine their efficiency and productivity. The weight of evidence provided by natural systems suggests that the tightening of nutrient cycles and increases in soil organic matter reserves can be achieved where disturbance frequency, residue decay rates, and nutrient imports are reduced. The relationship between plant diversity and these factors, as well as their influence on harvestable matter, is less firmly established and is likely to vary greatly among plant-soil systems. Agricultural intensification has suppressed the importance of biological mechanisms in arable systems by reducing diversity and increasing the supply of readily available nutrients in soils. The long- and short-term influences of such agricultural practices on gross- and net-harvestable productivity and resource status must be better articulated. By taking a more holistic approach to research we can reduce mistakes in management caused by reductionism by anticipating when and where feedbacks could enhance or compromise soils' productive potential and system efficiency. New information management and synthesis tools and a broadening of objectives to value resource condition equally with productivity will help us to develop principles of rational agriculture abandoned when the humus theory was deposed.

10.2 DUALISTIC VIEW OF SOILS

10.2.1 Soil as a Metaphor

The historical and spiritual association between humans and soils has shaped attitudes about management that can be traced through philosophical and scientific writing (Table 10.1). Soil reverence was widespread in early religions (Hillel 1991; Lines-Kelly 2004) and has its origins in early lithic mythologies that attributed mineral parentage to humans, wherein gods were thought to impart lifeless minerals with souls (Eliade 1978). The close association between humans and soil is also featured in the Old Testament (Holland and Carter 2005); where according to Davis (2002), the first 11 chapters of Genesis can be understood as a stewardship doctrine chronicling the story of Adam's progressive alienation from God and fertile soil. She asserts God named Adam after soil, *adamah,* to keep him humble. This juxtaposition of humility and value is echoed in concepts tied to *Humus,* which the ancients used to refer to the "fatness of the land" (Waksman 1936). Humus is the English translation of *ghôm,* which has an Indo-European root shared by the terms: *Human, Humility, Hubris,* and *Humor* (Kurtz and Ketcham 2002). The Talmud explains how Adam was created as a golem from dust. The need to care for the soil as a proxy for ourselves

TABLE 10.1
Historical Summary of Philosophical and Scientific Texts that Have Influenced Attitudes About Soils and Scientific Principles of Management

10th–5th century bc
Moses
"Genesis" 1:28 "In the beginning, there was heaven and earth and then God said, let there be life" to them be fruitful and multiply." Establishes view of man as land steward and despot.

4th century bce
Democritus
Fragments of the "Presocratics" introduce atomism and mechanical determinism wherein the workings of the universe are cast as entirely mechanical, driven by the vibrations, the velocities, and impacts of the constituent atoms.

4th century bce
Aristotle
"On Generation and Corruption" presents natural philosophy that defines nature as a purposeful agent. Text poses questions about the origins of mater and transformation that persist into the modern era. Animal and mineral kingdoms are seen to have separate divine origins.

29 bce
Virgil
"The Georgics" is didactic poem divided into four books, the first two of which are devoted to agriculture and include guidance on farming and stewardship; text includes the warning: "Not that all soils can all things bear alike."

200 ce
Tamlud (Misnah; Tractate Sanhedrin 38b)
Adam is formed as a golem when dust was "kneaded into a shapeless hunk" by god.

2nd–12th centuries ce
Unknown Alexandrian
"Physiologus" fables and bestiaries advance humanist concept with nature and its symbols used as moral guides; this vision of a human-centered cosmos advances a utilitarian philosophy of resource management. Cited in Harrison (1999).

1310
Latin Geber
"Summa perfectionis" translation attributed to eighth-century Islamic scholar Jabir ibn Hayyan brings knowledge from hermetic alchemy to the West and provides the basis for corpuscular theory. Advocacy of the pursuit of purity using a rational, systematic chemical analysis had a significant influence on European scientists including Robert Boyle. Cited in Newman (2006).

1563
Bernard Palissay
"Recette Veritable" presented an idealized conception of stewardship of an ideal garden that includes a manual of chemistry that describes the elements in context of their transformational nature; reasserts a natural philosophy where Earth, or nature, relies on recycling of substance whose nutritive value is enhanced as they pass through matter— founds humanist notion of natural systems with directed perfection. Cited in Jeanneret (2001).

1619
Daniel Sennert
"De chymicorum cum Aristotelicis et Galelinicis consensu ac dissensu" paves the way for Boyle's mechanical philosophy.

1661
Robert Boyle
"Skeptical Chemist" calls for chemists to assert their discipline as equal to alchemy and medicine and argues that experimental proof is needed before theory is accepted as truth.

Continued

TABLE 10.1 (*Continued*)
Historical Summary of Philosophical and Scientific Texts that Have Influenced Attitudes About Soils and Scientific Principles of Management

1687

Isaac Newton

"Mathematical Principles of Natural Philosophy" proposed the laws of motion, and introduced the use of calculus. Proposed a religious basis for mechanistic sciences as he ushers in the Age of Reason.

1697

George Stahl

"Zymotechnia fundamenlalis sive fermentalionis theoria generalis" renamed terra pinguis, one of Aristotiles elements, as phlogiston from the ancient phlogios for "fiery" building on the notion that burning substances lose phlogiston, the invisible, life-giving material.

1777

Antoine Lavoisier

"Memoir on Combustion in General" asserts that pure air is the true combustible body and that fire is not fixed in bodies, particularly metals. Instead, matter of free fire exists in substances by virtue of properties manifest as they come into equilibrium with neighboring bodies.

1788

Joseph Randall

"Semi-Virgilian Husbandry" concluded that atmospheric water absorbed exhalations from Earth and returned them to soil. Cited by Klein et al. (1988).

1790s

Jan Ingen-Housz

"Experiments upon Vegetables, Discovering Their Great Power of Purifying the Common Air in the Sun-shine, and of Injuring It in the Shade and at Night to Which Is Joined, A New Method of Examining the Accurate Degree of Salubrity of the Atmosphere" advanced knowledge of organic matter mineralization.

1794–1796

Erasmus Darwin

"Zoonomia" espoused vitalist notions with life arising out of decay to postulate a Lamarckian theory of evolution.

1804

Nicolas de Saussure

"Chemical Researches on Vegetation" provides early insights into nature of photosynthesis. This publication marks the end of philogistic period and the beginning of modern agricultural science.

1809

Albrect Thaer

"Principles of Rationale Agriculture" theory formally outlined the humus theory and proposed a quantitative fertility scale that was based on soil properties, plant nutrient requirements, and on the cropping systems themselves. Cited in Feller et al. (2003).

1813

Sir Humphrey Davy

"Elements of Agricultural Chemistry" suggested soils provide physical medium to hold up plant and permit uptake of humic substances.

1826

Carl Sprengel

"About Plant Humus, Humic Acids and Salts of Humic Acids" (*Archiv fuer die Gesammte Naturlehere*, 8, 145–220, 1826) disproves humus theory and provides evidence of mineral theory.

1828

Frederich Wöhler

"On the Artificial Formation of Urea" ([Über künstliche Bildung des Harnstoffs.] *Ann Phys Chem*, 12, 253–256, 1828). Disproved vitalism, the unique biotic origin of organic compounds, when he synthesized urea.

TABLE 10.1 (*Continued*)
Historical Summary of Philosophical and Scientific Texts that Have Influenced
Attitudes About Soils and Scientific Principles of Management

1838

Liebig, J. "Sur les phénomenes de la fermentation et de la putréfaction, et sur les causes qui les provoquent" (*Annales de Chimie et de Physique*, 2e Serie LXXI, 178, 1838.)

1840

Justis von Liebig

Organic Chemistry in Its Application to Agriculture and Physiology

---

1860

Louis Pasteur

"Expériences relatives aux générations dites spontanées" cited in Hebd, C. B., Séances (*Acad. Sci. Paris*, 303–307, 1860) argues effectively that "life" does not originate spontaneously but develops.

1862

Pasteur, L. Note remise au Ministère de l'Instruction publique et des cultes, sur sa demande, Avril 1862. In *L'oeuvre de Pasteur*, T.VII, p. 3. Noted the "immense role of infinitely small bodies in the general economy of nature."

1881

Charles Darwin

"The Formation of Vegetable Mould through the Action of Worms with Observations on Their Habits" associated decay and the role of biota with sustained fertility.

1910

Cyril Hopkins

"Soil Fertility and Permanent Agriculture" proposes a means for permanent maintenance of soil fertility, warns against extractive practices that mine the soil.

1911

F. H. King

Farming of Forty Centuries in China, Korea, and Japan extols the virtues of traditional farming systems capable of supporting large populations for long time periods. "Again, the great movement of cargoes of feeding stuffs and mineral fertilizers to western Europe and to the eastern United States began less than a century ago and has never been possible as a means of maintaining soil fertility in China, Korea or Japan, nor can it be continued indefinitely in either Europe or America. These importations are for the time making tolerable the waste of plant food materials through our modern systems of sewage disposal and other faulty practices; but the Mongolian races have held all such wastes, both urban and rural, and many others which we ignore, sacred to agriculture, applying them to their fields." (C. B. King, Courier Dover Publications, Mineola, NY.)

1924

Rudolf Steiner

Spiritual Foundations for the Renewal of Agriculture. A Course of Eight Lectures" held at Koberwitz Silesia, founds Anthroposophy with his effort to relate cosmic and earthy forces, describing the farm as an individuality standing on its head, with the soil serving as a diaphragm to facilitate exchange between above- and belowground forces. In this he reasserts the need for ethereal origins for life. (First published by S. Steiner Verlag, Dornach, Switzerland, 1984. Translated into English and published by the Bio-Dynamic Farming and Gardening Association, 1993, Kimberton, PA.)

1925

Sir Ronald Aylmer Fisher

Statistical Methods for Research Workers pioneers modern statistical methods and lays the mathematical foundation for hypothesis testing through controlled experimentation, application of probability theory, maximum likelihood, randomization, and blocking. (Originally published by Oliver and Boyd in Edinburgh, U.K.)

Continued

TABLE 10.1 (*Continued*)
Historical Summary of Philosophical and Scientific Texts that Have Influenced Attitudes About Soils and Scientific Principles of Management

1936

Selman Waksman

Humus: Origin, Chemical Composition, and Importance in Nature is a foundational work initiating scientific inquiry into soil microbiology and biochemistry; Waksman's work on antibiotics resulted in his receipt of a Nobel prize in 1952. (Williams & Wilkins, Baltimore, MD.)

1938

W. A. Albrecht

"Loss of Organic Matter and Its Restoration" in *Soils and Men* provides scientific evidence of soil degradation resulting from agricultural intensification and points to organic matter management as key to stewardship. In this he asserts: "Organic matter thus supplies the 'life of the Soil' in the strictest sense." (U.S. Department of Agriculture, U.S. Printing Office.)

1940

Lord Nortbourne

Look to the Land. "The soil is the whole world by itself. But as a world it is an entity, and one with which as an entity, every farmer and gardener is intimately concerned." (2nd ed. Sophia Perennis Press, Hillsdale, NY.)

1942

Evelyn Balfour, Lady

The Living Soil presents the first systematic comparison of organic and conventional systems and outlines new humus theory by outlining the separate physical, chemical, and biological contributions of organic matter to plant health. (Faber and Faber, London.)

1945

Jerome Irving Rodale

Pay Dirt: Farming and Gardening with Composts initiates the organic gardening movement in the United States with a soil-centered approach. (Devin-Adair, New York.)

1947

Ehrenfried Pfieffer

Soil Fertility, Renewal, and Preservation: Bio-Dynamic Farming and Gardening. Outlines principles of biodynamic practices and preparations; evaluates their influence on yields. (Faber and Faber, London.)

1949

Aldo Leopold

A Sand County Almanac and Sketches Here and There. Articulates the land ethic thus: "The land ethic simply enlarges the boundaries of the community to include soils, waters, plants, and animals, or collectively: the land." (Oxford University Press, New York.)

1968

Eugene Odum

"Strategy for Ecosystem Development." (*Science.*)

1976

H. H. Koepf et al.

Biodynamic Agriculture: An Introduction. Asserts "The good farmer can tell how biologically-active the soil is, not by counting bacteria, but by the tilth and the workability of the soil." (Anthroposophic Press, Spring Valley, NY.)

1977

Wendell Berry

The Unsettling of America: Culture and Agriculture. Offers a critique of American agriculture where traditional systems are uprooted by fortune seekers, he asserts: "By separating ourselves from the production and processing of food—mainly through the replacement of agriculture with agriscience and agribusiness—we are suffering, and not just in terms of our health but, more importantly, morally, culturally, as communities of people." (University of California Press, Berkeley.)

TABLE 10.1 (*Continued*)
Historical Summary of Philosophical and Scientific Texts that Have Influenced Attitudes About Soils and Scientific Principles of Management

1978

Bill Mollison and David Holmgren

Permaculture One. Begins a series of publications advancing theory of permanent agriculture based on use of natural perennial systems. (Transworld Publishers, Australia.)

1981

Wendell (Ed) Berry

The Gift of Good Land: Further Essays Cultural and Agricultural. An edited volume that revisits the theme that traditional practices sustain the land and culture; extols the virtues of stewardship, and looks to religious origins for this proviso. (North Point Press, San Francisco.)

1984

Richard Lowrance, Benjamin R. Stinner, and Garfield J. House, Eds.

Agricultural Ecosystems; Unifying Concepts. Early summary of papers introducing agroecology as a science based on the 1982 symposium held during the Ecological Society of America meetings, State College, PA. (John Wiley & Sons, New York.)

1985

Masanobu Fukuoka

The Natural Way of Farming: The Theory and Practice of Green Philosophy. Compares the merits of the "Natural" or the philosopher's way of farming, with "Scientific" farming which he asserted was designed to support dialectical materialism. (Japan Publications, Tokyo.)

is echoed in the works of Virgil, Palissay, and countless others who have associated wise management to an idealized agrarian culture. The view of man as a despot, potentially using and abusing Earth's resources also dates to early Greek traditions. A dualistic view of man was expressed through the medieval bestiaries where humans were instructed to tame their destructive nature (Conley 1998).

This tension between soil stewardship and exploitation recurs through history. Harrison (1999) argues that a biblical imperative to dominate the Earth played an important role in the rise of modern science and that by linking resource exploitation to the Fall, Christianity promoted restoration and by doing so made "despot" and "steward" parts of a future tradition. This dualistic view of the human endeavor has defined the agenda and character of science (Brown 2003) and, so, the fate of the soil resource. Exploitative practices have been tempered by the pervasive use of soil a metaphor for humanity and the belief that cultural fortunes are tied to its stewardship. Metaphors are widely used in both religion and science to link concepts in a way that allows abstract ideas to be understood in terms of deeply grounded physical or social experiences (Gerhart and Russell 2004). Our legacy of metaphorical reasoning has left us with a view of soil as either something to be revered and protected or something dirty to be equated with waste that is to be used (Lines-Kelly 2004). These two views of soil can be traced through our science and our attitudes about management.

10.2.2 Vitalism and Humus Theory

Before the nineteenth century, knowledge about soils and agriculture was pursued along with medicine and chemistry as part of natural philosophy with theories developed and applied in a holistic fashion. Science was not yet separated from philosophy and agricultural questions were frequently at the center of philosophic debates. Key points of contention addressed by scholars interested in plant–soil interactions can be traced to Aristotle's thoughts about *the origins of life and mass, and the basis for residue transformation* (Manlay et al. 2007). Scientists in the sixteenth and seventeenth centuries struggled to understand the nature of organic and inorganic matter and to discover how living and nonliving matter, including noncombustible elements, were created and destroyed. It was during this period that modern chemistry and experimental methods grew up out of natural philosophy as alchemists, atomists, and advocates of corpuscular theories attempted to elucidate the nature of matter (Newman 2006). Plant growth and organic matter decay were important models system for investigators interested in these topics. A letter written by a French friar in 1679 provides insight into the state of confusion that existed as he wrote if one were to "take a pot ... with seven to eight pounds of earth and grow in it any plant you choose; the plant will find in this earth and in the rainwater which has fallen on it, the principles of which it is composed in its mature state" (Klein et al. 1988). For most of the seventeenth century it was believed that water and air absorbed nutrients, including carbon, released by plant and animal respiration and by putrification, and that these nutrients were then returned to the soil to be recycled by plants. There was no mechanistic understanding of respiration, photosynthesis, or decomposition. Using a mint plant in 1780, Joseph Priestley discovered that plants could "restore air which has been injured by the burning of candles." This discovery informed Lavoisier's research and led him to conclude that respiration was a sort of combustion. This understanding was followed by the discoveries of Senebier and de Saussure who, at the turn of the nineteenth century, revealed the basics of photosynthesis and showed that increases in plant mass were primarily due to the uptake of CO_2 and water (Pennazio 2005).

Histories of this period often cite the "humus theory" as a prime example of outdated romantic science that was disproved by modern methods. Carl Sprengel (1826, 1839) and Justis von Liebig (1840) developed the mineral theory of plant nutrition (Wendt 1950). The demise of the humus theory receives a great deal of attention because the argument was about more than plant mineral nutrition. By the turn of the nineteenth century, natural philosophers struggled to transition from a chemicotheological to chemicophysical models of organic metamorphosis (Marald 2002). The humus theory, which was put into print by Albrect Thaër in 1809, asserted plants took up C, N, O, and H from soil and transformed them through a "vital" life force; additionally; the theory also suggested addition of nonessential salts and lime would benefit plants by aiding decomposition (van der Ploeg et al. 1999). The idea that plants absorbed humus originated with Wallerius in 1761 and was restated in Davy's *Agricultural Chemistry* published in 1813 (Rossiter 1975). This widely used textbook propagated misconceptions about plant C uptake that should have been corrected by de Saussure's work that outlined the fundamentals of photosynthesis more than a decade earlier. The humus theory considered more than plant C and nutrient acquisition. It also addressed the nature of matter, asking whether and how organic and inorganic materials and living and nonliving matter were transformed into one another. *During the 50 years leading up to Davy's book, atomistic and corpuscular theories proposed mechanical models for matter transformation that were vitalist in nature.* Elements continued to be divided into organic and inorganic categories, with organic elements distinguished by possession of a vital force. It was still believed that organic elements could not be synthesized from inorganic matter without some divine addition. Vitalism officially, or at least technically, died in 1828, when Frederick Wohler synthesized urea, an organic molecule, with inorganic precursors. This caused him to write of "the great tragedy of science, the slaying of a beautiful hypothesis by an ugly fact."

The origins of vitalist thinking can be traced all the way back to Democratus. According to Aristotle, Democratus believed the soul was composed of fire atoms and these provided the essence of life (Kinne-Saffran 1999). The notion of fire atoms was embedded in Phlogiston theory which argued that flammable matter lost "phlogiston" elements when combusted. Lavoisier's work on combustion brought an end to the phlogiston period in 1777. The demise of the humus and phlogiston theories are frequently connected (van der Ploeg et al. 1999); this is probably because both were holdovers of chemicotheology.

The official end of vitalism did little to resolve vague and incorrect notions about the contributions of soils to plant productivity and resolve uncertainty about nutrient transformations. The notion that elements, and even matter, could be created spontaneously within plants through *biological transmutation persisted into the middle of the nineteenth century*. Albrect Thaer subscribed to this notion and believed that calcium could be changed into silicon within the plant and that calcium was formed from potassium and that humus was the original source of lime (Ihde 1984). De Saussure had rejected the notion of transmutation and proved that the presence of silicates in the plant were not due to the life force, as asserted by Lampadius in 1832, but instead that silicates were a product of the amount of silicon in the soil. Even though de Saussure had argued that certain minerals were essential for plant growth in 1804, uncertainty about this and the origins of mineral nutrients persisted. Sprengel and Liebig both pursued the identity of the so-called indispensable mineral salts, while Boussingault and Ville sought to understand nitrogen fixation (Aulie 1970). The matter of transformation was addressed definitively in 1842, when A. F. Wiegmann and L. Polstroff won the Royal Goettingen Society of Science's prize for an experiment comparing plant growth in a synthetic soil or in sand alone and allowed them to conclude: "The inorganic constituents of plants can in no respect be regarded as products of their vital activity either as formations from unknown elements or as peculiar derivations of the 4 elements known to make up organic substances" (Browne 1977).

Liebig receives a great deal of attention in the historical recounting of this era for several reasons including his aggressive style and advocacy for the use of inorganic fertilizers (Werner and Holmes 2002; Pennazio 2005). Liebig was an active participant in the discussion of organic compounds with interests in both their origin and fate. He agreed with several of the core ideas advanced by the humus theory, including the notion that a life force gave rise to living plants and that there was a close association between soil organic matter, soil physical condition, and overall plant productivity that had nothing to do with nutrient supply. He engaged in several high-visibility arguments with the English school of Lawes and Gilbert about the need for nitrogen inputs to soils, which he did not deem necessary, and clashed with Louis Pasteur about biotic contributions to decay. Pasteur's work on fermentation silenced Liebig's assertions that decomposition obeyed the same physical laws that applied to mechanical systems and that transformations occurred through purely physical, thermodynamic processes (Rosenfeld 2003). In 1882, Pasteur articulated how putrefaction played a major role in the recycling of elements between the living and mineral worlds (Schwartz 2001).

Liebig's vitriol and advocacy for chemistry are often blamed for divisions between biologists and chemists that formed in the nineteenth century (Rosenfeld 2003). His writings attacked plant physiologists, incorrectly asserting in 1840 that most physiologists continued to believe that humus was the chief nutritive source for plants (Werner and Holmes 2002). Further, Liebig charged that physiologists' ignorance stemmed from flawed methodology (Werner and Holmes 2002; Marald 2002). Plant physiologists responded in kind, asserting Liebig was ignorant of plant physiology and was unaware of advances in cell biology being made with improved microscopic methods (Holmes 1989). Divisions that formed between the biological and physical sciences and distinctions between preferred methods persist to this day even within the soils discipline where chemistry and physics have received the greatest emphasis. The taint of their chemicotheological past and incorrectly perceived lack of quantitative methods explain the second class status that soil biology and biochemistry have held during the modern period.

10.2.3 RISE OF MODERN AGRICULTURE AND SPECIALIZATION

Scientific discoveries made in the mid- to late nineteenth century had a profound influence on how agricultural science was practiced in the West. Major advances in plant physiology and chemistry had already been made using methodology credited to Bacon that was informed by the gravimetric methods of eastern alchemists (Newman 2006). Eastern alchemy had already contributed to a medieval green revolution in the Mideast where it also fostered development of early scientific principles of resource conservation (Watson 1974, 1981). In the West, however, modern agricultural research developed with a more narrow focus on production. According to (Holmes 1989), it was Liebig's development of reproducible quantitative methods performed in an assembly-line fashion that played the founding role in the development of modern agricultural chemistry and our approach to education in agricultural science. Experimental methods that were mechanical and mathematical began to suppress more theoretical and spiritual approaches in the 1700s. Newtonian science was disseminated to the public through *experimentum crucis* that demonstrated "key principles" and this approach become backbone of university instruction in the natural sciences (Velho and Velho 1997). The nineteenth century culminated in the founding of disciplinary divisions, reductionism, and an abiotic bias that lingers in agricultural chemistry to this day. Science historians argue that mechanistic sciences that are focused on quantification principally service politicians and economic ends (Schaffer 2005). Liebig is probably most criticized by proponents of alternative agriculture for modern agriculture's emphasis on production and use of "chemical manures" (see Kirschenmann, Chapter 5).

10.2.4 SUSTAINABLE AGRICULTURE AND AGROECOLOGY

10.2.4.1 Alternative Agriculture Movements

Adoption of modern agricultural research methods may have spawned reductionism and suppressed the use of alternative experimental approaches within the discipline but did not extinguish more integrative approaches entirely. The systems thinking that had previously been an integral part of agricultural research found a variety of outlets. Alternative agriculture movements grew in responses to problems in sustainability that resulted from perceived perturbations of nutrient cycles caused by modern agriculture (Marald 2002). For a time, this area of inquiry retained vitalist notions to the extent that nutrient recycling by organic life cycles was seen to be by divine origin (Johnston 1853). The rift between modern and alternative approaches to agriculture is tied to the adoption of "chemical manures" developed by Liebig and others that upset this divine order. Notions of God's economy or *oeconomy,* which referred to the divine government of the natural world, saw a system that matched needs in an efficient if not perfect manner (Worster 1977). Chemicotheological ideas that had made an explicit connection between recycling, agricultural production, and societal well-being were rapidly displaced in the late nineteenth century by waste management practices that emphasized municipal economic and technological theory (Marald 2002). The replacement of manure by chemical fertilizers and ensuing development of sewer systems decoupled the city from the farm and sparked debate that was taken up by luminaries of the day. Victor Hugo decried the waste of fertilizer when he wrote "the sewer is the conscience of the city" (1896). It was this kind of debate that made waste recycling and composting central tenets of sustainable agriculture (Blum 1992; Conford 1995).

By the turn of the twentieth century soil scientists were divided in their views about the ability of soils and modern agricultural practices to support agriculture indefinitely. Interest in the benefits of humus remained as many of the specific benefits of organic matter to production systems remained poorly understood. The U.S. Bureau of Soils denied that soil exhaustion was causing crop yield decline, asserting instead that this was due to the "toxic excretia" of roots that could be offset through the addition of manures (Whitney and Cameron 1903; and Whitney 1908, cited in Schulman 1999). Both Cyril Hopkins (1910) and F. H. King (1911) argued soil exhaustion was the chief cause for observed yield decline. Hopkins focused his attentions on rotation and the tendency of modern practices to "mine the soil," while King advanced a systems view of the problem and its solution, writing in 1910 that "the husbandman is an industrial biologist and as such is compelled to shape his operations so as to conform with the time requirements of his crops." William Albrecht (1938) provided early evidence of soil exhaustion of fertile prairie soils by tracking declines in soil nitrate concentrations observed during the cropping season under corn.

These works, and in particular those of King, influenced Lord Northbone, who proposed the concept of "organic agriculture" in 1940 as a system of management centered on soil stewardship to promote nutrient cycling to support the whole (Scofield 1986). His organic ideal was a system "having a complex but necessary interrelationship of parts, similar to that in living things" (Heckman 2006). Northborne's romantic vision of this integrated whole influenced the works of Sir Albert Howard and Ehrenfried Pfeiffer, among others. Both King and Howard studied traditional farming systems that had sustained productivity without degrading soil resources as models of success. Howard's work addressed issues of land use, culture, and values; these sentiments were expanded by Aldo Leopold, Wendell Berry, and Wes Jackson and continue to hold sway to this day. Contemporary literary collections by Freyfogel (2001) and Jackson and Jackson (2002) continue to articulate agrarian and arcadian visions of a culture founded on land stewardship principles and thus perpetuate a reverence for soil.

Rudolf Steiner, the founder of anthroposophy and biodynamic agriculture, was famously inspired by Goethe and his approach to science. Goethe had advocated an approach to science that was self-conscious, holistic, and empirical (Zemplén 2003). It is likely relevant that before turning to agriculture, Steiner studied the relationship between life, health, and growth. The biodynamic movement he founded merged spiritualism and science using an approach to agriculture that related the earth organism to the cosmos. His conception recalls alchemal works that place humans and resource stewardship at the center of well-being (Figure 10.1). Biodynamic agriculture includes an explicit focus on soils and organic matter management that was later articulated in scientific terms through the works of Ehrenfried Pfeiffer and Herbert Koepft. Suggestions of elemental transmutation as part of the composting procedure recall vitalist thinking.

This merger of spiritualism, systems thinking, and soil reverence was also present in Masanobu Fukuoka's work (1985). Fukuoka proposed a strategy for natural agriculture that married scientific understanding with practical and philosophical dimensions (Figure 10.2). Fukuoka's comparison of natural and scientific farming systems summarizes his beliefs about differences in their productive potential. He wrote that science-based farming "is just so much wasted effort. Yields that improve on nature can never be achieved." This trust in nature's inherent wisdom helps explain the conservatism in alternative agriculture and the emphasis on stewardship instead of manipulation through inputs.

Alternative agriculture movements were unified in the view that it is the system of practice that is the mechanistic agent, causing either positive or negative outcomes. This is a medieval view of science wherein humans are seen as responsible agents that act as system engineers (Grant 1996). The degradation of soils that followed conversion to "modern" methods reinforced the perception that production agriculture was not sustainable. Most if not all of the alternative agriculture visions used natural systems as their model for perfection. This is the case for natural systems or perennial systems agriculture and permaculture movements advocated by Wes Jackson (1980, and Chapter 6, this volume) and Bill Mollison and David Holmgren (1978). Figure 10.3 provides a visual summary of permaculture principles. Both movements argued that for agriculture to be sustainable, it must mimic natural systems and rely principally on perennial plants. Agricultural intensification was seen as artificial and energetically inefficient. The organic, permaculture, and perennial systems movements were more utilitarian and less spiritual in their perspective than some of the other movements.

The emphasis placed on soil organic matter by all of these alternative movements is notable. The more pragmatic focus of the organic and perennial systems movements has made them more accessible to, and informed by, the agricultural research community. Wes Jackson left an academic position to found the Land Institute along with Dana Jackson. Jerome I. Rodale's conception of organic agriculture was strongly influenced by the research of Sir Albert Howard and Lady Eve Balfour (Lotter 2003). Howard and Balfour's emphasis on biology and organic matter has been credited to Charles Darwin's studies of earthworms (Lotter 2003). Howard's work attracted interest in compost and effective waste management as means to solve societal ills and instilled the belief that traditional agricultural systems contained an ecological wisdom. He also championed the need to integrate livestock, asserting in 1956 that "no permanent or effective system of agriculture has ever been devised without the animal." Lady Eve Balfour articulated what has become the modern humus theory, when she speculated in 1943: "Humus benefits the soil in three ways: mechanically,

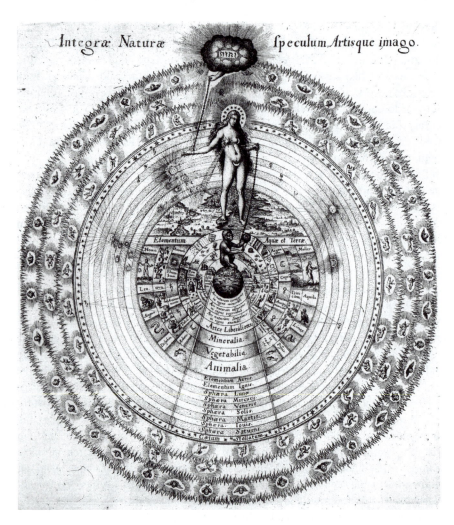

FIGURE 10.1   Frontispiece, titled "The Macrocosm and the Microcosm," from Robert Fludd's Utriusque cosmi historia (1617) released in five parts during the period 1617–1621. In this series, Fludd attempted to synthesize Western esoteric ideas on the relationship between the Cosmos and Man while supplanting the ideas of the Greek philosophers with alchemal concepts of nature that were based on Christian principles.

as a plant food, and by fundamentally modifying the soil bionomics. Of the three, this last, hitherto largely ignored, is probably the most important."

Interest in soil biology, biochemistry, and plant health was also rekindled within academic institutions during the early part of the twentieth century. Selma Waksman, and other agronomic scientists worked to elucidate the contributions of humus to soil productivity in biochemical terms (1936). In 1938, William Albrecht wrote: "Even though we can now feed plants on diets that produce excellent growth without the use of any soil whatever, yet the decaying remains of preceding plant generations, resolved by bacterial wrecking crews into simpler, varied nutrients for rebuilding into new generations, must still be the most effective basis for extensive crop production by farmers." He championed theories of balanced nutrition and discussed implications for animal health in the early 1940s. Albrecht was among the first to document yield reductions caused by overapplication of macronutrients. In addition, he pointed out the trade-offs between production and quality, observing that hybrids could be developed where nutrient content per unit-mass was reduced. By the end of the twentieth century alternative agriculture movements and associated scientific efforts had begun to formalize theories about what a sustainable agriculture would look like.

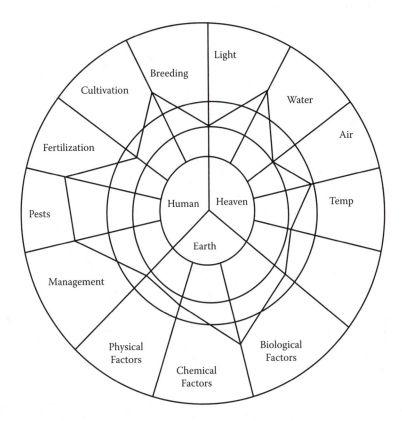

FIGURE 10.2 Adapted from "The Natural Way of Farming" (Fig 2.10) compares harvest components. In this scheme Fukuoka describes trade-offs among farming systems with harvest components in Mahayana Natural Farming (1) depicted by the outermost circle, where the human spirit and life blend and humans devote themselves to service of nature in an idealized state of understanding and enlightenment; in Himayana Natural Farming (2) depicted by the second-largest circle, where humans aspire to the goals of Mahayana but use natural farming or organic techniques to receive nature's bounty; in Scientific Farming (3) depicted by the irregular shape which "represents the distortions and imperfections arising from narrow research findings" with humans estranged from nature, attempts to draw as much as it can from nature and apply knowledge to produce results that eclipse nature; and finally; the smallest circle depicts harvested benefits based on application of Liebig's laws alone (4).

10.2.4.2   Research Traditions

Advocates for sustainable or alternative agriculture criticized the goals and the experimental methods used by production agriculture. The wariness about reductionism held by alternative agriculture movements can be traced to Goethe, the famous poet, who is best known to scientists for his contributions to optics (Smith 2000). His commentary on the scientific process is as or more important. Goethe objected to "the stretching of nature upon the rack" in the seventeenth century when he foresaw how science and the methods used to pursue it would create and shape reality (Sepper 2005). He argued, even before the Age of Reason was in full swing, that controlled experimentation could not be used as the sole means to capture truth and feared mechanistic studies might miss the intrinsic importance of phenomena (Zemplén 2003). Our struggle to capture information and describe phenomena using empirical and exploratory methods remains an important and difficult issue, particularly where emergent properties are concerned. Howard (1945) articulated important dimensions of this critique of methods and perspective when he reflected on what he considered to be the irrelevance of Rothamsted, noting "in an evil moment were invented the replicated and randomized experimental plots, by means of which the statisticians can be furnished with all the data

FIGURE 10.3 Permaculture mandala by Graham Burnett illustrating the principles of permaculture. (Burnett, G. 2008. *Permaculture: A Beginner's Guide.* Spiralseed, Westcliff on Sea, Essex, UK. www.spiral seed.co.uk. With permission.)

needed for their esoteric and fastidious ministrations." Howard's objection was not to the work of R. A. Fisher, who pioneered experimental design while working with data from Rothamsted's classical experiments, but to the questions being evaluated by component-style research. He charged that "authority has abandoned the task of illuminating the laws of Nature, has forfeited the position of the friendly judge, scarcely now ventures even to adopt the tone of the earnest advocate: it has sunk to the inferior and petty work of photographing the corpse—a truly menial and depressing task."

Despite such criticism, experimentation, with its focus on hypothesis testing of simple cause-and-effect relationships, has been regarded as superior, to the extent that it is the standard approach, with exploratory experimentation devalued despite its proven power (Stephenson 1995; Steinle 2002). With our emphasis on controlled experimentation designed around a central hypothesis, we abandoned an early tradition of applied empiricism. The earlier tradition of observational science underpinned the theory of "rational husbandry" championed by Wallerius and the Thaer School in Germany in the later part of the eighteenth century. This tradition lost sway because it was thought to lack theoretical and methodological maturity (Velho and Velho 1997). Even though the school had uncovered practical factors influencing yield and devised an integrative framework for management (Feller et al. 2003), it was criticized because results were difficult to transfer to other sites and the practices being used degraded the soil (Krohn and Schafer 1976). Many of the tools pioneered by precision agriculture, and now precision conservation (e.g., Lerch et al. 2005; Massey et al. 2008), could be argued to be a return to the objectives and applied empiricism used by rational agriculture. When combined with advances in observational and statistical methodology, these and related approaches will help us evaluate questions not effectively addressed by the strict positivist framing of traditional field plot studies. Although component-style research and "restricted empiricism" imposed in many field trials effectively evaluates cause-and-effect relationships between factors, this approach frequently fails to capture complex interactions needed to understand systems (Tilman 1991; Anderson 1992; Laughlin 2005).

The experimental methods favored by agronomists and ecologists reflect disciplinary values, with agronomists valuing production and ecologists prizing conservation (Hess et al. 2000; Chan et al. 2007). Worster (1977) ties the conservation tradition embraced by ecologists to religious traditions. He explains the very word "ecology," which is derived from the Greek *oikos* for house, emerged in the nineteenth century as a more scientific replacement for the earlier term *oeconomy*, which was used by priests to connect economy to natural balance and so reassert the chemicotheological view of nature. This notion of nature's balance informs the ecosystem concept (Golly 1993) and systems theory (von Bertalanffy 1969). The assumed perfection of nature is also integral to agroecology and many other disciplines including industrial ecology (Egerton 1973; Cooper 2001; Ehrenfeld 2004; Fiksel 2006). By adopting nature as a standard, ecology retains an Aristotelian view of causality wherein factors are understood to reinforce or dampen positive feedbacks that optimize and conserve system integrity; this conception is in contrast to the linear or positivistic view embraced by traditional experimentalists, including many agronomists (Ulanowicz 1990). Ecologists are heirs to the vitalist and romantic traditions, wherein they revere soil and tend to presume that agricultural land use is despotic and likely to degrade soil. The agronomy tradition, which is more mechanistic and utilitarian, tends to see humans as benign stewards where productivity is seen as a moral and measurable value to be prized even over conservation (Lowry 2005; Holland and Carter 2005). This focus on production, however, does not mean that agronomists accept soil degradation as a condition of production agriculture. Practitioners do not necessarily assume systems are self-optimizing and instead focus efforts to improve system performance, or production, through the use of improved technologies and/or imported resources. Some contemporary agronomists worry that net productivity achieved by extensive systems will not satisfy demand and champion a vision where intensive management maximizes productivity while sustaining soil quality (Cassman et al. 2003). Determining where and how this vision can or cannot be achieved is a critically important task that cannot be effectively undertaken by separate agronomy or ecology traditions.

Norton (2005) argues conservation policy is now distorted by two competing ideologies with the moralists who value conservation struggling against utilitarians who prize economic efficiency above all. He asserts that the "preexperiential" attachment of parties engaged in the debate limits our ability to improve environmental policy and suggests improved communication among social and natural sciences might allow us to use science to improve management. The current framing of modern agriculture has resulted in the kind of "socio-technical lock-in" described by Callon (2005) wherein investment in favored technologies has set out trajectories for investment and valuation that favor some enterprises over others. We assume profit so drives decision making that rational farmers must be compensated to adopt conservation practices in proportion to any economic loss they suffer as a consequence of yield reductions (e.g., Kurkalova et al. 2006). This pairing suggests ecology and economy must be traded against one another and possibly, by inference, that alternative systems can survive by subsidy alone. According to Callon (2005), disenfranchised groups, in this case advocates for alternative agriculture, mobilize to form competencies that promote goods and services that are undervalued by the mainstream. This is certainly the case for advocates for sustainability who have sought to describe the benefits of an alternative system and the ecosystem services provided by agricultural landscapes.

In theory, adoption of sustainable agricultural practices that substitute natural processes for industrially produced inputs should allow farmers to regain control over production processes otherwise lost to companies that supply those inputs (Pfeffer 1992). Advocates for sustainable agriculture have decried the disproportionate attention paid by land grant universities to commercial and corporate goods (Warner 2007). These institutions and disciplines therein are currently hard-pressed to meet the needs of society as we seek to reconcile and then monetize cultural and biophysical values. Some see progress reflected in the growing consensus that agronomic metrics need to include a broader suite of services (Miller 2008). Proposed metrics include those that value critical support services provided by land and productive soil (Millennium Ecosystem Assessment 2005). Efforts to implement soil quality standards that include a standard of care for the soil now being explored by the Natural Resources Conservation Service (formerly the Soil and Water Conservation Society) might provide an important opportunity (Andrews et al. 2004; Tugel et al. 2005).

The solution to our sustainability challenge cannot be predicated entirely on the soil and its stewardship. Still, its use as a metaphor for humanity that reminds us to tame an exploitative nature remains helpful. A renewed awareness of resource limits is causing us to ask whether we can structure agricultural systems so that they provide the needed balance of services. Pursuit of this question will be aided by the formalization of theories about ecological soil management and how they apply to agricultural systems. Early proponents of agroecology began this task with an effort to improve communication between agronomists and ecologists (Lowrance et al. 1984). There are opportunities to expand an agroecology tradition that applies ecological concepts to the design of agricultural systems (Altieri 1989; Gleissman 1998) and take advantage of appropriate tools and technologies developed by mainstream agriculture and other research traditions as we move forward to pursue a hybrid tradition to develop a new version of the humus theory.

10.3 SOIL BIOLOGY AS A SYSTEMS DRIVER

10.3.1 Ecosystem Development and the Tenets of Agroecology

10.3.1.1 Ecology and Nature as Conceptual Frameworks

The vision of sustainable agriculture advanced by early advocates holds that humans can act as system engineers, structuring agricultural systems in such a way that production is optimized without loss or degradation of soil; key elements of optimized systems include nutrient recycling where nutrients are supplied in proportion to the system within which they reside (livestock or food systems that recycle appropriate quantities of waste), so that proper balance between inputs and outputs are achieved; plant soil interactions and organic matter reserves that develop as a characteristic of systems are thought to determine the degree to which some theoretical optimum is achieved. This understanding of agroecosystem development has been informed by ecological theories that have sought to explain succession and stability in plant communities (e.g., May 1974; Holling 1974). Most of these theories have focused on competition for resources and reproductive rates of community members as the important drivers with these factors seen to interact with the environment, stress, and disturbance. Although no common theory for ecosystem organization is accepted, Grime and Tilman have made notable attempts to unify theories that relate the attributes of plants and their interactions with the environment to plant community structure (Craine 2005). Of particular relevance are ecological "maximization theories" that have highlighted the importance of plant soil feedbacks as drivers for nutrient cycles that develop within systems and become causal agents (Loreau 1998). Maximization theories posit ecosystem properties are the indirect result of selection for different functional or demographic traits in organisms shaped by competition for nonliving resources and that through this competition, resource use intensity and

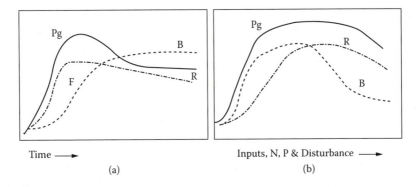

FIGURE 10.4 Do high-input, short-rotation agroecosystems function like young systems and are low-input diversified agroecosystems mature systems? (a) Adapted from Eugene Odum, "Strategy for Ecosystem Development" (1969). (b) Schema applied to agricultural input intensity; Pg, gross production; R, respiration; B, total living biomass and organic matter. Energy flow = P/R. Younger (high intensity) systems have high P/R and P/B ratios, mature (low intensity) have high B/P and B/E efficiency.

productivity are maximized (Loreau 1998). Related theories have sought to explain why productivity and biomass tend to increase with system succession and are accompanied by a declining productivity-to-biomass ratio, and increased nutrient recycling (Finn 1982; Ulanowick and Hannon 1987; Bergon et al. 1990; Schneider and Kay 1994). The influence of plant competition and associated facilitation mechanisms on succession appears to be most important in severe environments (Bertness and Callaway 1994; Callaway and Walker 1997). The potential for positive or negative facilitation by neighbor species is increasingly accepted to vary with environmental factors (Brooker and Callaghan 1998; Travis et al. 2005).

The basic elements of maximization theory originally stated by Eugene Odum's 1969 "Strategy of Ecosystem Development" are that system succession is orderly and directional, that the physical environment influences the pattern, rate, and limits of development, but the process is community controlled, and that development culminates in a stabilized ecosystem where both biomass and symbiotic function between organisms are maximized per unit available energy flow. Odum's framework is relevant to sustainability assessments because it integrates time, matter, and energy flows and evaluates their interactions with biotic communities in a way that might allow us to evaluate paradigms that have been articulated for agricultural systems.

Odum proposed energy flow as an integrative measure and defined this as the ratio between productivity and respiration (Figure 10.4). Young systems, which are characterized by early successional stages dominated by annual plants, tend to have high productivity to respiration and productivity to biomass ratios where, according to his definition, biomass includes above- and belowground biomass plus organic matter stocks. Odum also attributed the observed shift from systems that support high net aboveground productivity to those that invest more C to belowground reserves to declining soil nutrient status and increased reliance on symbiotic associations. This view has been borne out for a number of natural systems. Notable examples include work on a Hawaiian chronosequence that provides evidence of positive feedback between plant growth and soil organic matter reserves with interactions between nutrient stocks and symbionts evolving over geological timescales (Vitousek and Hobbie 2000; Pearson and Vitousek 2002). Odum's depiction charts the course of the system to its maximum and does not reflect the natural tendency of the system to degrade as soils weather and nutrients and organic biomass decline in concert (Jenny 1940; Torn et al. 1997).

According to Odum, and other maximization theories, systems gain efficiency as they mature because they invest more in biological associations. Plant facilitation studies suggest plant diversity and interspecies benefits are more important in extreme environments (Walker and del Moral 2003). Plant investment in nutrient acquisition increases as substrates weather. This is more important for geotrophic nutrients which include those supplied principally by substrate weathering. Accumulation of biotrophic nutrients (C and N) recruited by living organisms results from shift in biomass allocation and reductions in decay rates. Odum argued that as systems matured, they increased biomass to productivity and biomass to energy flow ratios. The idea that energy flux and net productivity were maximized with system development was introduced by Lotka in 1922, and H. T. Odum and Pinkerton articulated this in the form of systems theory in 1955. The mechanistic basis for these concepts and relevance to agronomic systems has yet to be established.

10.3.1.2 Applications of Theory to Agriculture

Researchers pioneering the field of agroecology have sought to apply ecosystem principles to agriculture and assumed that managed systems that should emulate natural ones to be sustainable and, in addition to this, agroecosystems needed to meet the demands for crop production. Working in Georgia at Horseshoe Bend on a loamy forest soil, Stinner et al. (1984) carried out nutrient budgets, comparing conventional, no-tillage, and old-field systems to evaluate the hypothesis that by reducing disturbance, no-tillage systems would better approximate the natural system (here represented by the old-field) and achieve greater productivity, biodiversity, and efficiency (Figure 10.5). Plant removal was the largest nutrient sink. The no-tillage and old-field systems maintained larger litter stores than the conventionally tilled system and this was tied to nutrient retention within the system. House and Stinner (1983) investigated the relationship between nutrient cycling and food web structure, linking nutrient status to pest pressure with results indicating a possible causal connection between soil tillage and insect herbivory rates. Work published by House et al. in 1984 unified theories advanced by Odum and Balfour with its assertion that reduced tillage would lead to fungally dominated systems and this would accrue to slower organic matter turnover rates within no-tillage systems and thus nitrogen retention (Figure 10.6). Results from Horseshoe Bend, the long-term study site in Georgia, were consistent with the idea that organic matter status could serve as an earmark for system performance. While few subsequent works have carried out the kind of detailed assessment conducted by these workers, many have used organic matter as a proxy for ecosystem health.

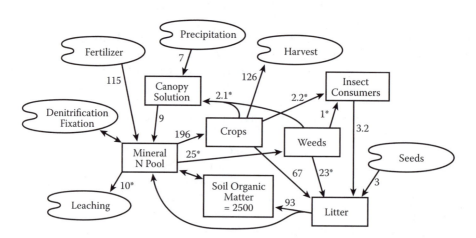

FIGURE 10.5 Nutrient Budgets from Stinner et al. (1984) developed for conventional tillage, no-tillage, and old-field systems. Emphasis was on input–output patterns as well as internal cycling between the major component processes (primary production, decomposition, and consumption). The major hypothesis of the study was that no-tillage systems, because of minimal soil disturbance, are more nutrient conservative than are conventional tillage systems.

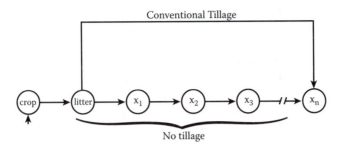

FIGURE 10.6 Depiction of how management influences the decomposition pathway in conventional and no-tillage systems first printed in House et al. (1984). The original legend asserted "tillage by-passes many biologically mediated transition steps occurring in the no-tillage system."

There is contradictory evidence relating agricultural intensity to soil organic matter reserves. As is true for natural systems, organic matter status in agricultural fields is the result of plant–soil–microbe interactions that are influenced by the chemical, biological, and physical environment. Our understanding of cause-and-effect relationships that determine equilibrium levels in agronomic systems is significantly informed by studies that have manipulated inputs to simplified systems. The direct effect of manure and/or organic waste additions to soil organic matter stocks is well described in modeling terms (Paustian et al. 1997). Increases in organic matter and nutrient stocks caused by importation of organic mass can result in stockpiling or, the "myth of manure," as Schlesinger et al. (2000) put it, and do not necessarily provide evidence of positive feedbacks that enhance nutrient cycling. Fertilizer, lime, and water additions that prompt increases in productivity also contribute to direct effects and are consistent with movement from left to right, along the X axis of Odum's diagram, with gross productivity shifting upward as leachable nutrients are replenished. Feedbacks between plant productivity and soil condition are difficult to evaluate in agricultural systems because genetic improvement of crops can move gross productivity upward and alter biomass partitioning to decouple above- and belowground productivity relationships. Odum's schema suggests there might be a trade-off in productivity as systems mature, with gains in stored biomass occurring at the expense of net productivity that could be harvested. Agronomic systems differ in that we enforce the community structure. In conventional and alternative agronomic situations crop species are selected and edaphic factors are managed to maximize productivity. This includes fertilization, tillage practices, and water management that move us along Odum's axis to the left to poise the system at some "optimum." Whether and how to achieve this optimum and whether organic matter in the soil must be traded off against harvested production are at issue.

We must determine how productivity and soil quality can best be maintained and determine whether both can be optimized through exploitation of beneficial feedbacks. Comparison of the performance of intensive and extensively managed farming systems, particularly those shoehorned into factorial-type experimental designs, may not provide fully satisfactory information for a variety of reasons; still, data from this kind of comparison is what we have to draw on for now. The ability of diversified systems to sequester organic matter varies with systems and climate regime. Often, more diversified cereal-based production systems receiving lower levels of fertilizer additions contain higher levels of organic matter than intensively managed and/or manure-amended counterparts (Wander et al. 1994; Rasmussen 1998; Willson et al. 2002; Wander and Nissen 2004; Marriott and Wander 2006). Increases in organic matter observed in most livestock-based systems are largely attributed to direct effects of waste application (Stockdale et al. 2002) and, along with legume-based diversified rotations, their support of greater plant cover and thus opportunity for solar capture during the year. Greater diversity and intensity of plant cover fosters competition for nutrients that can dampen mineralization in rain-fed systems and thus conserve organic matter.

Research has pointed to N limitation in legume-based production systems as a mechanism for C sequestration (Cormack et al. 2003). Potential gains in organic matter might be offset by reductions in crop productivity if not compensated for by belowground inputs. There are feedbacks to moisture that occur as well. By growing plants for a larger proportion of the time, diversified systems maintain greater season-long evapotranspiration that dries the soil, reduces water drainage, and increases uptake and/or immobilization of N (Randall et al. 1997; Schulte et al. 2006). Alterations in structure and habitat characteristics can also occur. Studies of farming system effects on soil physical properties emphasize tillage regime, and it is well known that adoption of practices that reduce physical disturbance can result in increased organic matter levels by reducing decay rates. Macilwain (2004) and others suggest that increases in organic matter derived from additions or crop diversification can be undercut in organic farming systems and similar enterprises that rely heavily on tillage for weed control. This assertion has been disproved by the performance of many long-term trials (e.g., Marriott and Wander 2006).

The direct positive and negative effects of fertilizer addition and tillage have been disproportionately studied and may be overemphasized in models, particularly in situations where feedbacks occur. Negative feedbacks on organic matter stocks are increasingly observed in agricultural systems; these include anything that accelerates decay by changing biotic composition, physiological status, enzymatic activity, plant tissue composition, and/or the edaphic factors that influence heterotrophic activity. Positive feedbacks would have the reverse effect, prompting increases in organic matter reserves that are due to slowed decay. Evidence of positive feedbacks leading to organic matter accumulation is principally correlative (Ehrenfeld et al. 2005). This is because system comparisons cannot separate direct effects from the positive feedbacks likely to contribute to frequently observed differences in organic matter reserves of diversified and conventional annual cropped systems.

To adapt Odum's framework to assess plant–soil paradigms in agriculture, we consider how imposed plant communities and attendant technologies influence soil processes and then quantify resulting changes in soil function and its resistance and resilience to stress. Feedbacks between nutrient status and biotic responses and their interactions with the physical environment are likely to be important but receive far less attention than direct effects. Feedbacks, which appear to differ among high and low input/disturbance systems, may underpin the central paradigms for ecologically sound soil management.

10.3.2 Systems, Intensity, and Feedback

10.3.2.1 Natural Systems; Biology, and C and N Cycling

In natural systems, changes in nutrient availability are accompanied by changes in plant populations that are consistent with Odum's conception of ecosystem development. Open areas with high nitrification rates favor nonmycrorrhizal annuals with high seed production, growth rates, and N demand (Kottke 2002). In natural systems, increased aboveground productivity is associated with reduced immobilization and lower retention of nutrients within soils (Aerts and deCaluwe 1997). This is evident at the evolutionary scale as species from high-fertility habitats promote N mineralization and nitrification to a greater extent than do species from low-fertility habitats (van der Krift and Berendse 2001). Management can influence this. The benefit of microbial partnerships to plants diminishes when the availability of mineral nutrients is high; both nitrogen fixation (Elgersma et al. 2000; Rastetter et al. 2001) and mycorrhizal infection (Grant et al. 2005) are suppressed in soils where nutrients are abundant. Influences of N fertility status on soil organic matter levels vary among in natural systems subject to N deposition, and this has been attributed to the variable effects of N availability on microbial production of extracellular enzymes (Waldrop et al. 2004). Nutrient additions can alter enzymatic abundance (e.g., Sinsabaugh et al. 2005) and promote or suppress organic matter decay (Fog et al. 1998). There is growing evidence that high N availability can stimulate cellulose degradation but this is challenged by evidence that N saturation can inhibit lignin degradation (Berg and Matzner 1997; Berg et al. 1998). Interactions between N and decomposition rates appear to vary with litter biochemistry (Sinsabaugh et al. 2002), which varies with N nutrition. Evaluation of N response of the root development and nutrient uptake of 55 plant species from the Great Plains, Levang-Brilz and Biondini (2002) found that reductions in the supply of N increased the root-ratio in 62 percent of the species. In addition to litter quality, N can have direct effects on microbes. In a sort of chronosequence of grassland ecosystems, Zeglin et al. (2007) found that soil microbial responses to N enrichment were constrained by pH, and that this prevented negative feedbacks that could accelerated organic matter decay. Fraterrigo et al. (2006) linked reduced mineralization rates in a recovering forest to reductions in fungal communities that were the legacy of disturbance. It is these kinds of interactions and feedbacks that we need to learn to manage in our production systems.

10.3.2.2 Agricultural Systems; Biology, and C and N Cycling

Cultural practices promote mineralization and nitrogen fertilizer addition increases nitrification rates (Shi et al. 2004). In these situations, soil N is supplied largely through heterotrophic mineralization (Schimel and Bennett 2004) with plants acquiring inorganic N from microbial waste (Hodge et al. 2000). Feedbacks between plants and soil N cycling differ notably when inorganic nutrient stocks are low. The notion that nutrient limitation fosters tighter nutrient cycles is widely held. Under conditions of limited nutrient availability, plants and microbes rely heavily on the depolymerization of organic N, and both N fixation and mycorrhizal-associations are favored (Vitousek 2004). Root residues, residue quality, and rhizodeposition can be altered to slow mineralization and nitrification rates when nutrients are limiting (Allmaras 2004; Booth et al. 2005). Under these conditions, microbes and roots are thought to compete for amino acids (Owen and Jones 2001) and this is thought to reduce available nutrient levels and associated decay rates of organic matter (Moorhead et al. 1998; Hamel, 2004). Accumulation of organic matter by this mechanism would result in movement along Odum's axis to the right. Under these conditions, NO_3^- assimilation is promoted and nitrification is suppressed (Booth et al. 2005). Direct effects include changes in resource condition (increased N availability and reduced organic matter retention) and result from the acceleration of biotic processing of matter and energy (Coleman 1994).

There is growing evidence that fertilization can alter biotic composition and physiologic status in agricultural systems. Subtle interactions between resident organic matter quality and inputs appear to influence interactions between nutrient abundance and organic matter turnover rates. Losses of organic matter and thus soil quality in more intensive systems are frequently associated with N additions but the mechanistic basis for this is not fully explored (e.g., Russell et al. 2006). The point at which N additions, biomass (and presumably residues returned), and organic matter stocks are decoupled, with organic matter levels failing to respond to increased C inputs, appears to vary among sites (Vanotti et al. 1997; Moran et al. 2005). This phenomenon is often attributed to priming effects, that might include changes in biotic composition, plant and microbial physiological responses, and physical quality that manifest in altered decay rates. Priming occurs in alternative and conventional systems when the addition of a nutrient, which could be C or N, causes a disproportionately large influence on decay rates (Kuzyakov 2002). Soil quality is likely to determine whether or not a system will have a tendency toward positive or negative priming. For example, high soil N, and presumably P, levels inherited from a history of manure addition appear to have prevented organic matter accumulation in an organic livestock system receiving manure additions where this and higher plant productivity should have caused organic matter levels to exceed conventional and legume-based organic systems that added less C to the system (Wander et al. 2007). Accelerated mineralization accounted for the failure of that system to retain or accumulate C-rich fractions, including particulate organic matter, in that soil. Results in the Morrow Plots, which compare rotation length and fertility source in the oldest study of agricultural intensity in North America, show clearly how C inputs and organic matter stocks can be decoupled. Despite a fourfold increase in grain yield experienced in the fully fertilized continuous corn plots since 1955, soil organic carbon (SOC) levels have failed to increase in concert and may even have declined (Aref and Wander 1997; Wander and Nissen 2004; Khan et al. 2007). Organic matter levels increase with rotation length and manure additions but the degree to which this occurs is not explained by aboveground biomass. Feedbacks between the crop mix, nutrient abundance, and edaphic factors manifest in the organic matter record point to positive priming in the 1- and 2-year systems and negative priming in the 3-year rotation. The notion that diversified systems retain organic matter better because decay rates are slowed appears to be contradicted by the frequent observation that soil biological activity and mineralization potentials are increased (Poudel et al. 2002; Kramer et al. 2002). This is true for Rodale's Farming Systems Trial, where the legume-based system has retained more organic matter than the conventional counterpart even though it supports greater biological activity and receives similar amounts of aboveground organic inputs (Wander et al. 1994; Drinkwater et al. 1998). Feedbacks in root–soil interactions and changes in community characteristics previously mentioned might contribute to this in a variety of ways. Biotic feedbacks tend to receive the most attention.

The role of biota remains a key point of interest; improved analysis and new molecular techniques may help shed light on feedbacks and interactions. Evidence from fertilized grasslands suggests that N additions shift populations toward bacterial domination (Moore and Hunt 1988) and increased nutrient export (de Vriesa et al. 2006), soil food webs are thought to differ in extensive and intensive production systems (Moore et al. 2005; Six et al. 2006). Changes seen in the soil resource where bacterial and fungal proportions differ (e.g., tilled versus no-tilled systems and grassland comparisons) have been attributed to the shorter generation times and reduced metabolic efficiency of bacteria (Moore and de Ruiter 1997). Girvan et al. (2004) used rRNA to show that fertilizer addition can shift the biomass toward the bacterial community and reduced overall diversity. Jumpponen et al. (2005) found that N enrichment prompted minimal changes in arbuscular mycorrhizal colonization but altered community composition. Other agronomic factors are likely to have equal or greater influence on biotic community. Tillage practices and plant species composition have substantial influence on fungal communities wherein bacteria seem to be more sensitive to fertilizer additions (Kennedy et al. 2005; Garcia et al. 2007). Zhang et al. (2005) evaluated the influences of deintensification in a large-scale comparison of organic and conventional cash-grain farming systems with a system that included animals, a successional field, and a plantation woodlot and found that the proportion of fungi and microbial C:N ratios increased with deintensification as the contributions of total bacterial and Gram-positive bacterial diminished. A study of a similar transect of management intensity carried out in Kansas found that increased fungal activity was correlated with C sequestration (Bailey et al. 2002).

Plants and plant roots also play a defining role in that they control the belowground C cycle by influencing substrates and thus the microbial communities that regulate matter and energy flow. It is these feedbacks from plants to biota and soils that one might hope to manage. For example, microbes can influence the flow of C and N through the excretion of products to stimulate root exudation (Phillips et al. 2000). Roots and associated exudates can stimulate mineralization and/or prompt immobilization depending on nutrient balance (Castells et al. 2004). Allmaras (2004) found that rhizodeposition was more than doubled in N-fertilized plots. Rhizodeposition has been shown to have the ability to cause positive priming of organic C under conditions of high background N (Gerzabek et al. 1997; Kuzyakov and Bol 2006). In a container study, Cheng et al. (2003) found that root-driven respiration accelerated organic matter turnover by two- to fourfold, depending on the crop species considered, but did not find that N additions increased net CO_2 losses. Other short-term efforts have failed to observe changes in organic matter decay rates with N addition (Baer et al. 2002; Torn et al. 2005). Shifts in biotic or abiotic attributes that ultimately cause accelerated mineralization and organic matter loss might be the result of cumulative feedbacks that take time to manifest. The physical influences of roots on habitat deserve closer investigation. Root–soil interactions can protect organic matter from microbes by promoting aggregation while enhancing the habitat that supports the biotic community (Rasse et al. 2005). Feedbacks that alter the microbial and faunal environment might explain why extensive systems seem to support a larger microbial biomass that is characterized by reduced metabolic activity and/or reduced respiratory quotients (House and Stinner 1983; Wander et al. 1994). Understanding of this and how interactions between management and habitat quality vary among soil types (Yoo and Wander 2008) will help us tailor management for different environments.

10.4 CONCLUSIONS

The use of soils as a metaphor for humanity has led us to see ourselves as ecosystem stewards. A resulting reverence for soils has had profound influence on soil stewardship and related sciences. By the start of the twentieth century the romantic notions of natural philosophy were a thing of the past. Chemistry and physics had been elevated over biology to become the dominant disciplines in natural science. Modern agricultural research relied on techniques that emphasize cause-and-effect relationships characterized through component-style research. These reductionist methods and emphasis on crop production left systems-level questions unaddressed. Agroecology responded to the critique of production agriculture levied by organic and alternative agriculture movements in the early twentieth century and looked to nature to provide a model for sustainability. Central tenets of agroecology, which is informed by agronomy and ecology traditions, hold that the farming system and nutrient regimes imposed on a landscape will influence soil biology and the resulting feedbacks on the system will determine its efficiency and productivity. The weight of evidence suggests that tightening of nutrient cycles and increases in soil organic matter reserves can be achieved where disturbance frequency, residue decay rates, and nutrient imports are reduced. The relationship between plant diversity and these factors and their influence on harvestable crops is less firmly established. Huge opportunities exist for combinations of plant functional traits and intelligent cultural practices. Agricultural intensification has suppressed the importance of biological mechanisms in arable systems by increasing the supply of readily available nutrients in soils. Despite an increased understanding of the physical, chemical, and biological factors that influence plant productivity and nutrient use efficiency, we have yet to devise and implement strategies to optimize biologically mediated processes to increase nutrient use efficiency and productivity in our major production systems. Pressures on the land and growing acceptance of resource limits are rekindling interest in these kinds of efficiencies. Advances being made in statistical and experimental methods and a broadening of our objectives to include resource conservation and stewardship will enable us to constructively pursue the goal to develop principles of rational agriculture that was abandoned when the humus theory was deposed.

REFERENCES

Aerts, R. and deCaluwe, H. 1997. Nutritional and plant-mediated controls on leaf litter decomposition of *Carex. Ecology*, 78, 244–260.

Albrecht, W. A. 1938. Loss of organic matter and its restoration, in *Soils and Men,* U.S. Department of Agriculture, Yearbook of Agriculture. U.S. Govt. Printing Office. pp. 347–360.

Allmaras, R. R. 2004. Corn-residue transformations into root and soil carbon as related to nitrogen, tillage, and stover management, *Soil Science Society of America Journal*, 68, 1366.

Altieri, M. 1989. Agroecology: A new research and development paradigm for world agriculture, *Agriculture, Ecosystems, and Environment*, 27, 37–46.

Anderson P. 1972. More is different. *Science,* 177, 393–396.

Andrews, S. S., Karlen, D. L., and Cambardella, C. A. 2004. The soil management assessment framework: A quantitative soil quality evaluation method with case studies. *Soil Science Society of America Journal,* 68, 1945–1962.

Aref, S. and Wander, M. M. 1997. Long-term trends of corn yield and soil organic matter in different crop sequences and soil fertility treatments. In *Advances in Agronomy* 62. Academic Press, San Diego, 153–197.

Aulie, R. P. 1970. Boussingault and the nitrogen cycle, *Proceedings of the American Philosophical Society,* 114, 435–479.

Baer, S. G. et al. 2002. Changes in ecosystem structure and function along a chronosequence of restored grasslands. *Ecological Applications,* 12, 688–710.

Bailey, V. L., Smith, J. L., and Bolton, H. 2002. Fungal-to-bacterial ratios in soils investigated for enhanced C sequestration, *Soil Biology and Biochemistry,* 34, 997–1007.

Berg, B. and Matzner, E. 1997. Effect of N deposition on decomposition of plant litter and soil organic matter in forest systems, *Environmental Review,* 5, 1–25.

Berg, B. et al. 1998. Long-term decomposition of successive organic strata in a nitrogen saturated Scots pine forest soil, *Forest Ecology and Management,* 107, 159–172.

Bergsma, T. T., Robertson G. P., and Ostrom, N. E. 2002. Influence of soil moisture and land use history on denitrification end-products, *Journal Environmental Quality,* 31, 711–717.

Binswanger, H. C. and Smith, K. R. 2000. Paracelsus and Goethe: Founding fathers of environmental health. *Bulletin of the World Health Organization,* 78, 1162–1164.

Blum, B. 1992. Composting and the roots of sustainable agriculture, *Agricultural History,* 66, 171–188.

Booth, M. S., Stark, J. M., and Rastetter, E. 2005. Controls on nitrogen cycling in terrestrial ecosystems: A synthetic analysis of literature data, *Ecological Monographs,* 75, 139–157.

Brooker, R.W. and Callaghan, F.V. 1998.The balance between positive and negative interactions and its relationship to environmental gradients: a model. *Oikos,* 81, 196–207.

Brown, T. L. 2003. *Making Truth: Metaphor in Science.* University of Illinois Press, Urbana.

Browne, C. A. 1977. *A Source Book of Agricultural Chemistry.* Arno Press, New York.

Burnett, G. 2008. *Permaculture: A Beginner's Guide.* Spiralseed, Westcliff on Sea, Essex, UK. www.spiralseed. co.uk.

Callaway, R. M. and Walker, L. R. 1997. Competition and facilitation: a synthetic approach to interactions in plant communities. *Ecology,* 78, 1958–1965.

Callon, M. 2005. Disabled persons of all countries, unite! In *Making Things Public, Atmospheres of Democracy*, Latour, B., and Weibel, P. Eds. MIT Press, Cambridge, MA, 308–313.

Cassman, K. G. et al. 2003. Meeting cereal demand while protecting natural resources and improving environmental quality. *Annual Review Environmental Resources,* 28, 315–358.

Castells, E., Penuelas, J., and Valentine, D. W. 2004: Are phenolic compounds released from the Mediterranean shrub *Cistus albidus* responsible for changes in N cycling in siliceous. *New Phytologist,* 162, 187–195.

Chan, K. M. A. et al. 2007. When agendas collide: Human welfare and biological conservation, *Conservation Biology,* 21, 59–68.

Cheng, W. D., Johnson, E., and Fu, S. 2003. Rhizosphere effects on decomposition: Controls of plant species, phenology, and fertilization, *Soil Science Society of America Journal,* 67, 1418–1427.

Coleman, D. D. 1994. The microbial loop concept as used in terrestrial soil ecology studies. *Microbial Ecology,* 28, 245–250.

Conford, P. 1995. The alchemy of waste: The impact of Asian farming on the British organic movement, *Rural History,* 6, 103–114.

Conley, T. 1998. A chaos of science, *Renaissance Quarterly,* 51, 934–941.

Cooper, G. 2001. Must there be a balance of nature? *Biology and Philosophy,* 16, 481–506.

Cormack, W. F., Shepherd, M., and Wilson, D. W. 2003. Legume species and management for stockless organic farming, *Biological Agriculture & Horticulture,* 21, 383–398.

Craine, J. M. 2005. Reconciling plant strategy theories of Grime and Tilman. *Journal of Ecology,* 93, 1041–1052.

Davis, E. 2002. The Bible and our topsoil. *CropChoice,* from *Prairie Writers Circle* http://www.cropchoice. com/leadstryc76a.html?recid=1105 (accessed 11 June 2008).

de Vriesa F. T. et al. 2006. Fungal/bacterial ratios in grasslands with contrasting nitrogen management, *Soil Biology & Biochemistry,* 38, 2092–2103.

Drinkwater, L. E., Wagoner, P., and Sarrantonio, M. 1998. Legume-based cropping systems have reduced carbon and nitrogen losses, *Nature,* 396, 262–265.

Egerton, F. N. 1973. Changing concepts of the balance of nature, *Quarterly Review of Biology,* 48, 322–350.

Ehrenfeld, J. 2004. Industrial ecology: A new field or only a metaphor? *Journal of Cleaner Production,* 12, 825–831.

Ehrenfeld, J. G., Ravit, B., and Ekgersma, K. 2005. Feedback in the plant soil system, *Annual Review Environmental Resources*, 30, 75–115.

Elgersma, A., Schlepers, H., and Nassiri, M. 2000. Interactions between perennial ryegrass (*Lolium perenne* L.) and white clover (*Trifolium repens* L.) under contrasting nitrogen availability: Productivity, seasonal patterns of species composition, N_2 fixation, N transfer, and N recovery, *Plant & Soil*, 221, 281–299.

Eliade, M. 1978. *The Forege and the Crucible: The Origins and Structures of Alchemy.* Chicago, University of Chicago Press.

Eltun, R. and Bordheim, O. 1999. Yield results during the first eight years crop rotation of the Apelsvoll cropping system experiment. In *Designing and Testing Crop Rotations for Organic Farming*, Olesen, J. E. et al. Eds. DARCOF Report No. 1, 79–89.

Fraterrigo, J. M., Balser, T. C., and Turner, M. G. 2006. Microbial community variations and its relationship with nitrogen mineralization in historically altered forests, *Ecology*, 87, 570–579.

Feller, C. L. et al. 2003. The principles of rational agriculture by Albrecht Daniel Thaer (1752–1828). An approach to the sustainability of cropping systems at the beginning of the 19th century, *Journal of Plant Nutrition and Soil Science*, 166, 687–698.

Fiksel, J. 2006. Sustainability and resilience: towards a systems approach, *Sustainability: Science, Practice, & Policy*, 2, 1–8.

Finn, J. T. 1980. Flow analysis of models of the Hubbard Brook ecosystem, *Ecology*, 61, 562–571.

Fog, K. 1988. The effect of added nitrogen on the rate of decomposition of organic matter, *Biology Review*, 63, 433–462.

Freyfogel, E. T. 2001. *The New Agrarianism: Land, Culture, and the Community of Life.* Island Press, Washington, D.C.

Fukuoka, M. 1985. *The Natural Way of Farming.* Japan Publications, Tokyo.

Garcia, J. P. et al. 2007. One-time tillage of no-till: Effects on nutrients, mycorrhizae, and phosphorus update, *Agronomy Journal,* 99, 1093–1103.

Gerhart, M. and Russell, A. M. 2004. Metaphor and thinking in science and religion. *Zygon, Journal of Religion & Science*, 39, 13–38.

Gerzabek, M. H. et al. 1997. The response of soil organic matter to manure amendments in a long term experiment at the Ultuna Sweden, *European Journal of Soil Science,* 48, 273–282.

Girvan, M. S. et al. 2004. Responses of active bacterial and fungal communities in soils under winter wheat to different fertilizer and pesticide regimens. *Applied Environmental Microbiology*, 70, 2692–2701.

Gliessman, S. 1998. *Agroecology; Ecological Processes in Sustainable Agriculture.* Ann Arbor Press, Ann Arbor, MI.

Golly, F. B. 1993. *A History of the Ecosystem Concept in Ecology: More Than the Sum of the Parts.* Yale University Press, New Haven, CT.

Grant, C. et al. 2005. Soil and fertilizer phosphorus: Effects on plant P supply and mycorrhizal development, *Canadian Journal Plant Science*, 85, 3–14.

Grant, E. 1996. *The Foundations of Modern Science in the Middle Ages: Their Religious Origins in a Medieval World.* Cambridge University Press, Cambridge.

Hamel, C. 2004. Impact of arbuscular mycorrhizal fungi on N and P cycling in the root zone, *Canadian Journal of Soil Science,* 84, 383–395.

Harrison, P. 1999. Subduing the Earth: Genesis 1, early modern science, and the exploitation of nature, *Journal of Religion,* 79, 86–109.

Heckman, J. 2006. A history of organic farming: Transitions from Sir Albert Howard's War in the soil to USDA National Organic Program, *Renewable Agriculture and Food Systems*, 3, 143–150.

Heggenstaller, A. H. and Liebman, M. 2006. Demography of *Abutilon theophrasti* and *Setaria faberi* in three crop rotation systems, *Weed Research*, 46, 138–151.

Hess, G. R. et al. 2000. A conceptual model and indicators for assessing the ecological condition of agricultural lands. *Journal of Environmental Quality*, 29, 728–737.

Hillel, D. 1991. *Ours of the Earth: Civilization and the Life of the Soil*, Berkeley, University of California Press.

Hodge, A., Robinson, D., and Fitter, A. H. *2000*. Are *microbes* more effective than plants at competing for nitrogen? *Trends in Plant Science*, 5, 304–308.

Holland, L. and Carter, J. S. 2005. Words v. deeds: A comparison of religious belief and environmental action, *Sociological Spectrum*, 25, 739–753.

Holling, C. S. 1973. Resilience and stability of ecological systems, *Annual Review of Ecology and Systematics,* 4, 1–23.

Holmes, F. 1989. The complementarily of teaching and research in Liebig's laboratory, *Osiris,* 5, 121–161.

Hopkins, C. 1910. *Soil Fertility and Permanent Agriculture.* Ginn and Co . Boston.

House, G. J. and Stinner, B. R. 1983. Arthropods in no-tillage soybean agroecosystems—Community composition and ecosystem interactions, *Environmental Management,* 7, 23–28.

House, G. J., Stinner, B. R., Crossley, Jr., D. A., and Odum, E. T. 1984. Nitrogen cycling in conventional and no-tillage agro-ecosystems: Analysis of pathways and processes, *Journal of Applied Ecology,* 21, 991–1012.

Howard, S. A. 1945. *Farming and Gardening for Health and Disease.* Faber and Faber, London.

Hunter, R. G. and Faulkner, S. P. 2001. Denitrification potentials in restored and natural bottomland hardwood wetlands, *Soil Science Society of America Journal,* 65, 1865–1872.

Ihde, A. J. 1984. *The Development of Modern Chemistry.* Courier, Dover, NY.

Jackson, D. L. and Jackson, L. L. 2002. *The Farm as a Natural Habitat; Reconnecting Food Systems with Ecosystems.* Island Press, Washington, D.C.

Jeannerete, M. 2002. *Perpetual Motion: Transforming Shapes in the Renaissance from da Vinci to Montaigne.* Johns Hopkins University Press, Baltimore.

Jenny, H. 1940. *Factors of Soil Formation: A System of Quantitative Pedology,* Dover Press, NY.

Johnston, J. F. W. 1853. The circulation of matter, *Blackwood's Edinburgh Magazine,* 73, 550–560.

Jumpponen, A. J. et al. 2005. Nitrogen enrichment causes minimal changes in arbuscular mycorrhizal colonization but shifts community composition-evidence from rDNA data, *Biology and Fertility of Soils,* 41, 217–224.

Kennedy, N., Connolly, J., and Clipson, N. 2005. Impact of lime, nitrogen, and plant species on fungal community structure in grassland microcosms, *Environmental Microbiology,* 7, 780–788.

Khan et al. 2007. The myth of nitrogen fertilization for soil carbon sequestration. *Journal of Environmental Quality,* 36, 1821–1832.

King, F. H. 1911. *Farming for Forty Centuries in China, Korea and Japan.* Courier Dover Press. NY.

Kinne-Saffran, E. and Kinne, R. K. H. 1999. Vitalism and synthesis of urea from Friedrich Wöhler to Hans A. Krebs. *American Journal Nephrology.* 19, 290–294.

Klein, R. M. et al. 1988. Precipitation as a source of assimilable nitrogen: A historical survey, *American Journal of Botany,* 75, 928–937.

Kottke, I. 2002. Mycorrhizae—Rhizosphere determinants of plant communities. In *Plant Roots the Hidden Half,* 3rd ed., Waisel, Y. et al., Eds. Marcel Dekker, New York, 919–932.

Kramer, A. W. et al. 2002. Short-term nitrogen-15 recovery vs. long-term total soil N gains in conventional and alternative cropping systems, *Soil Biology and Biochemistry,* 34, 43–50.

Krohn, W. and Schafer, W. 1976. The origins and structure of agricultural chemistry. In *Perspectives on the Emergence of Scientific Disciplines,* Lemaine, G. et al. Eds. The Hague, Paris, 27–52.

Kurkalova, L., Kling, C., and Zhao, J. H. 2006, Green subsidies in agriculture: Estimating the adoption costs of conservation tillage from observed behavior. *Canadian Journal of Agricultural Economics–Revue Canadienne D Agroeconomie,* 54, 247–267.

Kurtz, E. and Ketcham, K. 2002. *Spirituality of Imperfection: Storytelling and the Journey Toward Wholeness.* Bantam Books, New York.

Kuzyakov, Y. 2002. Review: Factors affecting rhizosphere priming effects, *Journal of Plant Nutrition and Soil Science,* 165, 382–396.

Kuzyakov, Y. and Bol, R. 2006. Sources and mechanisms of priming effects induced in two grassland soils amended with slurry and sugar, *Soil Biology & Biochemistry,* 38, 747–758.

Lerch, R. N., Kitchen, N. R., Kremer, R .J. Donald, W.W., Alberts, E. E., Sadler, E. J., Sudduth, K. A., Myers, D. B., and Ghidey, F. 2005. Development of a conservation- oriented precision agriculture system, *Water and Soil Quality Assessment,* Nov–Dec, 411–420.

Loreau, M. 1998. Ecosystem development explained by competition within and between material cycles, *Proceedings Royal Society London,* Biological Sciences Series, 265, 33–38.

Laughlin, R. B. 2005. *A Different Universe: Reinventing Physics from the Bottom Down.* Basic Books, New York.

Levang-Brilz, N. and Biondini, M. E. 2002. Growth rate, root development, and nutrient uptake of 55 plant species from the Great Plains Grasslands, USA, *Plant Ecology,* 165, 117–144.

Liebig, J. V. 1840. *Chemistry in Its Application to Agriculture and Physiology.* British Association, London.

Lines-Kelly, R. 2004. Soil: our common ground—A humanities perspective. http://www.regional.org.au/au/asssi/supersoil2004/keynote/lineskelly.htm# (accessed 11 June 2008).

Lotka, A. J. 1922. Contributions to the evolution of energetics, *Proceedings of the National Academy of Science,* 8, 147–155.

Lotter, D. W. 2003. Organic agriculture, *Journal of Sustainable Agriculture,* 21, 59–128.

Lowrance, R., Stinner, B. R., and House, G. J. (Eds.) *Agricultural Ecosystems; Unifying Concepts*, John Wiley & Sons, New York.

Lowry, R. C. 2005. Explaining the variation in organized civil society across states and time. *The Journal of Politics*, 67, 574–594.

Macilwain, C. 2004. Organic: Is it the future of farming? *Nature,* 428, 792–793.

Manlay, R. J., Feller, C., and Swift, M. J. 2007. Historical evolution of soil organic matter concepts and their relationships with the fertility and sustainability of cropping systems. *Agriculture Ecosystems & Environment.* 119, 217–233.

Marald, E. 2002. Everything circulates: Agricultural chemistry and recycling theories in the second half of the nineteenth century, *Environment and History*, 8, 65–84.

Marriott, M. E. and Wander, M. M. 2006. Total and labile soil organic matter in organic farming systems, *Soil Science Society of America,* 70, 950–959.

Massey, R. E. et al. 2008. Profitability maps as an input for site-specific management decision making, *Agronomy Journal*, 100, 52–59.

May, R. M. 1972. Will a large complex system be stable? *Nature,* 238, 413–414.

May, R. M. 1974. *Stability and Complexity in Model Ecosystems*, Princeton University Press, Princeton.

Millennium Ecosystem Assessment, 2005. Island Press, Washington, D.C.

Miller, F. P. 2008. After 10,000 years of agriculture, whither agronomy? *Agronomy Journal,* 100, 22–34.

Mollison, B. and Holmgren, D. 1978. *Permaculture One.* Transworld, London.

Moore, J. C. and de Ruiter, P. C. 1997. Compartmentalization or resource utilization within soil ecosystems. In *Multitrophic Interactions in Terrestrial Ecosystems.* Grange, A. and Brown, V., Eds. Blackwell Science, Oxford, 375–393.

Moore, J. C. and Hunt, H. W. 1988. Resource compartmentalization and the stability of real ecosystems, *Nature,* 333, 261–261.

Moore, J. C., McCann, K., and de Ruiter, P. C. 2005. Modeling trophic pathways, nutrient cycling, and dynamic stability in soils, *Pedobiologia,* 49, 499–510.

Moorhead, D. L., Westerfield, M. M., and Zak, J. C. 1988. Plants retard litter decay in a nutrient-limited soil: A case of exploitative competition? *Oecologia*, 113, 530–536.

Moran, K. K. et al. 2005. Role of mineral-nitrogen in residue decomposition and stable soil organic matter formation, *Soil Science Society of America Journal,* 69, 1730–1736.

Neff, J. C. et al. 2002. Variable effects of nitrogen additions on the stability and turnover of soil carbon, *Nature,* 419, 915–917.

Newman, W. R. 2006. *Atoms and Alchemy; Chemistry & the Experimental Origins of the Scientific Revolution.* University of Chicago Press, Chicago, 1–250.

Nissen, T. M. and Wander, M. M. 2003. Management and soil-quality effects on leaching and nitrogen-use efficiency, *Soil Science Society of America Journal*, 67, 1524–1532.

Norton, B. G. 2005. *Sustainability: A Philosophy of Adaptive Ecosystem Management.* University of Chicago Press, Chicago.

Odum, E. P. 1969. Strategy for ecosystem development, *Science,* 126, 262–270.

Owen, A. G. and Jones, D. L. 2001. Competition for amino acids between wheat roots and rhizosphere microorganisms and the role of amino acids in plant N acquisition, *Soil Biology Biochemistry,* 33, 651–657.

Paustian, K., Collins, H. P., and Paul, E. A. 1997. Management controls on soil carbon. In *Soil Organic Matter in Temperate Agroecosystems: Long-Term Experiments in North America,* Paul, E. A. et al., Eds. CRC Press, Boca Raton, FL, 15–49.

Pearson, H. L. and Vitousek, P. M. 2002. Soil phosphorus fractions and symbiotic nitrogen fixation across substrate-age gradient in Hawaii, *Ecosystems*, 5, 587–596.

Pennazio, S. 2005. Mineral nutrition of plants: A short history of plant physiology, *Rivista di biologia*, 98, 215–236.

Pfieffer, E. 1947. *Soil Fertility, Renewal and Preservation: Bio-Dynamic Farming and Gardening*, Faber and Faber, London.

Phillips, C. J. et al. 2000. Effects of agronomic treatments on structure and function of ammonia-oxidizing communities, *Applied and Environmental Microbiology,* 66, 5410–5418.

Poudel, D. D., Horwath, W. R., Mitchell, J. P., and Temple, S. R. 2001. Impacts of cropping systems on soil nitrogen storage and loss, *Agric. Syst.* 68, 253–268.

Randall, G. W. et al. 1997. Nitrate losses through subsurface tile drainage in Conservation Reserve Program, alfalfa, and row crop systems, *Journal of Environmental Quality*, 26, 1240–1247.

Rasse, D. P., Rumpel, C., and Dignac, M. F. 2005. Is soil carbon mostly root carbon? Mechanisms for a specific stabilization, *Plant & Soil*, 269, 341–356.

Rasmussen, P. E. et al. 1998. Long-term agroecosystem experiments: assessing agricultural sustainability and global change, *Science*, 282, 893–896.

Rastetter, E. B. et al. 2001. Resource optimization and symbiotic nitrogen fixation. *Ecosystems*, 4, 369–388.

Ribe, N. and Steinle, F. 2002. Exploratory experimentation: Goethe, land, and color theory. *Physicstoday.org.* 55, (7): 43. http://scitation.aip.org/journals/doc/PHTOAD-ft/vol_55/iss_7/43_1.shtml (accessed 11 June 2008).

Rosenfeld, L. 2003. Justus Liebig and animal chemistry, *Clinical Chemistry*, 49, 1696–1707.

Rossiter, M. 1995. *The Emergence of Agricultural Science.* Yale University Press, New Haven, CT.

Russell, A. E., Laird, D. A., and Mallarino, A. P. 2006. Nitrogen fertilization and cropping system impacts on soil quality in midwestern mollisols, *Soil Science Society of America Journal*, 70, 249–255.

Schaffer, S. 2005. Public experiments. In *Making Things Public*, Latour, B. and Weibel, P., Eds. MIT Press, Cambridge, MA, 298–307.

Schimel, J. P. and Bennett, J. 2004. Nitrogen mineralization: Challenges of a changing paradigm. *Ecology*, 85, 591–602.

Schimel, J. P. and Gulledge, J. 1998. Microbial community structure and global trace gases, *Global Change Biology*, 4, 745–758.

Schlesinger, W. H. 2000. Carbon sequestration in soils: Some cautions amidst optimism. *Agriculture, Ecosystems, and Environment*, 82, 121–127.

Schneider, E.D. and Kay, J. J. 1994. Life as a manifestation of the second law of thermodynamics, *Mathematical and Computer Modelling*, 19(6-8): 25–48.

Schulman, S. W. 1999. The business of soil fertility: A convergence of urbanagrarian concern in the early 20th century, *Organization & Environment*, 12, 401–424.

Schulte, L. A. et al. 2006. Agroecosystem restoration through strategic integration of perennials, *Journal of Soil and Water Conservation*, 6, 164–169.

Schwartz, M. 2001. The life and works of Louis Pasteur, *Journal of Applied Microbiology*, 91, 597–601.

Scofield, A. M. 1986. Organic farming—The origin of the name, *Biological Floriculture and Horticulture*, 4, 15.

Sepper, D. L. 2005. Goethe and the poetics of science, *Janus Head*, 8, 207–227.

Shi, W. et al. 2004. Microbial nitrogen transformations in response to treated dairy waste in agricultural soils, *Soil Science Society of America Journal*, 68, 1867–1874.

Sinsabaugh, R. L. et al. 2005. Extracellular enzyme activities and soil organic matter dynamics for northern hardwood forests receiving simulated nitrogen deposition. *Biogeochemistry*, 75, 201–215.

Sprengel, C. 1826. Ueber Pflanzenhumus,Humussaure und humussaure Salze (About plant humus, humic acids and salts of humic acids). *Archiv furdie Gesammte Naturlehre* 8, 145–220.

Sprengel, C. 1839. *Die Lehre vom Du¨ nger* (Fertilizer science). Immanuel Muller Publishing Co., Leipzig, Germany.

Six, J. et al. 2006. Bacterial and fungal contributions to carbon sequestration in agroecosystems, *Soil Science Society of America Journal*, 70, 555–569.

Steenwerth, K. L. et al. 2002. Soil microbial community composition and land use history in cultivated and grassland ecosystems of coastal California, *Soil Biology and Biochemistry*, 34, 1599–1611.

Steinle F. 2002. Experiments in history and philosophy of science, *Perspectives on Science*, 10, 408–432.

Stephenson, R. H. E. 1995. *Goethe's Conception of Knowledge and Science.* Edinburgh University Press, Edinburgh.

Stinner, B.R., Crossley, Jr., D. A., Odum, E. P., and Todd, R. L. 1984. Nutrient budgets and internal cycling of N, P, K, Ca, and Mg in conventional tillage, no-tillage, and old field systems on the Georgia Piedmont, *Ecology*, 65, 354–369.

Stockdale, E. A. et al. 2002. Soil fertility in organic farming systems—Fundamentally different? *Soil Use Management*, 18, 301–308.

Stuedler, P. A. et al. 1996. Microbial controls of methane oxidation in temperate forest and agricultural soils. In *Microbiology of Atmospheric Trace Gases*, Murrell, J. C. and Kelly, D. P., Eds. Springer, Berlin, 69–81.

Thieta, R. K., Frey, S. D., and Six, J. 2006. Do growth yield efficiencies differ between soil microbial communities differing in fungal:bacterial ratios? Reality check and methodological issues, *Soil Biology & Biochemistry*, 38, 837–844.

Tilman, D. 1991. The schism between theory and ardent empiricism: A reply to Shipley and Peters, *American Naturalist,* 138, 1283–1286.

Torn, M. S. et al. 1997. Mineral control of soil organic carbon storage and turnover. *Nature,* 389, 170–173.

Torn, M. S., Vitousek, P. M., and Trumbore, S. E. 2005. The influence of nutrient availability on soil organic matter turnover estimated by incubations and radiocarbon modeling, *Ecosystems,* 8, 352–372.

Travis, J. M. J., Brooker, R. W., and Dytham, C. 2005. The interplay of positive and negative species interactions across an environmental gradient: insights from an individual-based simulation model, *Biology Letters,* 1, 5–8.

Tugel, A. J. et al. 2005. Soil change, soil survey, and natural resources decision making: A blueprint for action. *Soil Science Society of America Journal,* 69, 738–747.

Ulanowicz, R. E. 1990 Aristotelean causalities in ecosystem development, *Oikos,* 57, 42–48.

Valentine, D. W., Holland, E. A., and Schimel, D. S. 1994. Ecosystem and physiological controls over methane production in northern wetlands, *Journal Geophysical Research Atmosphere,* 99, 1563–1571.

van der Krift, T. A. J. and Berendse, F. 2001. The effect of plant species on soil nitrogen mineralization, *Journal of Ecology,* 89, 555–561.

van der Ploeg, R. R., Böhm, W., and Kirkham, M. B. 1999. On the origin of the theory of mineral nutrition of plants and the law of the minimum, *Soil Science Society of America Journal,* 63, 1055–1062.

Vanotti, M. B., Bundy, L. G., and Peterson, A. E. 1997. Nitrogen fertilizer and legume-cereal rotation effects on soil productivity and organic matter dynamics in Wisconsin, In *Soil Organic Matter in Temperate Agroecosystems: Long-Term Experiments in North America,* Paul, E. A. et al. Eds. CRC Press, Boca Raton, FL, 105–120.

Velho, L. and Velho, P. 1997. The emergence and institutionalization of agricultural science, *Cadernos de Ciencia and Technologia Brasilia,* 14, 205–233.

Vitousek, P. 2004. *Nutrient Cycling and Limitation, Hawaii as a Model System.* Princeton University Press, Princeton, NJ.

Vitousek, P. M. and Hobbie, S. 2000. Heterotrophic nitrogen fixation in decomposing litter: patterns and regulation, *Ecology,* 81, 2366–2376.

von Bertalanffy, L. 1968. *General System Theory: Foundations, Development, Applications.* George Braziller, New York.

Waksman, S. A. 1936. Humus, Origin, Composition, and Importance in Nature. Williams & Wilkins, Baltimore, MD.

Waldrop, M. et al. 2004. Nitrogen deposition modifies soil carbon storage through changes in microbial enzymatic activity, *Ecological Applications,* 14, 1172–1177.

Walker, L. R. and del Moral, R. 2003. *Primary Succession and Ecosystem Rehabilitation,* Cambridge University Press, Cambridge, UK.

Wander, M. M. and Nissen, T. M. 2004. Value of soil organic carbon in agricultural lands, *Mitigation and Adaptation for Global Climate Change,* 9, 417–431.

Wander, M. M. et al. 1994. Organic and conventional management effects on biologically active soil organic matter pools, *Soil Science Society of America Journal,* 58, 1130–1139.

Wander, M. M. et al. 2002. Soil quality: Science and process, *Agronomy Journal,* 96, 23–33.

Warner, K. D. 2007. *Agroecology in Action: Extending Alternative Agriculture through Social Networks.* MIT Press, Cambridge, MA.

Watson, A. M. 1974. The Arab agricultural revolution and its diffusion: 700–1100, *Journal of Economic History,* 34, 8–35.

Watson, M. 1981. A Medieval green revolution: New crops and farming techniques in the early Islamic world. In *The Islamic Middle East: 700–1900: Studies in Economic and Social History,* Udovitch, A.L., Ed. Darwin Press, Princeton, NJ.

Wendt, G. 1950. *Carl Sprengel und die von ihm geschaffene Mineraltheorie als Fundament der neuen Pflanzenernährungslehre* (CarlSprengel and his mineral theory as foundation of the modern scienceof plant nutrition). Ernst Fischer Publishing Co.,Wolfenbuttel, Germany.

Werner, P. and Holmes, F. L. 2002. Justus Liebig and the plant physiologists, *Journal of the History of Biology,* 35, 421–441.

Willson, T. C., Paul, E. A., and Harwood, R. R. 2001. Biologically active soil organic matter fractions in sustainable cropping systems, *Applied Soil Ecology,* 16, 63–76.

Worster, D. 1977. *Natures Economy: The Roots of Ecology.* Anchor Books, Garden City, NY.

Yoo, G. and Wander, M. M. 2008. Tillage effects on aggregate turnover and sequestration of particulate and humified soil organic carbon, *Soil Science Society of America Journal,* 72, 670–676.

Zhang, W. J. et al. 2005. Responses of soil microbial community structure and diversity to agricultural deinten-sification, *Pedosphere,* 15, 440–447.

Zeglin, L. H. et al. 2007. Microbial responses to nitrogen addition in three contrasting grassland ecosystems, *Oecologia,* 154, 349–359.

Zemplén, G. A. 2003. The Janus faces of Goethe: Goethe on the nature, aim, and limit of scientific investiga-tion, *Periodica Polytechnica: Social and Management Sciences,* 11, 259–278.

11 The Role of Biodiversity in Agronomy and Agroecosystem Management in the Coming Century

Daniel Hillel and Cynthia Rosenzweig

CONTENTS

11.1 INTRODUCTION

Human beings have gradually increased in numbers and expanded the extent and range of their activity, eventually gaining dominance over, and drastically modifying, entire terrestrial and marine biomes for food production throughout the world. Consequently, numerous other species have been endangered or even eradicated. Recent calculations suggest that rates of species extinctions are now on the order of 100 to 1000 times those before humans dominated the Earth. For some well-documented groups, extinctions are even greater. Over the past 2000 years, humans have driven to extinction as many as one-quarter of Earth's bird species (Steadman 1995). Unless checked, the continued increase of human population and the intensified manipulation of the environment for short-term advantage threaten to turn human success into eventual failure. Having tampered with nature in hopes of gaining control over it, humans are now more dependent than ever on its complex workings, in which the diversity and intrinsic mutuality of all life forms are essential factors.

A crucial imperative is to ensure the adequate production and supply of food for a growing population in a world in which biotic, terrestrial, and aquatic resources have already been seriously degraded or depleted. Despite the lower fertility levels projected and the increased mortality risks to which some populations are subject, the population of the world is expected to increase by some 3 billion in the coming decades, from the current 6.3 billion to a total of some 9 billion before it stabilizes (United Nations Population Division 2003). The yearly addition of an estimated 70 to 80 million people will impose greater demands for food, housing, healthcare, education, political organization, public order, and employment. The world's average population density of 45 people per square kilometer is projected to rise to 66 people per square kilometer by 2050. Since only about 10 percent of land is arable, population densities per unit of arable land are roughly 10 times higher (Cohen 2003). Given the poverty and famine that prevail in some regions and the foreseen change of the Earth's climate, it is an open question whether, and how, humanity can provide for itself while avoiding irreversible damage to natural ecosystems and their biodiversity. Biodiversity is here defined as the range of species that exist in an agroecosystem and the genetic resources within a species. Increasing awareness of the issue and the development of modern methods of conservation management offer hope for some progress in this difficult task. Utilizing the promise inherent in such methods must, however, be constrained by an understanding of the potential problems and hazards they pose (Hillel et al. 2002).

11.2   HISTORICAL BACKGROUND

For the greater part of their career as a species, humans roamed over the landscape in small bands subsisting as hunters, gatherers, and occasional scavengers. Being omnivorous, humans availed themselves of a variety of food sources opportunistically and eclectically, gathering edible plant products and killing some animals for their meat (as well as for their skin, bones, antlers, and other usable parts). In time, humans learned to manipulate their environment, initially through the creation of fires (Caldararo 2002). Although their lives were physically rigorous, they were venturesome and adaptable enough to spread out from their native African savanna into all the habitable continents. Relying on their ingenuity and tool-making ability, they adapted to widely varying environments—from icy northern Eurasia to arid Australia.

A dramatic change in human lifestyle began toward the end of the era known by geologists as the Pleistocene and the beginning of the current era, called the Holocene. That change evidently took place earliest in the Near East, some 10 to 12 millennia ago, during what archaeologists call the Neolithic Age. As the last ice age ended, the warming trend gave rise to a profusion of plant and animal life in that region that afforded humans an abundance of food sources and favorable sites for regular, and eventually permanent, habitation (Hillel 1991).

As groups of humans shifted from nomadic to sedentary living and began to form settlements, they also learned, after collecting seeds of wheat and barley, to domesticate selected plants (Kislev et al. 2004). Thus, agriculture began. At first it was in the form of rain-fed farming in relatively humid areas, and later it was in the additional form of irrigated farming in the main river valleys. Simultaneously, animal husbandry developed based on the herding of livestock (sheep, goats, cattle, etc.) both in conjunction with village-based farming and in the context of an alternative lifestyle, namely, seminomadic pastoralism.

Of the many plants with edible products, relatively few were found suitable for early domestication. Prominent among these were selected species of the *Graminea* family (the cereal grains of wheat, barley, oats, rye, and sorghum), the *Leguminosa* family (peas, lentils, chickpeas, and several types of beans), vegetables of various genera, and a number of fruit-bearing woody plants or trees (olives, grapes, almonds, pomegranates, figs, and dates). Similarly, only a limited number of animals lent themselves conveniently to domestication. Breeding programs, along with natural hybridization, played a pivotal role in shaping the genetic and evolutionary trajectories of domesticated animal species (Arnold 2004).

Consequently, human societies abandoned their prior lifestyle as roaming hunter–gatherers, and, as they became sedentary producers of food, came to depend on their managed crops and livestock for subsistence. Agriculture has created plants and animals (e.g., wheat, rice, maize, cattle, swine, and poultry) that are currently some of the most prevalent and widespread organisms. Thanks to these, humans have become the dominant species on Earth. An inextricable mutual dependency thus developed between humans and their domesticates.

The same processes of transition to an agricultural or pastoral economy, which first took place in the Near East, appeared in at least seven independent centers (Smith 1995) and rapidly spread from these places as well. The latter centers included southern and eastern Asia, central Africa, and Central America, each with its own indigenous selection of domesticable plants and animals. In all those locations, the agricultural transformation improved food security and thereby set in motion a progressive increase of population density. So productive has been the enterprise of agriculture that an ever-decreasing number of farm workers have been able to feed an ever-increasing number of people. Urban centers then developed, in which people engaged in a variety of other occupations (industry, art, science, medicine, instituted religion, the military, and other societal functions), thus creating the basis for complex civilizations.

A less auspicious consequence of those same developments was a narrowing of the array of foods that served to sustain the population. The domesticated lifestyle provided only a limited number of tended species and strains instead of the wide selection of types and sources of food that humans had previously been able to collect or hunt in the wild. As the variety of foods was reduced, so were the nutritional balance and quality of the diet. The study of archaeological remains from around the world reveals that the shift from hunting and gathering to increased nutritional focus on domesticated grains (~10,000 years ago) coincided with a decline in health, including increased evidence of morbidity related to dental abnormalities, iron-deficiency anemia, infection, and bone loss (Larsen 2003). Moreover, reliance on a small number of crops and animals, maintained in managed sites, made societies vulnerable to failures of production resulting from the vagaries of weather as well as from pests and diseases of crops and livestock. People living in close communities, and eventually in cities, themselves became more vulnerable to communicable diseases.

So the great advantages of domestication were not without attendant disadvantages. However, with the increase of population density made possible by the initial success of agriculture, there could no longer be a return from permanent husbandry to the lifestyle and economy of nomadic hunting–gathering. Humans also changed biologically because of the selective pressures of living in built environments, decreased mobility, and changes in diet consistency associated with increasing sedentism (Leach 2003). The agricultural transformation thus became effectively irreversible (Hillel 1991).

As long as human exploitation of the land and its biotic resources was restricted to small enclaves, the surrounding expanses of relatively undisturbed natural ecosystems could remain intact, with their biodiversity preserved. But, as the extent and intensity of human exploitation of the terrestrial domain increased, along with the increase of population, natural habitats were reduced and fragmented. This process of human encroachment has continued and accelerated until nearly half of Earth's continental surface has come under direct human management. A similar process has occurred in Earth's aquatic (freshwater as well as oceanic) ecosystems. Even where humans have not intervened directly, the secondary effects of their activity (such as the chemical residues of industrial production) have caused indirect deleterious effects. Entire biomes are now threatened and numerous "wild" species have already been eliminated. Projections indicate that biodiversity loss will continue into the future, as expressed in declines in populations of wild species and reduction in area of wild habitats (Jenkins 2003).

Within the agricultural lands themselves, poor management practices have induced processes of degradation. Denudation of the vegetative cover, coupled with surface pulverization by tillage or by the trampling of livestock or machinery, has made the soil vulnerable to erosion by wind during dry periods and by water during rainstorms. In extreme cases, the fertile topsoil has been completely scoured away, and the less fertile subsoil (or even the sterile bedrock) has been exposed. Soil productivity is thus greatly impaired, as is its capacity to support various forms of life (Hillel, 1991).

Quite another process of soil degradation occurs in irrigated lands, particularly in river valleys located in arid regions. There, the traditional practice of flood irrigation with large volumes of water causes much percolation through the soil, which tends to raise the water table, to saturate the soil excessively (a phenomenon called "waterlogging"), and to accumulate salts at or near the soil surface (Hillel 1998). The result is soil salination, a process that destroys soil productivity.

Fortunately, the picture is far from entirely bleak. Many of the ills described can be prevented or alleviated. New trends and opportunities offer hope that further threats to biodiversity can be avoided. Human population growth seems to be slowing. Moreover, agriculture has already begun to develop and adopt better methods of production coupled with biological control and conservation aimed at preserving, even enhancing, the diversity of life on Earth (Edwards et al. 1993; Smith et al. 1995). The new approaches are impelled by a growing recognition of the indispensable importance of biodiversity to agriculture.

11.3 DEPENDENCE OF AGRICULTURE ON BIODIVERSITY

All the plants whose products are utilized by humans, either directly or indirectly (via plant-consuming animals), were derived originally from biological diversity; that is to say, from wild ancestors. So were all domesticated animals. Those domesticates were selected and bred for their desirable traits. As environmental circumstances and stresses change, as the requirements and preferences of humans change, and as domesticated organisms themselves are vulnerable to diseases and pests, the need arises repeatedly to breed new varieties.

Traditionally, agricultural breeding has been done with the close genetic relatives (either wild genotypes or domesticated varieties and strains) of the relevant organisms. *In situ* genetic diversity is often considered a resource for future crop improvement (Ladizinsky 1989). Different strains may contain different genes, including perhaps genes that impart resistance to certain pests and environmental stresses. Recently, new possibilities have arisen to transfer desired traits (genes) not just between strains of the same species, but even from one species to another, thus greatly enlarging the range of potential genetic resources available to agriculture (though the new techniques also present new hazards). Either way, breeding plants and animals for agricultural purposes was and remains dependent on nature's rich array of life forms, that is, on natural biodiversity.

Of all the myriad species of plants or animals whose products can be useful to humans, agriculture utilizes directly only a few hundreds. Among those, just 80 crop plants and 50 animal species provide most of the world's foods. However, what is not generally appreciated is that those relatively few species depend vitally for their productivity on hundreds of thousands of other species. Among the latter are insects and birds that pollinate crop flowers and feed on crop pests.

Even more numerous and varied are the microbial species that live on plants and animals and are especially abundant in the soil. They help decompose residues (including pathogenic and toxic agents), transmute them into nutrients for the continual regeneration of life, and create soil structure. Agricultural productivity and sustainability benefit from microorganisms in many ways, including the conversion by bacteria of elemental nitrogen from the atmosphere into soluble ammonium and nitrates that serve as essential nutrients for plants. Nitrogen-fixing bacteria may be either symbiotic (e.g., *Rhizobium* bacteria that attach themselves to the roots of legumes) or nonsymbiotic (free-living). Quite another function is fulfilled by mycorrhizal fungi, which live in association with crop roots and facilitate the uptake of phosphorus and other relatively immobile nutrients (Hartel 2005).

Biological control agents (so-called because they prey on insects and other kinds of pests), as well as pollinators, generally reside in natural or seminatural ecosystems (Vandermeer 1997).

Clearing away such ecosystems in the belief that such action prevents the invasion of pest species into fields and orchards may actually do more harm than good by depriving agriculture of beneficial organisms.

In ways both visible and invisible, agriculture thus depends on nature's biodiversity. Biodiversity operates not only on a present functional level, but also provides standing insurance against future extinctions, as well as evolutionary flexibility in regard to future climate change (Lande 1988). Genetic diversity in wild populations is a substrate for both natural and directed selection. Hence, diminution of that diversity endangers agriculture just as it endangers all the processes of life on Earth, which are inherently interdependent.

Growing conditions differ from place to place (due to differences in soil, water regime, temperature, exposure to sun and winds, daylength, prevalence of diseases and pests, etc.), and also differ from season to season (due to the variability of climate). Therefore, pure stands of genetically similar, or essentially identical, plants are at greater risk than are genetically diverse stands. Conversely, genetically diverse crops can better survive in environments in which conditions fluctuate. Though the latter may not provide yields that are as great during favorable or normal seasons, they are more likely to provide an adequate yield during unfavorable seasons, during which pure stands (lacking genetic variety and hence adaptability to changing conditions) may be devastated by inclement weather or other disruptive factors.

11.4 FUNCTIONS AND EFFECTS OF BIODIVERSITY IN AGRICULTURE

Specific functions that biodiversity fulfills in agriculture include pollination, insect pest protection, and disease control. Recent work has brought to light the interactive role of agricultural biodiversity in maintaining healthy assemblages of bird species.

11.4.1 POLLINATORS

Declines in pollination have been reported in every continent except Antarctica (Kearns et al. 1998), and underpollination for some crops caused by pollinator limitation already reaches 70 percent in some places (Reddi 1987). This is significant because pollinators play a key role in agricultural productivity (Buchman and Nabhan 1996). Although the majority of the world's staple crops (wheat, rice, maize, potato, yam, and cassava) are either wind-pollinated or self-pollinated, or are propagated vegetatively (Richards 2001), many other important agricultural species do rely on pollinators. For instance, over 80 percent of the 264 species grown as crops in the European Union (EU) are dependent on insect pollination (Corbet et al. 1991; Williams 1996). In addition, the yield of tomatoes, sunflowers, olives, grapes, and soybean—all major crops—is optimized by regular pollination (Richards 2001). Fruit trees and legumes may be particularly hard hit by loss of pollinators, especially because they are grown intensively.

When compared to wind-pollinated plants, or plants pollinated by a wide taxonomic group, plants that have specialist animal pollination (a 1:1 species relationship such as figs and fig wasps) have the lowest risk of pollen wastage during animal transport. They also have the lowest risk of pollen clogging and allelopathy because of heterospecific pollination. These same plants, though, have the highest risk of pollination failure if their pollinators are lost (Wilcock and Neiland 2002). For this reason, decline in biodiversity may have cascading effects on species survivorship because it may disrupt these close-knit, highly efficient relationships. Just as a high diversity of pollinators may help increase the diversity of plants, a high diversity of plants supports more pollinators.

An experiment demonstrating the effect of habitat isolation (which often occurs in agricultural regions as native areas are converted to cultivation) on pollination has come from isolated "islands" of radish and mustard plants. The areas were set up in an agricultural landscape at varying distances from a species-rich grassland (Steffan-Dewenter and Tscharntke 1999). Increasing isolation resulted in fewer bee visits per hour to the islands and also reductions in the taxonomic diversity of the visitors. Also, fruit and seed set declined with increasing isolation from the grasslands.

Similarly, Albrecht et al. (2007) looked at the effects of managing meadows according to the prescriptions of ecological compensation areas (ECA) in Switzerland on both pollinator species richness and abundance, and the reproductive success of plants in nearby intensively managed meadows (IM). Species richness and abundance of small-sized pollinators in IM declined significantly with increasing distance from ECA (Figure 11.1). The number of fruits and seeds per plant of *Raphanus sativus* (radish) in the IM decreased with increasing distance from ECA. The authors concluded that establishing ECA is an effective method of enhancing both pollinator species richness and abundance and pollinator services to nearby intensely managed farmland. In another study, the amount of woody border had a significant positive effect on the overall richness of insects at the family level in agricultural fields (Mänd et al. 2002).

Research in human-dominated tropical landscapes in Sulawesi, Indonesia has shown that forests annually provide pollination services worth $63 per hectare through pollination of coffee plants. Spatially explicit land use simulations demonstrate that depending on the magnitude and location of ongoing forest conversion, pollination services are expected to decline continuously and thus directly reduce coffee yields by up to 18 percent, and net revenues per hectare up to 14 percent within the next two decades (Priess 2007).

11.4.2 Insect Pests

Small-scale farmers in the tropics have long used crop diversification as a way of minimizing the risk of crop failure. Vegetation or crop diversity has been frequently recommended as a way of reducing pest problems, and the lack of it has been blamed for infestations (Tonhasca and Byrne 1994). Experimental studies and theoretical arguments suggest that the differences in pest abundance between diverse and simple systems can be accounted for by the response of herbivore host-finding behavior to patterns of resource availability (Risch et al. 1983).

The so-called resource concentration hypothesis (Root 1973) applies to specialist herbivores and suggests that the presence of nonhost species disrupts the ability of pests to attack their main host effectively. There have been several mechanisms elucidated that interfere with an insect's host-seeking behavior: camouflage—the host plant is guarded from insect pests by the presence of other plants; crop background—certain pests prefer certain backgrounds of a particular color and/or texture; masking or dilution of attraction stimuli—the presence of the nonhost plant masks or dilutes the attractant stimuli of the host plant, leading to a breakdown or reorientation of feeding and reproduction; and repellent chemical stimuli—aromatic odors of certain plants disrupt the insect's host-finding ability.

Some mechanisms interfere with pest populations as a whole. These include mechanical barriers, such as companion crops that block the dispersal of herbivores across the polyculture; lack of stimuli that causes the herbivore landing on a nonhost to leave the plot quickly; and microclimate influences. Other field studies have supported the hypothesis that increasing crop diversity will decrease pest abundance. For instance, lepidopterous stem-borers constitute one of the major constraints to efficient maize and sorghum production in the developing world (Ampofo 1986). Ogol et al. (1999) investigated the effect of using an agroforestry system involving maize and the tree legume *Leucaena leucocephala* in western Kenya. Abundance of adult and larval/pupal stages of maize stem-borers, oviposition preference, foliar damage, borer entry/exit holes, maize plant mortality, and stem breakage because of borer damage were significantly greater in the maize monocrop than in the maize–*leucaena* intercrop. The reduced pests in the maize–leucaena plots were associated with reduced yield loss per plant, and the 3.0 m leucaena hedgerow spacing plots yielded more than the maize monocropped plants even though they had 25 percent fewer maize plants to begin with.

During a good part of the twentieth century, farmers throughout the world have relied heavily on chemical pesticides (Pimentel 1997). But often these pesticides kill natural enemies and provoke

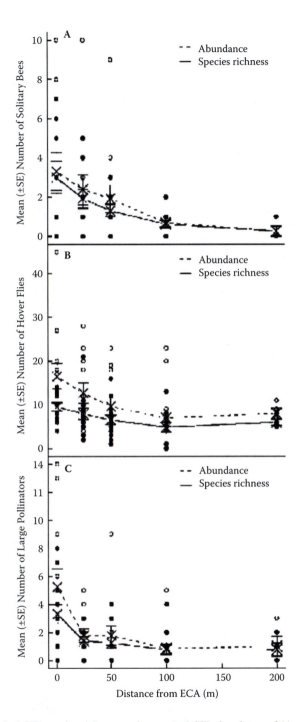

FIGURE 11.1 Mean (± 1 SE) species richness and mean (± 1 SE) abundance of (a) solitary bees, (b) hover-flies, and (c) large-sized pollinators on three species of phytometers in ecological compensation areas (ECA) (0 m) and at distances of 25 m, 50 m, 100 m (n = 13), and 200 m (n = 4) from the ECA within intensively managed meadows (IM). Untransformed, pooled data of the three phytometer species are shown. (From Albrect et al. 2006. With permission.)

GENETIC DIVERSITY AND DISEASE CONTROL IN RICE

Zhu et al. (2000) studied genetically diversified rice crops that were planted in fields in Yunnan Province, China, to test the effect of such a planting on rice-blast disease. *Magnaphorthe grisea*, the fungus that causes blast disease, interacts on a gene-for-gene basis (Staskawicz 1995; Baker 1997) with its host and has a very varied pathogenesis (Ou 1980). It exists as a mixture of genetic variants that attack host genotypes with different resistant genes. For this reason, host-resistance genes often remain effective for only a few years in agricultural production before succumbing to new pathogenic races (Ou 1980). Yunnan Province, China, has a cool, wet climate that fosters the development of the blast. To control it, farmers make multiple foliar fungicide applications.

When mixtures of rice varieties were planted, blast was controlled so well that only one foliar fungicide spray was applied. When these mixtures were compared to monocultures, the researchers found that the diversification had a substantial impact on rice-blast severity. Panicle-blast severity on the valuable glutinous varieties averaged 20 percent in monocultures, but was reduced to 1 percent when dispersed within the mixed population. Blast severity also decreased, to a smaller extent, among the hybrid varieties.

Canopy microclimate data collected at one survey site in 1999 indicated that height differences between the taller glutinous and shorter hybrid varieties resulted in temperature, humidity, and light conditions that were less conducive for glutinous blast than those in the uniform crop heights of the monocultures. Induced resistance (which occurs when inoculation with a virulent pathogen race or races induces a plant defense response that is effective against pathogen races that would normally be virulent on that host genotype) may have also played a role in reducing the blast occurrence in the mixed rice fields (Zhu et al. 2000) (Figure 11.2).

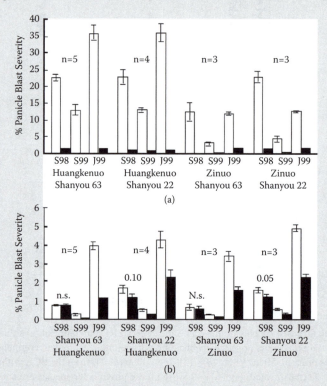

FIGURE 11.2

resistance in the pest they are intended to kill. The absence of natural enemies may allow benign insects to increase their population to such an extent that not only do they become pests, but they are also able to acquire resistance to pesticides. This pattern is known as the "pesticide treadmill" (Vandermeer 1997). In Central America, for instance, a host of predatory and parasitic arthropods was removed from the agricultural system, and its loss resulted in greater problems, to the point that the cotton industry of Guatemala, El Salvador, and Nicaragua was severely damaged.

In the last decades of the twentieth century, an increasing awareness of the limitations and damages associated with chemical pesticides has led to the development of sophisticated techniques of "integrated pest management" (IPM). Such methods are based on the judicious combination of biological controls with sparing applications of chemicals only when absolutely necessary. The biological control component of IPM, in turn, depends on ecosystem biodiversity. For example, spiders are one of the orders that show great potential as biological control agents (Nyffeler 2003).

11.4.3 DISEASE CONTROL

Genetic diversity is likely to reduce the odds of crop failure and to contribute to greater stability of production. Similar benefits may be inherent in mixed-species and multispecies cropping systems such as are common in subsistence farm units. As shown in a rice-blast study (see box), favorable aspects of microclimate conditions may become highly fractionated, causing insects to experience difficulty in locating and remaining in suitable microhabitats. In contrast, uniform monoculture crops, standing like battalions of identical soldiers in close formation, may be susceptible to pest and disease outbreaks. Pathogens spread more readily, and epidemics tend to be more severe when the host plants (or animals) are more uniform, numerous, and crowded. Owing to their high densities and the large areas over which they are grown, both crop plants and livestock are repeatedly threatened by ever-new infestations of pests and diseases. Existing pests and diseases are continually evolving strains that overcome the innate defenses of particular strains or breeds, as well as of chemical treatments applied by farmers.

Many historical examples can be cited to prove that monoculture stands or concentrations of crops and livestock with uniform genetic traits, though they may be more productive in the short run, entail the risk of succumbing, sooner or later, to changing conditions. Catastrophic outbreaks of disease, invasions of insects, and climatic anomalies have caused many wholesale crop and animal destructions in the past. Such episodes have contributed to famine, especially where, in the absence of sufficient diversity, no varieties or breeds were present that could withstand the destructive conditions.

Among the notable examples of disastrous outbreaks are the infestation of red rust on wheat in Roman times, the mass poisoning from ergot-tainted rye during the Middle Ages in Europe, the failure of the vaunted vineyards of France in the late nineteenth century, and the potato famine that hit Ireland in the 1840s and 1850s. The last was caused by the fungus *Phytophthora infestans*, which arrived accidentally from North America and attacked the genetically uniform potato stock

FIGURE 11.2 (See previous page) Panicle blast severity (mean percentage of panicle branches that were necrotic due to infection by *Magnaporthe grisea*) of rice varieties planted in monocultures and mixtures. (a) The susceptible, glutinous varieties Huangkenuo and Zinuo. (b) The resistant, hybrid varieties Shanyuo22 and Shanyuo63. S98, Shiping County, 1998; S99, Shiping County, 1999; J99, Jianshui County, 1999; open bar, blast severity for a variety grown in monoculture control plots; black bar, blast severity of the same variety when grown in mixed culture plots in the same fields. Error bars are one s.e.m.; *n*, number of plot means that contribute to individual bars for each of the four combinations of susceptible and resistant varieties. All differences between pairs of monoculture and mixture bars are significant at *P*, 0.01 based on a one-tailed *t*-test, unless indicated by 0.05 (significant at *P*, 0.05), 0.10 (significant at *P*, 0.10), or n.s. (not significant at $P = 0.10$). (From Zhu, Y. et al. *Nature*, 406, 718–722, 2000. With permission.)

that served as the mainstay of Irish farms. As a result, about 1,100,000 people died from starvation or typhus and famine-related diseases, and 1.5 million people emigrated to North America in just the famine years (Mokyr 2004).

There is reason to be concerned especially over the massive concentration of agricultural production (and of food consumption) on three primary crops—wheat, rice, and maize—which together account for more than half of the total nutritional energy derived from plants in the world at large. In principle, such a concentration creates vulnerability. One example of the vulnerability of wheat is the recent outbreak of scab—*Fusarium* head blight—on wheat and barley in Minnesota and the Dakotas. Many farmers in areas where scab has been severe are forced to abandon farming for lack of alternative crops to grow profitably (McMullen et al. 1997). Ultimately, the best insurance against the future failure of crops is the enhancement of biodiversity, both to allow improvement of current crops and to discover appropriate future substitutes for them.

11.4.4   BIRDS

Donald et al. (2001) investigated the relationship between agricultural intensification and the collapse of Europe's farmland bird population, and found that population declines and range contractions were significantly greater in countries with more intensive agriculture. The effects are discernible at a continental level, making them comparable in scale to deforestation and global climate change as major anthropogenic threats to biodiversity.

The effects of different farming methods on bird communities were also compared in northern Italy (Genghini et al. 2006). In particular, the study surveyed the frequency and number of bird species, the number of individuals and bird diversity found in organic (nonuse of synthetic pesticides), integrated (reduced use of pesticides on the basis of economic threshold), and conventional orchards. Bird diversity was greater and insectivorous species were more abundant on organic and integrated farms ($p < 0.05$) than conventional ones.

Sinclair et al. (2002) studied the affects of agriculture on avifauna in the Serengeti by comparing agricultural fields with native savannah and grasslands. The authors documented a substantial but previously unnoted decline in avian biodiversity in the agricultural lands. The abundance of species found in agriculture was only 28 percent of that for the same species in native savannah (Figure 11.3). Insectivorous species feeding in the grass layer or in the trees were the most reduced. Some 50 percent of both insectivorous and granivorous species were not recorded in the agricultural sites, with ground-feeding and tree species the most affected. Although there was a concurrent decline in insects in the agricultural regions, the authors noted that the great reduction in insectivorous birds would be likely to affect farmers' ability to control insect-pest outbreaks. Also, the lack of raptors in the agricultural sites, particularly those that consume rodents (e.g., black-shouldered kite, *Elanus caeruleus,* and long-crested hawk eagle, *Spizaetus ayrestii*) and that are abundant in the savannah, may contribute to the frequent outbreaks of rodents such as the Natal multimammate mouse (*Mastomys natalensis*) in the agricultural zones.

11.5   CULTIVATED PLANTS AND WILD RELATIVES

Though clear benefits do exist in planting agricultural lands near wild ones, the action is not without potential consequences. One often-problematic process that has gained attention occurs when cultivated crops breed with their wild relatives (Ellstrand and Schierenbeck 2000; Ellstrand 2003). For example, although France's sugar beet seed production fields are many kilometers from the wild sea beets growing along the Golfe du Lion, the sea beets have been able to pollinate the sugar beets being used in seed production. By the mid-1970s, northwestern Europe's sugar beet fields had become pocked with beets that were flowering prematurely, or "bolting," a trait of the wild relatives. Investigation found that these bolters were the result of cultivated beets being pollinated by their wild beet relatives.

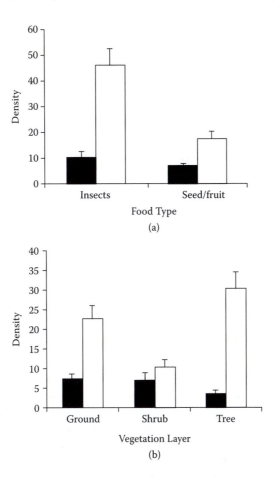

FIGURE 11.3   Index of bird density per square kilometer estimated on 5 km transects in Serengeti savannah (open bar) and adjacent smallholding agricultural land (solid bar). (a) There was a fourfold decline in density in agriculture for insectivores and a threefold decline in seed and fruit feeders for the same species. (b) In terms of vegetation structure, there was a 10-fold decline in tree dwellers, a threefold decline in ground dwellers, but only a 40 percent decline in shrub dwellers. Vertical lines indicate one s.e.m. (From Sinclair, A. R. E., Mduma, S. A. R., and Arcese, P. 2002. *Proceedings of the Royal Society of London B,* 267, 2401–2405, 2002. With permission.)

   The problem is not unidirectional. Just as wild alleles can move into domesticated crops, alleles from domesticated crops can move into wild populations. Individual crop plants typically contain less genetic variation than individual populations of their wild relatives (Ladizinsky 1985). The evolutionary result of continued substantial gene flow from a single cultivar to a wild population would be a decrease in genetic diversity. Some wild species might become extinct because of assimilation with the crop species. Also, the hybridized wild species may suffer from sterility, and thereby wild populations may be reduced.

   A literature review (Ellstrand and Schierenbeck 2000) found 28 well-documented examples in which interspecific hybridization preceded the evolution of new lineages that became either weeds in managed ecosystems or invasive species in unmanaged ones. Ellstrand also found that spontaneous hybridization between a given crop and at least one wild relative somewhere in the world is the rule rather then the exception (Ellstrand 2003). For the 25 most important food crops, all but 3 have some evidence for hybridization with one or more wild relatives, causing a wide array of effects.

   For instance, natural hybridization with cultivated rice has caused the near extinction of the endemic Taiwanese taxon *Oryza rufipogon* spp. *formosana* (Oka 1992). Collections of this wild rice over the last century show a shift toward characteristics of the cultivated species and a decline in fertility. Throughout Asia, typical specimens of other subspecies of *O. rufipogen* and the wild *O. nivara* are rarely found because of extensive hybridization with the crop (Chang 1995). Also, hybridization with maize may have played a role in the extinction of the populations that were maize's progenitors (Small 1984).

Although growing plants near natural areas does have many benefits, the dangers of gene flow need to be considered. Surrounding a field with plants that would interfere with the pollen can help reduce gene flow. Such plantings are often used to enhance the effect of isolation distance. Saeglitz et al. (2000) found that planting hemp around a crop was extremely effective in preventing contamination from plants outside the crop space. Forest borders for field crops, or "trap crops," might not only be beneficial in preventing gene flow, but also offer other benefits such as pest management.

11.6   GENETIC BASES OF AGRICULTURAL CROPS

Genetic diversity within each species of crop, that is, among its wild progenitors or relatives, as well as among its cultivated varieties and strains, has long been a foundation of agriculture. Traditional methods of plant breeding, based on the selection and cross-breeding or hybridization of genetically distinct strains, are still the most commonly used. They have been and continue to be used in the effort to improve crop immunity or resistance toward such yield-reducing factors as fungal diseases or insect infestations, as well as to improve crop adaptation to environmental stresses such as heat spells, dry spells, or salinity.

The preservation of genetic diversity among wild plants can best be achieved in the natural setting, within native habitats and living ecosystems. The preservation of agricultural cultivars can be accomplished in designated fields and greenhouses. Both of these are termed "*in situ* preservation."

Where such methods of living-plant preservation are not practical or sufficient, further efforts must be made to preserve the seed stocks of numerous species and varieties *ex situ*, in specially organized and carefully maintained collections. Such collections can serve as genetic pools from which plant breeders may draw genes that can impart to new varieties superior tolerance or resistance to pests, diseases, or weather anomalies. The need for improved varieties arises repeatedly, as new pests appear or as old pests themselves acquire immunity to prior modes of control.

Large seed-storage facilities (called seed banks) have been organized and are maintained by such agencies as the U.S. Department of Agriculture, the various units of the Consultative Group on International Agricultural Research, and many other national and international organizations. (The Food and Agriculture Organization of the United Nations maintains a global listing of crop varieties.) The seed banks hold large collections of "landraces" (farmers' indigenous cultivars) and wild relatives of crop species, as well as modern crop varieties and special breeding stock. They are intended to be preserved indefinitely as sources of genetic diversity for future breeding work. The recently opened "Doomsday" Seed Vault, created by the Global Crop Diversity Trust, will keep millions of food crops safely stored in the Norwegian archipelago of Svalbard in the case of losses from climate change, wars, and natural disasters. Such progress in preservation should be commended as well as encouraged.

11.7   BIODIVERSITY IN THE SOIL

Soils are among the most species-rich habitats on the planet (Brussaard et al. 1997; Wall and Virginia 2000; Wall 2004). Almost every phylum known aboveground is represented in soil, and each with a wealth of species diversity (Figure 11.4). Yet, it is estimated that few of these species, perhaps fewer than 10 percent, have been identified and described (Groombridge 1992).

Life in soils includes vertebrates (e.g., prairie dogs, gophers, lizards, pack rats), macrofauna (large invertebrates up to several centimeters long, e.g., ants, termites, millipedes, spiders, centipedes,

FIGURE 11.4   The living soil. (*National Geographic,* September 1984. With permission.)

earthworms, enchytraeids, isopods, snails), micro- and mesofauna (microscopic invertebrates less than a millimeter in length such as the tardigrades, rotifers, nematodes, and mites), as well as algae, lichens, protozoa, fungi, bacteria, and viruses (Wall Freckman et al. 1997). The abundance of these organisms is astounding. A cubic meter of a grassland soil can harbor millions of organisms—10 million nematodes, 45,000 oligochaetes, 48,000 mites and *Collembola*, and thousands to millions of microorganisms (Overgaard-Nielsen 1955).

Numerous species in soil are directly involved in ecosystem processes and ecological services that contribute to sustaining agriculture (Pankhurst and Lynch 1994; Daily et al. 1997; UNEP 2001; Wall et al. 2001; Wall 2005). These include enhancement of soil fertility through the decomposition of organic matter and the cycling of nitrogen and carbon; maintenance of soil structure and hydrological cycles through aggregation of soil particles and increased soil aeration; movement and transfer of soil organic matter and other microscopic biota throughout the soil; increase in plant community diversity and plant fitness through symbiotic, mutualistic, and parasitic associations (Hendrix et al. 1990; Bardgett et al. 2001); soil carbon sequestration and trace gas flux; and air and water purification by degrading pollutants (Coleman and Crossley 1996). Through these many connections, soil biota are an essential and intimate link to ecosystem functioning in aboveground terrestrial systems, and to freshwater and marine sediments (Wall Freckman et al. 1997; Wagener et al. 1998).

Soil biodiversity is determined by multiple factors: vegetation (chemical quality and quantity, biomass, plant species, community composition), soil physical and chemical properties, climate, and the interactions among soil organisms (Anderson 1995; Giller et al. 1997). In natural systems, these factors have been integrated over time, resulting in associations among soils, aboveground biota, and climate (Hooper et al. 2000). Disturbances affecting soils can influence ecosystem functioning, ecosystem services, and test our ability to manage soils sustainably.

EUTROPHICATION

Inappropriate agronomic management may affect biodiversity in freshwater aquatic systems associated with agricultural lands. Eutrophication of lakes, ponds, rivers, and estuaries due to excessive nutrient deposition derived from soil erosion, fertilizer residues, and animal manure can have adverse affects on aquatic ecosystems by deprivation of oxygen and by promotion of toxin-producing algal blooms. These blooms alter marine habitats through shading and over-growth and adversely affect fish and other marine organisms. Animals, including humans, who consume fish and shellfish contaminated by harmful algal bloom toxins may develop paralytic, neurotoxic, or amnesic shellfish poisoning. In the southeastern United States, the emergence of the dinoflagellate *Pfiesteria piscicida* in the 1990s has resulted in the death of tens of thousands of fish in estuarine waters and rivers. Human health effects were also reported. Discharges of swine and poultry waste into rivers have been implicated in creating the conditions for such toxic algal blooms in coastal areas. Aquaculture activities may also play a role in the outbreaks. In the past, only a few regions of the United States were affected by harmful algal blooms. Now, virtually every U.S. coastal state has reported serious blooms, which may be responsible for more than $1 billion in losses in the last two decades through direct impacts on coastal resources and communities (NOAA 2004).

Land use change is the major global change driver affecting soils. Conversion of natural systems to agriculture diminishes the diversity both of plant species as well as of microbes (Wardle et al. 1999), mycorrhizae (Thompson 1987), nematodes (Freckman and Ettema 1993; Wasilewska 1997), termites (Eggleton et al. 1997), beetles (Nestel et al. 1993), and ants (Perfecto and Snelling 1995). It can also impact nearby aquatic systems and lead to eutrophication. Land use change that increases soil compaction and texture affects the diversity and abundance of vertebrates and larger inverte-brates because their habitat is often dependent on specific soil conditions (Andersen 1987).

There is considerable evidence indicating that disturbance to the soil habitat in natural ecosystems affects soil biodiversity (Freckman and Ettema 1993; Wardle and Lavelle 1997; Wardle et al. 1998), with cascading effects on other soil properties such as decreases in water infiltration, carbon and nutrient content, oxygen levels, salinity, and erosion (Pimentel and Kounang 1998; Wolters et al. 2000). Land use change, atmospheric deposition (e.g., acid rain, nitrogen), pollution (sewage, excess fertilizer, toxic chemicals), and invasive species (plants, animals) can alter the plant species, distributional pattern, chemical quality of the plant, root abundance and architecture, soil microclimate, and resulting food base for the soil and aquatic communities (see box). For example, a sudden change in plant composition that results in nonhost plants has immediate effects on the decline in diversity of primary consumers, root pathogens (e.g., obligate parasites-fungi, bacteria, plant-parasitic nematodes), and symbionts (mycorrhizae, rhizobia) (Sasser 1972; Rovira 1994). These changes to plants can, over time, affect plant fitness and community composition.

In general, it is easier to sustain soils and prevent degradation than to try to restore the soil community and functioning of degraded soils. Efforts at soil reclamation across large scales, whether from disturbances due to intensive chemical use in conventional agriculture, agricultural forestry, fire, or pollution, have focused generally on supplying a sufficient amount of organic matter in the form of plants or plant litter as a substrate base for "reclaiming" the soil community. In all cases, the objective is not necessarily to recreate the original soil species diversity of the preaffected natural soil but to restore the functioning of the soil community (for example, to enhance vegetation growth, decrease toxicity of chemicals, and promote soil structure).

In agricultural soils, the set of practices included in the term "conservation tillage" incorporates plant organic matter residues to soils. Over time, the enhanced soil food web mimics the functions of natural systems (Hendrix et al. 1986; Freckman and Ettema 1993). Earthworms present with no-till systems, for example, generally improve water infiltration and provide channels that facilitate root penetration (Edwards and Lofty 1980), although increased nutrient leaching can occur. Soil moisture is enhanced, soil carbon is sequestered, and soil quality and structure are improved (Rovira 1994) in comparison to soils that are intensively managed through tillage.

11.8 GENETIC MODIFICATION OF FOOD SPECIES

Modern biotechnology, including the generation of genetically modified (GM) species of crops, has increased awareness of the value of biodiversity, both within and among species. The prospect of transferring useful genes to completely unrelated plants greatly enlarges the pool of genes potentially available to crop breeders (Garcia Olmedo 1998). As examples, we may cite the transfer of a gene conferring protection toward insects from bacteria to maize, cotton, and potatoes. This might serve as a further inducement to preserve the full panoply of biodiversity for the benefit of human society, in addition to the fundamental ecological and ethical reasons for biodiversity conservation. However, great care is needed in the development of such genetic engineering.

11.8.1 Opportunities and Potential Risks

Among the successes cited for biotechnology are the insertion of *Bt* genes (from strains of the insect pathogen *Bacillum thuringiensis*) into maize, potato, and cotton to impart to these crops an inherent resistance to certain insect pests. Rice has also been modified: in one case, beta-carotene was produced and in another an undesirable component for sake brewing (glutelin) was reduced (FAO/ WHO 2000). The use of GM organisms has been increasing (Figure 11.5).

However, the development of biotechnology and genetic engineering is not problem-free. Behind the hoped-for benefits lurk potential pitfalls. Like all plant-breeding techniques, there can be unexpected effects. The decrease in glutelin levels in rice, for instance, was associated with an unintended increase in levels of prolamines, which could affect nutritional quality and allergenic potential (FAO/WHO 2000). Modified organisms may escape from greenhouses and fields into natural, or quasi-natural, ecosystems, and disrupt their biodiversity. Such an invasion of "alien" species of fish has already been noticed in the context of mariculture (FAO 2000). Also possible, as discussed above, is that pollen from transgenic crops can fertilize wild relatives, and thereby transfer the transgenes outside human control.

Another insidious possibility is that large commercial corporations, under the patent laws and the protection of "intellectual property rights," will appropriate the benefits to themselves. Many consider it unfair that the culminated work of generations of scientists, researching and publishing openly and cooperatively, should now be sanctified as the commercial property of exclusive groups,

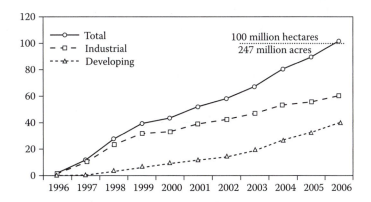

FIGURE 11.5 Area of biotechnology crops in industrial and developing countries, 1996–2006 (million hectares). (From James 2006. With permission.)

theirs to grant or withhold according to their profit interests. The concentration of vital scientific knowledge and its exclusive application to the benefit of a few enterprises should be prevented (Thompson 1998).

Apart from the general ethical question this arrangement raises, there is the specific conflict of interests between large, multinational corporations and people of the developing nations who are most in need of assistance. It is often the developing nations from whose territory the useful genetic material had been extracted in the first place. Now these nations may well find themselves unable to pay for the same genetic material after it has been put into directly useful form (i.e., incorporated into products or into new varieties of crops), and commercialized. Investments should be made at the source to stimulate the national capacity of developing countries to realize the potentialities of their own varieties and seed systems (FAO 2004).

The application of genetic transformation techniques to crop plants raises an important question relevant to the current trend toward loss of biodiversity: Does this technology offer the potential for mitigating the problem; or—contrariwise—does it pose a danger of exacerbating it? Proponents of the new technology contend that it can help to intensify production in favorable lands, thus alleviating the pressure on, and preventing the further degradation of, agriculturally marginal lands and their natural ecosystems. Opponents of the same technology fear that it can damage biodiversity in various ways, such as by permitting the greater use of pesticides and by introducing exotic genes and organisms that may disrupt natural plant communities. Other objections pertain to the exclusive commercial appropriation and exploitation of the technology, which may indeed hinder the free exchange of information and ideas that has always been the hallmark of science, to the special detriment of the poorer countries. Hence, the entire issue must be approached with discernment and caution.

Even prior to the birth of transgenic technology, the traditional plant-breeding methods that had evolved over the past 10 millennia have allowed extensive genetic alterations of the genomes of crop plants. Plant genomes, having about 25,000 genes each, have undergone numerous changes in the course of the improvement of their (both multigenic and oligogenic) agronomic traits. Classical breeding techniques are still the most effective approach to dealing with traits that are dependent on multiple genes distributed over the entire genome. On the other hand, genetic engineering appears to be useful in manipulating traits that depend on one or a few genes.

In the process of acquiring their agronomic traits, crop species ceased to be "natural" in that they lost the ability to survive by themselves in the open environment. Indeed, the genetic alterations achieved through domestication have been profound. Recombinant DNA technology, however, is opening possibilities that were out of reach by the traditional methods of breeding. Prominent among these possibilities are the production of various biodegradable polymeric compounds (plastics), oils of industrial uses, as well as pharmaceuticals.

The attainment of higher yields and the development of more environmentally sustainable practices have been and will continue to be the main challenges of agriculture. In 2025, the world's ~8 billion people will require an average world cereal yield of about 4 metric tons/ha. There will also have to be an approximate doubling of global use of synthetic nitrogen to produce the 3 billion tons of grain needed (Dyson 1999). Owing to demographic growth, available agricultural land per capita has steadily diminished in recent decades from about 0.5 ha to half that figure, and in the next 20 years it is projected to be further reduced to perhaps 0.15 ha. On a global scale, therefore, there is no option but to increase the per-hectare yields of all the main crops. The intensification of production on the most favorable lands should obviate the necessity to expand farming by further invading, and destroying, the remaining natural habitats (many of which are marginal for farming in any case).

Traditional plant breeding objectives have included the enhancement of traits that are directly related to increased yields, as well as the improvement of quality and other traits that are considered economically desirable. Most of the transgenic crop plants that have been approved for cultivation so far also address the same objectives of higher yields in the context of a "cleaner" agriculture. In effect, pest- and disease-resistant cultivars have the potential to increase yields by reducing losses (such as those due to pests) while reducing the need for various agrochemicals. For example, herbicide-resistant cultivars permit minimum tillage, a practice that in turn serves to protect the soil against erosion and to maintain the biota and fertility of the topsoil. However, the consequent increase in the use of herbicides may also disrupt biodiversity, both within and outside the target cropland. The question of whether or not the advantages outweigh the potential disadvantages requires careful case-by-case examination.

The effects of genetically modified herbicide-tolerant (GMHT) and conventional crop management on invertebrates (herbivores, detritovores, pollinators, predators, and parasitoids) have been compared in beet, maize, and spring oilseed rape sites throughout the United Kingdom. In general, the biomass of weeds was reduced under GMHT management in beet and spring oilseed rape and increased in maize compared with conventional treatments (Hawes et al. 2003). In maize, the weeds increased probably because of the higher persistence and greater efficacy of herbicides used in the conventional treatment. Herbivores, pollinators, and natural enemies changed in abundance in the same directions as their resources, and detritivores increased in abundance under GMHT management across all crops. The experiment suggests that the impact of GMHT cropping on invertebrate biodiversity acts primarily through changes in weed flora (Hawes et al. 2003).

In another study of the same crops, numbers of butterflies in beet and spring oilseed rape and of *Heteroptera* and bees in beet were small under the relevant GMHT crop management, whereas the abundance of Collembola—soil and litter dwellers—was consistently greater in all GMHT crops (Haughton et al. 2003). It is worth noting that experimental evidence from several countries has also implied that invertebrate populations could be affected in GMHT crops through reduced biomass and diversity of weeds. The most consistent effects appeared through the timing of the application of herbicides. The experiments with GMHT sugar and fodder beet in Europe showed that leaving weeds to be controlled later favored a range of invertebrates, including natural enemies of crop pests (Squire 2003).

Some biotechnology applications may not respond to the main challenges while addressing specific sectoral demands, from alterations of postharvest properties for food processing (such as delayed-ripening tomatoes) to those that meet certain nutritional requirements (such as the enhancement of provitamin A and iron in rice). The relevance of such applications must be weighed in relation to their specific merits or shortcomings in each case.

Transgenic technology has found an increasing number of applications, notwithstanding the objections and warnings of its critics. Its proponents claim that it has the potential to reconcile the needs of a growing population with the goal of conserving and promoting biodiversity. Over 40 million ha of transgenic crops are already grown in over a dozen countries. New techniques allow the development of transgenic plants whose purpose is to serve in the screening of new generations of agrochemicals that are intended to meet the requirements of higher specific activity (hence decreased application amounts per hectare), greater selectivity (so that they may affect only the target organism to be controlled), and higher biodegradability (so that they may not accumulate and persist in the environment).

Plant biotechnology also holds promise in the creation of medical products such as vaccines. In Africa, for instance, tobacco plants are being used to develop an affordable vaccine against the virus that causes cervical cancer. Making vaccines in plants may cut costs by orders of magnitude because one does not need a fermentation plant for yeast or bacteria, or a culture facility for human or animal tissue (Mthembu 2004). The use of transgenic plants to produce vaccines could translate into easier access, cheaper production, and alternative income generation (Royal Society of London 2000).

11.8.2 Biosafety Issues

The safety of biotechnology is a complex issue for which there is no simple answer, as it depends on the specific nature of the induced change and its actual performance in the environmental context.

This issue must therefore be treated on a case-by-case or application-by-application basis. In principle, however, no human endeavor is entirely devoid of risk, and that risk increases whenever humans act without knowing or fully understanding the potential consequences.

As in the approval of any technological innovation, the application of each proposal for transgenic application must be evaluated on the basis of benefits versus risks. Those risks must be monitored closely and continuously, retaining the possibility that the approval may be rescinded if the expectation derived from initial assessment is not borne out.

The concept of substantial equivalence, which implies a direct comparison of the proposed innovation with that currently in use, may offer a rational approach to the problem. Both intended and unintended effects must be monitored and assessed. Intended effects can be generally evaluated by well-established standard protocols, which are not restricted to transgenic products. In addition, new tools—high-throughput genomic proteomic and metabolomic methods—have been developed that allow the investigation of unintended effects. Those effects of genetic alteration, predictable or unpredictable, must be evaluated by methods independent of those used to achieve the respective alterations (whether by classical breeding or by genetic engineering).

Safety considerations concerning possible effects of transgenic products on human subjects should address possible toxic effects and potential allergenic effects. Standard toxicity tests can be performed with transgenic material, using the isogenic nontransgenic material as a control. Additionally, toxicity of new products differentially present in the transgenic line, whether intended or unintended, can be individually assessed as required.

Allergy problems evidently affect an increasing number of urban dwellers, for reasons that are not entirely known. A considerable number of items present in daily life—from rubber to peanuts—can cause allergic reactions in sensitive individuals. Molecules (allergens) or their parts (epitopes) that cause such reactions have been identified. Under the current legal framework, no gene encoding a known allergen may be expressed transgenically for commercial use and no gene from a particularly allergenic species may be transferred unless there is positive evidence that the gene product is not responsible for the observed allergy.

Once a gene has been transferred to a particular species, it becomes an integral part of the genetic makeup of the recipient species and will have the same fate as the remaining thousands of genes that constitute the genome. Interspecific genetic fluxes occur in nature to a very limited extent and are subject to barriers that are breached only rarely. A pollen grain must fly a certain distance, find an appropriate and mature recipient, pollinate, yield a viable seed capable of developing into a nonsterile mature plant, and finally, the progeny of this plant must be viable. For certain transformation events, accidental gene transfer to other species is highly unlikely, whereas for others there is a certain nonnegligible probability of occurrence. These cases should be subjected to experimental evaluation.

There is generally no danger of plant gene transfers into the genomes of humans or other animals. Humans and other animals species have daily consumed thousands of such genes, yet no evidence has been found in animal genomes of this type of horizontal transfer. The use of antibiotic resistance as a marker for selection in GM plants for human use, though, has also resulted in a fear that these genes may be transferred to those bacteria that cause disease in humans. There should be further research into alternatives to antibiotic resistance marker genes, as having antibiotic resistance genes present in new GM crops under development for potential foods is no longer acceptable (Royal Society of London 1998).

The transfer of genes from cultivated species to related wild species in the same habitat occurs with a probability that is different for each species–habitat combination. Whether the species is autogamous or allogamous, and whether or not there are closely related wild species in the vicinity of the cultivated one, determine the relevance of this issue in the approval procedure. Several million hectares of transgenic canola are currently cultivated in the North American continent, after initial approval, and the entire operation is under close monitoring. The results of these observations will be important to the future of this transgenic crop.

THE MONARCH BUTTERFLY AND *Bt* MAIZE

Concerns regarding the nontarget effects of transgenic crops containing transgenes from the organism *Bacillus thuringiensis* (*Bt*) arose after the publication by Losey et al. (1999) on the potential risk of maize pollen expressing the lepidopteron-active Cry protein to the monarch butterfly, *Danaus plexippus* L. A collaborative research effort by scientists in several states and in Canada, though, found that after 2 years of study that the impact of *Bt* maize pollen from commercial hybrids on monarch butterflies was negligible (Sears et al. 2001). The only transgenic maize that consistently affected monarchs was from a hybrid that is currently <2 percent of the maize planted in the United States and for which reregistration has not been applied (Hellmich et al. 2001). However, the *Bt* supportive studies assume that monarchs consume only pollen and not other maize tissues. Obrycki et al (2001) pointed out that the presence of anthers on milkweeds is of considerable importance because of the higher concentrations of *Bt* toxins that they contain. Besides their earlier research (Losey et al. 1999) that showed that pollen/anther mixtures were deleterious to monarch larvae, 2001 field observations and experimental evidence suggest that monarchs may be exposed to more maize anther material than previously assumed.

Induction of resistance in any target organism is to be expected sooner or later, whether it involves insecticides, antibiotics, or other agents. This induction can be either accelerated by malpractice or retarded by judicious management. In the case of transgenic plants that are resistant to insects, microbial diseases, or weeds, the implementation of proper management techniques (such as the provision of refuges, adequate doses, etc.) should be in the interest of all parties involved (farmers, authorities, and commercial enterprises).

Gene transfers to nontransgenic cultivars of the same species require only appropriate synchrony and are limited by distance. When GM imported maize was found growing in Oaxaca and the Tehuacán Valley, Mexico (2000–2002), concern arose about how GM might be affecting the diversity of maize at its center of origin. Also, this area contains many "landraces," which are "crop populations that have become adapted to farmers' conditions through natural and artificial selection" (Fitting 2006). Therefore, as GM maize makes its way into rural markets, it is important to store *ex situ* not only varieties, but also landraces.

The environmental safety of transgenic crops involves two additional aspects that require special attention. These are the possible effects on nontarget organisms and the possible induction of resistance in the target organisms. When genetic resistance to a particular target organism (bacteria, fungi, insects, and others) is introduced into a given cultivar, possible short-term and long-term effects on nontarget organisms must be evaluated. This evaluation should consider current practice as the control. Thus, for example, the effect of the introduction of *Bt* cotton on beneficial insects should be compared with the effect of the repeated insecticidal treatments that accompany the cultivation of nontransgenic cultivars. Transgenic crops that have been approved so far for plant protection are claimed to be more environmentally friendly than current technologies (which generally involve the use of pesticides) (Garcia Olmedo 1998).

11.9   CLIMATE CHANGE AND AGRICULTURAL BIODIVERSITY

The climatic consequences of increasing greenhouse gases are likely to include far-reaching changes in agriculture, as well as in natural ecosystems (Rosenzweig and Hillel 1998; IPCC 2001). Climate change will affect the regional patterns of temperature, precipitation, and evaporation, indeed the entire array of meteorological, hydrological, ecological, and agricultural relationships. Agricultural biodiversity will be affected by the magnitude and rate of climate change, and by its geographical and seasonal patterns. The functioning and productivity of agroecosystems in different countries

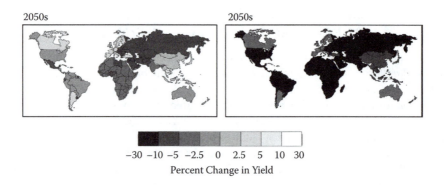

2050s   2050s

−30 −10 −5 −2.5 0 2.5 5 10 30
Percent Change in Yield

FIGURE 11.6   Potential changes (%) in national cereal yields for the 2050s (compared with the 1990s) under the HadCM3 SRES A1F1 climate change scenario with (left) and without (right) CO_2 effects. (From Parry et al., 2004. With permission.)

and regions will be altered (Figure 11.6). Some regions may benefit, while others suffer. Thus, threats to agricultural biodiversity are among the most serious of potential damages resulting from a change of climate.

The projected climate effects associated with increases in anthropogenic greenhouse gas emissions (including warmer temperatures, changed hydrological regimes, and altered frequencies and intensities of extreme climatic events) may inhibit crop production in some regions. Differential plant responses to higher concentrations of atmospheric CO_2 (Kimball et al. 2002) may also contribute to changes in biodiversity. Agricultural pests, overall, are likely to thrive under conditions of increasing atmospheric CO_2 concentrations and rapid climate change. A host of interactive changes in crop growth flow from these primary effects, some resulting in positive feedbacks and others in negative ones. All these changes, in concert, could have major impacts on the prospects for long-term food security.

11.9.1   SHIFTS IN AGROECOLOGICAL ZONES

As new areas become suitable for crop production while old agricultural areas become less so, the geographic shift of agriculture may encroach on natural ecosystems. This is more likely to occur in high-latitude and high-altitude regions as warming temperature prolongs growing periods. It may also occur in lower latitudes due to changes in hydrological regimes. Biodiversity in the affected ecosystems may be compromised.

Even apart from agriculture, climate change is likely to modify the zonation and bioproductivity of forests, grasslands, savannas, wetlands, tundras, and other biomes. A warmer regime might disrupt the prior adaptation of native plants and animals to their existing habitats. The flooding and waterlogging of some areas and the aridification of others could weaken currently vigorous biotic communities. For example, the thawing of permafrost could dry out tundras, just as the invasion of seawater can destroy freshwater wetlands (estuaries, deltas, marshes, lagoons) near coastlines.

The rate of climate change may be too rapid to allow some natural communities to adjust, and where evolving climate becomes increasingly unfavorable there could be a large-scale dieback of forests. Associated species that depend on these forest ecosystems may then be threatened with extinction. In this manner, climate change constitutes a threat to biodiversity in general, and to the survival of vulnerable or endangered species in particular. Conversely, some types of forests and other biomes may expand and become more vigorous as a consequence of the warming trend and enhance photosynthesis, demonstrating beneficial effects as well.

11.9.2 Pests and Climate Change

Climate affects not just agricultural crops but their associated pests as well. The major pests of crops include weeds, insects, and pathogens. The distribution and proliferation of weeds, fungi, and insects is determined to a large extent by climate (Rosenzweig et al. 2002). Organisms become pests when they compete with or prey upon crop plants to an extent that reduces productivity. Not only does climate affect the type of crops grown and the intensity of the pest problems, it affects the pesticides often used to control or prevent outbreaks (Coakley et al. 1999; Chen and McCarl 2001). The intensity of rainfall and its timing with respect to pesticide application are important factors in pesticide persistence and transport.

Because of the extremely large variation of pest species' responses to meteorological conditions, it is difficult to draw overarching conclusions about the relationships between pests and weather. In general, however, most pest species are favored by warm and humid conditions. But crop damages by pests are a consequence of the complex ecological dynamics between two or more organisms and therefore are very difficult to predict. For example, dry conditions are unfavorable for sporulation of fungi, but are also unfavorable for the crop; a weak crop during a drought is more likely to become infected by fungi than when it is not stressed. Pest infestations often coincide with changes in climatic conditions—such as early or late rains, drought, or increases in humidity—which in themselves can reduce yields. In these circumstances, accurately attributing losses to pests can be difficult.

Most analyses concur that in a changing climate, pests may become even more active than they are currently, thus posing the threat of greater economic losses to farmers (Rosenzweig et al. 2002). While the majority are invasive species from temperate zones, many of the worst weeds in temperate regions originated in tropical or subtropical regions, and in the current climate their distribution is limited by low temperature. Such geographical constraints will be removed under warm conditions. Warmer temperature regimes have been shown to increase the maximum biomass of grass weeds significantly. In crop monocultures, undesirable competition is controlled through a variety of means, including crop rotations, mechanical manipulations (e.g., hoeing), and chemical treatment (e.g., herbicides), which might increase in use under climate change.

With temperatures within their viable range, insects respond to higher temperature with increased rates of development and with less time between generations. (Very high temperatures reduce insect longevity.) Warmer winters will reduce winterkill, and consequently there may be increased insect populations in subsequent growing seasons. With warmer temperatures occurring earlier in the spring, pest populations will become established and thrive during earlier and more vulnerable crop growth stages. Additional insect generations and greater populations encouraged by higher temperatures and longer growing seasons will require greater efforts of pest management.

Warmer winter temperature will also affect those pests that currently cannot overwinter in high-latitude crop regions but do overwinter in lower-latitude regions and then migrate to the crops in the following spring and summer. Because warmer temperature will bring longer growing seasons in temperate regions, this should provide opportunity for increased insect damage. A longer growth period may allow additional generations of some insect pests and higher insect populations. Drought stress tends to bring increased insect pest outbreaks; insect damage may increase in regions destined to become more arid. If climate becomes warmer and drier as well, the population growth rates of small, sap-feeding pests may be favored.

Higher temperature and humidity and greater precipitation, on the other hand, are likely to result in the spread of some plant diseases, as wet vegetation promotes the germination of spores and the proliferation of bacteria and fungi, and influences the lifecycle of soil nematodes. In regions that suffer greater aridity, however, disease infestation may lessen, although some diseases (such as the powdery mildews) can thrive even in hot, dry conditions as long as there is dew formation at night.

11.9.3 Climate Change and Biotechnology

The prospect of a changing climate, caused by augmented atmospheric constituents, may provide motivation for the use of biotechnology (Rosenzweig 2001). There may be opportunities for optimizing photosynthetic and stomatal conductance responses to higher levels of atmospheric carbon dioxide. Biotechnology techniques may offer the potential for creating effective adaptations to changing climatic circumstances. Enhanced heat and drought tolerance of both crops and livestock are likely to be required, as are strategies to cope with shifting and newly emerging weeds, pests, and plant diseases. Finally, improved mitigation options could also be developed in regard to the ability of crops to sequester carbon, production of biofuels, reduction of methane emissions from rice-growing and ruminant livestock systems, and management of nitrous oxide emissions from nitrogen fertilization.

Several cautions are in order. GM organisms may not be able to cope with all of the effects of dynamic climate changes that occur in agricultural regions. For example, severe flooding may continue to be detrimental to crop production, regardless of genetic resources. Dissemination of new and severe crop pests may be so rapid as to bring large damages before development of appropriately modified crops. Finally, as emphasized in the previous section, much research and testing of GM crops is required, in any case, so that potential benefits and harms are more clearly understood.

11.10   CONSERVING BIODIVERSITY AND SUSTAINING FOOD PRODUCTION

Instead of the often-favored reductionist approach, which treats the production of food livestock without regard to ecological relationships, new agroecosystem approaches strive to integrate farming and food production units into the larger environmental domain, recognizing and preserving the role of native fauna and flora in their natural habitats. A more holistic approach to the integration of farming and ecology will better promote nutrient recycling, biological pest and disease control, pollination, soil quality maintenance, water-use efficiency, and carbon sequestration. Appropriate responses to weather anomalies (droughts and floods) and to off-site (along with on-site) effects of agricultural activities will also be engendered.

Differences among genomes can be of great value to agriculture, and the wider the spectrum of those differences, the greater their potential uses. Therefore, every effort must be made to preserve the full variety of genetic differences among species, as each species plays an ecological role in its own niche or habitat, and interacts with all other forms of life sharing the same community or ecosystem.

Conservation of genetic resources is thus a keystone of the agroecosystem approach. It ensures the broadest array of agricultural species and the myriad biota, such as soil organisms and pollinators, that provide services that enable food production and harvesting. Facilities such as germplasm banks need to be expanded and improved so that genetic resources for both crops and livestock are preserved. Germplasm collections should include the widest possible array of varieties and breeds, as well as their wild relatives. Both *in situ* and *ex situ* collections should be protected. All collections should be registered in a common, accessible database for the benefit of breeders and farmers everywhere. The various international agencies (with coordination and networking among such groups as the CGIAR, World Bank, USAID, NGOs, United Nations FAO, UNEP, and UNESCO), as well as national agricultural agencies, should be included. The rights of developing countries to their indigenous genetic resources should be respected and not be appropriated by outside commercial interests. Access to such resources should be freely available and fairness in rights assured.

Expanded knowledge of soil biota is also important because they are linked to critical ecosystem processes that sustain life. Research needs must include the role of soil biodiversity in plant health and ecosystem processes and as linkages with other terrestrial and aquatic systems if we are to understand how to manage ecosystems for food production sustainably for the long term. Globally, soil degradation has accelerated as human populations have expanded, threatening the stability of Earth's ecosystems, both natural and managed. Determining how soil species diversity will change under disturbances, such as with increasing land use, will help scientists, policymakers, and managers devise and implement strategies to preserve and maintain our terrestrial ecosystems and our food production base for the long term.

The enhanced greenhouse effect is expected to result in significant global warming during the course of this century. The potential impacts of climate change and climate variability on biodiversity need to more fully characterized, as both agricultural and natural ecosystems will thereby be affected. The zonation and adaptation of species will shift as the temperature and hydrological regimes change. Improved methods of assessing biodiversity in relation to climate change need to be developed.

Finally, the formulation and implementation of biodiversity policies is also a priority. National and international policies are needed to encourage the adoption of the agroecosystem paradigm on a wide scale and thus the conservation of biodiversity in food-producing systems. This will ensure nutritious food for the still-growing population, minimize exposure to agricultural chemicals, and promote both human and ecosystem health in an integrated way.

ACKNOWLEDGMENTS

We gratefully acknowledge our research assistant, Peter Neofotis, for his help in preparing this chapter. An earlier version appeared in *Advances in Agronomy*, 88, 2005.

REFERENCES

Albrecht, M. et al. 2007. The Swiss agri-environment scheme enhances pollinator diversity and plant reproductive success in nearby intensively managed farmland. *Journal of Applied Ecology*, 44, 813–822.

Ampofo, J. K. O. 1986. Maize stalk borer (Lepidoptera: Pyralidae) damage and plant resistance. *Environmental Entomology*, 15, 1124–1129.

Anderson, D. C. 1987. Below-ground herbivory in natural communities: A review emphasizing fossorial animals. *Quarterly Review of Biology*, 62, 261–286.

Anderson, J. 1995. The soil system. In *Global Biodiversity Assessment*, Haywood, V., Ed. Cambridge University Press, Cambridge, MA.

Arnold, M. J. 2004. Natural hybridization and the evolution of domesticated, pest and disease organisms. *Molecular Ecology*, 13, 997–1007.

Baker, B. et al. 1997. Signaling in plant–microbe interactions. *Science*, 276, 726–733.

Bardgett, R. D. et al. 2001. The influence of soil biodiversity on hydrological pathways and the transfer of materials between terrestrial and aquatic ecosystems. *Ecosystems*, 4, 421–429.

Brussaard, L. et al. 1997. Biodiversity and ecosystem functioning in soil. *Ambio*, 26, 563–570.

Buchmann, S. L. and Nabhan, G. P. 1996. *The Forgotten Pollinators*. Island Press, Washington, D.C.

Caldararo, N. 2002. Human ecological intervention and the role of forest fires in human ecology. *Science of the Total Environment*, 292, 141–165.

Chang, T. T. 1995. Rice. In *Evolution of Crop Plants,* 2nd ed., Smartt, J. and Simmonds, N. W., Eds. Longman, Harlow, U.K., 147–155.

Chen, C-C. and McCarl, B. A. 2001. An investigation of the relationship between pesticide usage and climate change. *Climatic Change*, 50, 475–487.

Coakley, S. M., Scherm, H., and Chakraborty, S. 1999. Climate change and plant disease management. *Annual Review of Phytopathology*, 37, 399–426.

Cohen, J. E. 2003. Human population: the next half century. *Science*, 302, 1172–1175.

Coleman, D. C. and Crossley, D. A. J. 1996. *Fundamentals of Soil Ecology*. Academic Press. Burlington, MA.

Corbet, S. A., Williams, I. H., and Osborne, J. L. 1991. Bees and the pollination of crops and wild flowers in the European Community. *Bee World*, 72, 47–59.

Daily, G. C., Matson, P. A., and Vitousek, P. M. 1997. Ecosystem services supplied by soil. In *Nature's Services: Societal Dependence on Natural Ecosystems*, Daily, G. C., Ed. Island Press, Washington, D.C., 113–132.

Donald, P. F., Green, R. E., and Heath, M. F. 2001. Agricultural intensification and the collapse of Europe's farmland bird populations. *Proceedings of the Royal Society of London B,* 268, 25–29.

Dyson, T. 1999. World food trends and prospects in 2025. *Proceedings of the National Academy of Sciences,* 96, 5929–5936.

Edwards, C. A. and Lofty, J. R. 1980. Effects of earthworm inoculation upon the root growth of direct drilled cereals. *Journal of Applied Ecology*, 17, 533–543.

Edwards, C. A. et al. 1993. The role of agroecology and integrated farming systems in agricultural sustainability. *Agriculture, Ecosystems, and Environment*, 46, 99–121.

Eggleton, P. et al. 1997. The species richness and composition of termites (Isoptera) in primary and regenerating lowland dipterocarp forest in Sabah, East Malaysia. *Ecotropica*, 3, 119–128.

Ellstrand, N. C. 2003. *Dangerous Liaisons? When Cultivated Plants Mate with Their Wild Relatives.* Johns Hopkins University Press, Baltimore, MD.

Ellstrand, N. C. and Schierenbeck, K. 2000. Hybridization as a stimulus for the evolution of invasiveness in plants? *Proceedings of the National Academy of Sciences*, 97, 7043–7050.

FAO. 2000. World review of fisheries and aquaculture: fisheries resources: trends in production, utilization, and trade. http://www.fao.org/DOCREP/003/X8002E/x8002e04.htm#TopOfPag (accessed 5/23/08).

FAO. 2004. Agricultural biotechnology; meeting the needs of the poor? The state of food and agriculture 2003–2004. Food and Agricultural Organization of the United Nations, Rome.

FAO/WHO. 2000. Safety aspects of genetically modified foods of plant origin. Report. June 2000. http://www.who.int/foodsafety/publications/biotech/en/ec_june2000_en.pdf.

Fitting, E. 2006. Importing corn, exporting labor: The neoliberal corn regime, GMOs, and the erosion of Mexican biodiversity. *Agriculture and Human Values*, 23, 15–26.

Freckman, D. W. and Ettema, C. H. 1993. Assessing nematode communities in agroecosystems of varying human intervention. *Agriculture, Ecosystems, and Environment*, 45, 239–261.

Garcia Olmedo, F. 1998. Tercera revolucion verde: plantas con Luz Propia. Editorial debate, Madrid, 220.

Genghini, M., Gellini, S., and Gustin, M. 2006. Organic and integrated agriculture: the effects on bird communities in orchard farms in northern Italy. *Biodiversity and Conservation*, 15, 3077–2094.

Giller, K. E. et al. 1997. Agricultural intensification, soil biodiversity, and agroecosystem function. *Applied Soil Ecology*, 6, 3–16.

Groombridge, B., Ed. 1992. *Global Biodiversity: Status of the Earth's Living Resources. World Conservation Monitoring Center.* Chapman & Hall, London.

Hartel, P. G. 2005. Microbial processes: environmental factors. In *Encyclopedia of Soils in the Environment*, Vol. 2, Hillel, D., Ed. Elsevier, Oxford, 448–455.

Haughton, A. J. et al. 2003. Invertebrate responses to the management of genetically modified herbicide-tolerant and conventional spring crops. II. Within-field epigeal and aerial arthropods. *Philosophical Transactions of the Royal Society of London B*, 358, 1863–1877.

Hawes, C. et al. 2003. Responses of plants and invertebrate trophic groups to contrasting herbicide regimes in the farm scale evaluations of genetically modified herbicide-tolerant crops. *Philosophical Transactions of the Royal Society of London B*, 358, 1899–1913.

Hellmich, R. L. et al. 2001. Monarch larvae sensitivity to *Bacillus thuringiensis*–purified proteins and pollen. *Proceedings of the National Academy of Sciences*, 98(21), 11925–11930.

Hendrix, P. et al. 1986. Detritus food webs in conventional and no-tillage agroecosystems. *BioScience*, 36, 374–380.

Hendrix, P. F. et al. 1990. Soil biota as components of sustainable agroecosystems. In *Sustainable Agricultural Systems*. Soil and Water Conservation Society, Ankeny, IA, 637–654.

Hillel, D. 1991. *Out of the Earth; Civilization and the Life of the Soul.* University of California Press, Berkeley.

Hillel, D. 1998. *Salinity Management for Sustainable Irrigation.* World Bank, Washington, D.C.

Hillel, D. et al. 2002. The role of biodiversity in world food production. Working Group 6. In Biodiversity: Its Importance to Human Health, Chivian, E, Ed. Report issued by the Center for Health and the Global Environment, Harvard Medical School, Boston.

Hooper, D. U. et al. 2000. Interactions between aboveground and belowground biodiversity in terrestrial ecosystems: Patterns, mechanisms, and feedbacks. *BioScience*, 50, 1049–1061.

IPCC. 2001. *Climate Change 2001: Impacts, Adaptation, and Vulnerability.* Intergovernmental Panel on Climate Change. Cambridge University Press, Cambridge, 1032.

James, C. 2005. Global Status of Commercialized Biotech/GM Crops:2005. ISAA Briefs No. 34. ISAAA, Ithaca, NY.

Jenkins, M. 2003. Prospects for biodiversity. *Science*, 302, 1175–1177.

Kearns, C. A. et al. 1998. Endangered mutualisms: The conservation of plant-pollinator interactions. *Annual Review of Ecology and Systematics,* 29, 83–112.

Kimball, B., Kobayashi, K., and Bindi, M. 2002. Responses of agricultural crops to free-air CO_2 enrichment. *Advances in Agronomy*, 77, 293–368.

Kislev, M. E., Weiss, E., and Hartmann, A. 2004. Impetus for sowing and the beginning of agriculture: ground collecting of wild cereals. *Proceedings of the National Academy of Sciences of the United States of America*, 101(9) 2692–2695.

Ladizinsky, G. 1985. Founder effect in crop-plant evolution. *Economic Botany*, 39, 191–199.

Ladizinsky, G. 1989. Ecological and genetic considerations in collecting and using wild relatives. In *The Use of Plant Genetic Resources,* Brown, A. H. D. et al., Eds. Cambridge University Press, Cambridge.

Lande, R. 1988. Genetic and demography in biological conservation. *Science*, 241, 1455–1459.

Larsen, C. S. 2003. Animal source foods and human health during evolution. *Journal of Nutrition*, 133, 11S-II.

Leach, H. M. 2003. Human domestication reconsidered. *Current Anthropology*, 44(3), 349–368.

Losey, J. E., Rayor, L. S., Carter, M. E. 1999. Transgenic pollen harms monarch larvae. *Nature,* 399, 214–215.

Mänd, M., Mänd, R., and Williams, I. H. 2002. Bumblebees in the agricultural landscape of Estonia. *Agriculture, Ecosystems, and Environment*, 89, 69–70.

McMullen, M., Jones, R., and Gallenberg, D. 1997. Scab of wheat and barley: A reemerging disease of devastating impact. *Plant Disease*, 81(12), 1340–1348.

Mokyr, J. 2004. Irish potato famine. *Encyclopædia Britannica Online* (accessed on 5/28/08).

Mthembu, N. 2004. Genetically altered tobacco fights cervical cancer. *Science in Africa*, March.

Nestel, D., Dickschen, F., and Altieri, M. A. 1993. Diversity patterns of soil macro-Coleoptera in Mexican shaded and unshaded coffee agroecosystems—An indication of habitat perturbation. *Biodiversity and Conservation*, 2, 70–78.

NOAA. 2004. http://www.hab.nos.noaa.gov/habfacts.html.

Nyffeler, M. and Sunderland, K. D. 2003. Composition, abundance, and pest control potential of spider communities in agroecosystems: A comparison of European and U.S. studies. *Agriculture, Ecosystems, and Environment*, 95, 579–612.

Obrycki, J. J. et al. 2001. Registration of *Bt* crops. *Public Information and Records Integrity Branch*, 11 September.

Ogol, C. K. P. O., Spence, J. R., and Keddie, A. 1999. Maize stem borer colonization, establishment, and crop damage levels in a maize–leucaena agroforestry system in Kenya. *Agriculture, Ecosystems, and Environment*, 76, 1–15.

Oka, H. I. 1992. Ecology of wild rice planted in Taiwan: II. Comparison of two populations with different genotypes. *Botanical Bulletin of Academia Sinica*, 33, 75–84.

Ou, S. H. 1980. Pathogenicity and host plant resistance in rice blast disease. *Annual Review of Phytopathology*, 18, 167–187.

Overgaard-Nielsen, C. 1955. Studies on enchytraeidae 2: Field studies. *Natura Jutlandica*, 4, 5–58.

Pankhurst, C. E. and Lynch, J. M. 1994. The role of the soil biota in sustainable agriculture. In *Soil Biota Management in Sustainable Farming Systems*, Pankhurst, C. E., et al., Eds. CSIRO, Melbourne.

Parry, M. L., Rosenzweig, C., Iglesias, A., Livermore, M., and Fischer, G. 2004. Effects of climate change on global food production under SRES emissions and socio-economic scenarios. *Global Environmental Change-Human and Policy Dimensions,* 14(1): 53–67.

Perfecto, V. and Snelling, R. 1995. Biodiversity and the transformation of a tropical agroecosystem: Ants in coffee plantations. *Ecological Applications*, 5, 1084–1097.

Pimentel, D. 1997. *Techniques for Reducing Pesticide Use. Economic and Environmental Benefits*. J. Wiley & Sons, New York.

Pimentel, D. and Kounang, N. 1998. Ecology of soil erosion in ecosystems. *Ecosystems*, 1, 416–426.

Priess, J. A. et al. 2007. Linking deforestation scenarios to pollination services and economic returns in coffee agroforestry systems. *Ecological Applications*, 17(2), 407–417.

Reddi, E. U. B. 1987. Under pollination a major constraint on cashewnut production. *Proceedings of the Indian Academy of Sciences*, B53, 249–252.

Richards, A. J. 2001. Does low biodiversity resulting from modern agricultural practice affect crop pollination and yield? *Annals of Botany*, 88, 165–172.

Risch, S. J., Andow, D., and Altieri, M. A., 1983. Agroecosystem diversity and pest control: Data, tentative conclusions, and new research directions. *Environmental Entomology*, 12, 625–629.

Root, R. B. 1973. Organization of plant–arthropod association in simple and diverse habitats: The fauna of collards (*Brassicae oleveceae*). *Ecological Monographs*, 43, 95–124.

Rosenzweig, C. 2001. Climate change and agriculture. In The Biobased Economy of the Twenty-First Century: Agriculture Expanding into Health, Energy, Chemicals, and Materials, Eaglesham, A. et al., Eds. National Agricultural Biotechnology Council Report 12. Boyce Thompson Institute, Ithaca, NY.

Rosenzweig, C. and Hillel, D. 1998. *Climate Change and the Global Harvest: Potential Impacts of the Greenhouse Effect on Agriculture.* Oxford University Press, New York, 324.

Rosenzweig, C. et al. 2002. Climate change and extreme weather events: Implications for food production, plant diseases, and pests. *Global Change and Human Health*, 2(2), 90–104.

Rovira, A. D. 1994. The effect of farming practices on the soil biota. In *Soil Biota: Management in Sustainable Farming Systems*, Pankhurst, C. E. et al., Eds. CSIRO, Melbourne, 81–87.

Royal Society of London. 1998. Genetically modified plants for food use. Statement. September.

Royal Society of London, U.S. National Academy of Sciences, the Brazilian Academy of Sciences, et al. 2000. Transgenic plants and world agriculture. *National Academy Press*, 1–40.

Saeglitz, C., Pohl, M., and Bartsch, D. 2000. Monitoring gene flow from transgenic sugar beet using cytoplasmic male-sterile bait plants. *Molecular Ecology*, 9, 2035–2040.

Sasser, J. N. 1972. Managing nematodes by plant breeding. *Proceedings Annual Tall Timbers Conference on Ecological Animal Control by Habitat Management.* February 24–25, 65–80.

Sears, M. K. et al. 2001. Impact of *Bt* corn pollen on monarch butterfly populations: A risk assessment. *Proceedings of the National Academy of Sciences*, 98(21), 11937–11942.

Sinclair, A. R. E., Mduma, S. A. R., and Arcese, P. 2002. Protected areas as biodiversity benchmarks for human impacts: Agriculture and the Serengeti avifauna. *Proceedings of the Royal Society of London B,* 267, 2401–2405.

Small, E. 1984. Hybridization in the domesticated-weed-wild complex. In *Plant Biosystematics,* Grant, W. F., Ed. Academic Press, Toronto, 195–210.

Smith, B. D. 1995. *The Emergence of Agriculture.* Scientific American Library, New York.

Smith, N. J. et al. 1995. Agroforestry developments and potential in the Brazilian Amazon. *Land Degradation & Rehabilitation*, 6, 251–263.

Squire, G. R. et al. 2003. On the rationale and interpretation of the farm scale evaluations of genetically modified herbicide-tolerant crops. *Philosophical Transactions of the Royal Society of London B,* 358, 1779–1799.

Staskawicz, B. J. et al. 1995. Molecular genetics of plant disease resistance. *Science*, 268, 661–667.

Steadman, D. W. 1995. Prehistoric extinctions of Pacific Island birds: Biodiversity meets zooarchaeology. *Science*, 267, 1123.

Steffan-Dewenter, I. and Tscharntke, T. 1999. Effects of habitat isolation on pollinator communities and seed set. *Oecologia*, 121, 432–440.

Thompson, J. P. 1987. Decline of vesicular-arbuscular mycorrhizas in long fallow disorder of field crops and its expression in phosphorus deficiency of sunflower. *Australian Journal of Agricultural Research*, 38, 847–867.

Thompson, P. B. 1998. *Agricultural Ethics: Research, Teaching, and Public Policy.* Iowa State University Press, Ames.

Tonhasca, Jr., A. and Byrne, D. N. 1994. The effects of crop diversification of herbivourous insects: A meta-analysis approach. *Ecological Entomology*, 19, 239–244.

UNEP. 2001. Soil biodiversity and sustainable agriculture. Convention of Biological Diversity Montreal, 12–16 November.

United Nations Populations Division. 2003. World Population Prospects; The 2002 Revision. United Nations ESA/P/WP.180.

Vandermeer, J. 1997. Biodiversity loss in and around agroecosystems. In *Biodiversity and Human Health,* Grifo, F. and Rosenthal, J., Eds. Island Press, Washington, D.C.

Wagener, S. M., Oswood, M. W., and Schimel, J. P. 1998. Rivers and soils: parallels in carbon and nutrient processing. *BioScience*, 48, 104–108.

Wall Freckman, D. et al. 1997. Linking biodiversity and ecosystem functioning of soils and sediments. *Ambio,* 26, 556–662.

Wall, D. H., Ed. 2004. *Sustaining Biodiversity and Ecosystem Services in Soils and Sediments.* Island Press, Washington, D.C.

Wall, D. H. 2005. Biodiversity. In *Encyclopedia of Soils in the Environment,* Vol. 1, Hillel, D., Ed. Elsevier, Oxford, 136–141.

Wall, D. H. and Virginia, R. A. 2000. The world beneath our feet: Soil biodiversity and ecosystem functioning. In *Nature and Human Society: The Quest for a Sustainable World*, Raven, P. R. and Williams, T., Eds. National Academy of Sciences and National Research Council, Washington, D.C., 225–241.

Wardle, D. A. and Lavelle, P. 1997. Linkages between soil biota, plant litter quality, and decomposition. In *Driven by Nature—Plant Litter Quality and Decomposition*, Cadisch, G. and Giller, K. E., Eds. CAB International, Wallingford, 107–124.

Wardle, D. A., Verhoef, H. A., and Clarholm, M. 1998. Trophic relationships in the soil microfood-web: Predicting the responses to a changing global environment. *Global Change Biology,* 4, 713–727.

Wardle, D. A., Giller, K. E., and Barker, G. M. 1999. The regulation and functional significance of soil biodiversity in agroecosystems. In *Agrobiodiversity,* Wood, D. and Lenne, J. Eds. CAB International, Wallingford, 87–121.

Wasilewska, L. 1997. The relationship between the diversity of soil nematode communities and the plant species richness of meadows. *Ekologia Polska,* 45, 719–732.

Wilcock, C. and Neiland, R. 2002. Pollination failure in plants: Why it happens and when it matters. *TRENDS in Plant Science,* 7, 6.

Williams, I. H. 1996. Aspects of bee diversity and crop pollination in the European Union. In *The Conservation of Bees,* Matheson, A. et al., Eds. Academic Press, London, 63–80.

Wolters, V. et al. 2000. Global change effects on above- and belowground biodiversity in terrestrial ecosystems: Interactions and implications for ecosystem functioning. *BioScience,* 50, 1089–1099.

Zhu, Y. et al. 2000. Genetic diversity and disease control in rice. *Nature,* 406, 718–722.

12 Long-Term Consequences of Biological and Biogeochemical Changes in the Horseshoe Bend Long-Term Agroecosystem Project

David C. Coleman, Mark D. Hunter,
Paul F. Hendrix, D. A. Crossley, Jr.,
Sofia Arce-Flores, Breana Simmons, and Kyle Wickings

CONTENTS

12.1 INTRODUCTION

We thank Patrick Bohlen and the organizers for inviting us to participate in this publication honoring the life and work of Ben Stinner, who was an early alumnus of the Horseshoe Bend project, where he conducted the research leading to his Ph.D. degree in 1984. Ben began his agroecosystem studies at the Horseshoe Bend facility at the University of Georgia (UGA). He was responsible for the initial setup and management of the well-known conventional tillage–no tillage experiment that generated much output. He helped in designing the study, clearing the fields, establishing the experimental plots, and getting crops into the ground. Ben was the lead student, with colleagues Andrea Yates and Garfield House under partial direction of Eugene P. Odum, Robert L. Todd, and D. A. Crossley, Jr. In this chapter we report on key findings from the Horseshoe Bend long-term project that Ben helped initiate, which built on the early work in his career on biogeochemistry and ecosystem ecology. Ben's progressive expansion into broader areas of agroecology is summarized beautifully in Chapter 2 by Deb Stinner, another UGA alumnus.

The maintenance of soil organic matter (SOM) is considered a desirable goal in agroecosystems (Coleman et al. 1984, 1994; Bossuyt et al. 2002). SOM maintains soil structural stability, enhances water-holding capacity, soil fertility, and crop production, ensuring long-term agricultural ecosystem stability (Dick et al. 1997; Hassink et al. 1997; Denef et al. 2004). Soil is estimated to be the largest terrestrial pool of carbon (C), containing 1500 Pg, twice that of the atmosphere (Schlesinger 1997).

Agroecosystems that utilize conservation tillage or no-tillage techniques are promising alternatives to deep moldboard plowing, because they enhance agricultural sustainability, and reduce losses to erosion (Six et al. 1999). Tillage practices disrupt soil aggregates, which may lead to increased aggregate turnover and increased decomposition of SOM (Six et al. 1998). Reduced aggregation leads to subsequent lower levels of SOM in conventional tillage (CT) than in no-tillage (NT) treatments (Paustian et al. 1998). Conversely, there is an increase in C content under NT which is attributed to a combination of slower litter decomposition and reduced soil disturbance under NT (Coleman et al. 1994; Paustian et al. 1997).

Six et al. (1999) developed a conceptual model of aggregate turnover, depicting organic C accumulation, mineral fraction, and particulate organic matter (POM). Faster macroaggregate turnover in CT than in NT results in (1) fewer macroaggregates being maintained, with more free microaggregates being present in CT, and (2) less fine POM and new microaggregates formed in CT. By the end of the cycle, fewer microaggregates contain crop-derived C in CT than in NT (Six et al. 1999).

Several soil scientists have used the natural signal from C_3-type vegetation ($\delta^{13}C \sim -26‰$) in the indigenous soil organic matter, and then followed the change that occurs from growing C_4-type vegetation ($\delta^{13}C \sim -12‰$) in experimental fields (Balesdent et al. 1987; Balesdent and Balabane 1992). Our studies have used a modified approach, converting from a regime of C_3 winter cover crops (wheat, rye, clover) and C_4 summer maize or grain sorghum crops, to all C_3 winter crops and summer (kenaf, cotton) crops. We are following changes in isotopic ratios of some soil C pools, measuring the long-term accretion of C in various soil fractions (Coleman et al. 1994; Six et al. 1998, 1999) in our agroecosystem. This research on a subtropical soil serves as a useful comparison and contrast to the results of Balesdent and Mariotti (1996), Balesdent et al. (1987, 2000) in temperate soils and Cerri et al. (1995) in tropical soils. We expect that C stabilization (i.e., slower C turnover) will be more pronounced over the long term in NT than in CT plots due to C protection within microaggregate fractions in the upper soil stratum of NT.

The studies presented here were conducted in the Horseshoe Bend Agroecosystem Long-term Research in Environmental Biology (LTREB) site, Athens, Georgia. Our objectives were (1) to measure the peak standing crop biomass of our summer crops across decadal time spans, and to relate those trends to other aspects of tillage management and winter cover crop; (2) to determine the influence of both tillage methods (CT and NT) on water-stable aggregate distribution and SOM dynamics; and (3) to investigate the changes in ^{13}C in both tillage practices. We hypothesized that the contrast between tillage treatments would change over time, with greater macroaggregation and organic matter fractions in NT versus CT, and faster C turnover in both macro- and microaggregates in CT versus NT.

12.2 MATERIALS AND METHODS

12.2.1 THE HORSESHOE BEND SITE

Horseshoe Bend is a 2-ha research site of the University of Georgia, situated in bottomland (fine loamy siliceous thermic Rhodic Kanhapludult in the Hiwassee series with 66 percent sand, 13 percent silt, and 21 percent clay) along the middle fork of the Oconee River, in Athens-Clarke County, GA. Mean annual precipitation is 1270 mm, and annual mean minimum and maximum temperatures are 8.3 and 19.3°C for CT and 9.5 and 17.5°C for NT plots (Hendrix et al. 2001). Soil pH is 6.0 in the surface 2 cm and 5.7 at depths of 5 to 10 cm. Research has been conducted continuously at Horseshoe Bend since Odum et al. (1974) set up old fields in the mid-1960s. From 1978 onward, four 0.1-ha plots have been managed with moldboard plowing (to 15 cm) followed by disking (CT), and another four 0.1-ha plots have been managed using a NT regime, with the only soil disturbance being direct seed drilling in these untilled plots. We sow winter cover crops of wheat and crimson clover, and various summer crops, including grain sorghum (*Sorghum bicolor*) corn (*Zea mays* L.), and beginning in 1999, cotton (*Gossypium hirsutum* L.), either engineered *Bt* (producing the Cry1Ac protein), or non-*Bt*. In spring and summer, we make topical applications of Roundup herbicide to control C4 weeds, such as Johnson grass (*Sorghum halepense*). The NT plots have built up a significant organic layer near the soil surface, and tend to be dominated by fungal tissues in the top 1 to 2 cm (Beare et al. 1992, 1994a, b). In contrast, the CT plots have a more uniform distribution of the organic carbon in the soil profile in the top 15 cm. The tillage and *Bt* treatments are set up in a split-split plot design.

12.2.2 Sampling and Analytical Procedures

For net primary production aboveground, we sampled quadruplicate ¼ m samples in late October. Thus four samples each were taken from rye and clover winter cover crop sites of each management plot, for a total of 64 samples. The biomass samples were oven-dried in a forced-air oven at 60°C. until fully dry, then weighed to determine total mass. At the time of sampling for crop biomass, samples for weeds and litter were taken in the same quadrats in some years, and also recorded.

For soil organic matter analyses, samples (5.8 cm dia, 0 to 5 and 5 to 15 cm) were taken from CT and NT plots in quadruplicate for each 0.1-ha plot. Samples were kept field moist at 4°C for no more than 24 h, and processed for macro- and microaggregates using the wet sieving technique of Beare et al. (1994a, b). Delta ^{13}C values were determined on ground samples of micro- and macroaggregates on a Finnigan MAT Isotope Ratio Spectrometer. All analyses were made in triplicate and analyzed by analysis of variance or nonparametric statistics.

The distribution of aggregates was measured using a modified Yoder (1936) wet-sieving apparatus that was designed to handle larger masses of soil on stacked sieves (21.6 cm dia), which allowed for complete recovery of all particles from individual samples (Beare and Bruce 1993). Soil was air-dried prior to sieving and suspended in distilled water at room temperature on the largest sieve 5 minutes before sieving. Each 50-g subsample was distributed on the surface of the top sieve of the three-sieve stack (2000 μm above the 250 μm sieve). The soil was wet-sieved oscillating 21 times per minute for 5 min, through the three sieves to obtain four aggregate size fractions: (1) 2000 μm (large macroaggregates), (2) 250 to 2000 μm, (3) 53 to 250 μm (microaggregates), and (4) <53 μm (silt- plus clay size particles). Following sieving, the sieves were lowered to the bottom of the stroke and the fresh organic matter from crop residue was aspirated from the surface before draining and placed into an aluminum cake pan. After wet sieving, the water columns were drained and the soil sieves were backwashed into cake pans and left to dry in a drying chamber (10°C). Subsamples of soil were taken to enable calculation of sample weight to oven-dry weights. The 53-μm pan was removed from the bucket and a stirrer was used to mix the water in the bucket. After mixing, a 200-ml subsample was taken and placed in its respective pan. Subsamples from each size fraction were ground and analyzed for total C and ^{13}C. Sub-samples were taken from the dry aggregate size fractions noted above to separate the POM-associated and mineral-associated C and N. Details of this method are described by Cambardella and Elliott (1992). Subsamples of intact aggregates were mixed with sodium hexametaphosphate and shaken for 12 h on a reciprocal shaker. The dispersed organic matter plus sand was collected on a 53-μm mesh sieve; the water in the soil slurry was evaporated in a forced-air oven at 50°C and the dried sample ground and analyzed by Dumas combustion on a Carlo/Erba analyzer. The difference between the C and N values for the evaporated soil slurry and those obtained from a nondispersed soil sample was considered to be equal to the C or N retained in the sieve.

12.2.3   Isotope Analysis

Total carbon content and ^{13}C content of aggregate size fractions was measured on a Finnigan Delta C mass spectrometer coupled to a Carlo Erba NA 1500 CHN combustion analyzer via Finnigan's Conflo II interface. The ^{13}C/^{12}C ratios are then reported as δ^{13}C values measured relative to a Peedee Belemnite (PDB) standard.

The fraction of new C was calculated using the equation:

$$F = (\delta t - \delta A)/(\delta B - \delta A)$$

where $\delta t = \delta^{13}$C at time t, $\delta A = \delta^{13}$C of the soil when a mix of C_3 and C_4 plants were grown (at time 0, 1997), δB = the average δ^{13}C of C_3 plants, and F = fraction of new C in the soil. F is used to estimate the turnover of soil C (Balesdent and Mariotti 1996).

12.2.4   Turnover and Net Inputs of Organic Carbon

Skjemstad et al. (1990) used the assumption of exponential decay as a means for comparing relative decay rates and turnover times in micro- and macroaggregates. We determined the first-order rate constants (k) as:

$$k = -\ln(\text{Cmix}_{\text{cotton}}/\text{Cmix}_{\text{time 0}})/t$$

where C_{mix} cotton is the concentration of C from the mixture of C_3 and C_4 plants remaining in each size fraction from the soil at the present time, C_{mix} time 0 is the concentration of C from the mixture at time zero, before the switch to only C_3 vegetation was made, and t is the time interval since that change took place (2005 – 1997 = 8 years). The net input rate was calculated as the concentration of new C in each size fraction divided by the length of time since only C_3 plants had been planted.

12.2.5   Statistical Analysis

The experiment was analyzed as a nested split plot design, and analyzed by the SAS statistical package for analysis of variance (ANOVA-PROC GLM, SAS Institute 1990). Tukey's HSD ($p < 0.05$) was used for mean separation.

12.3   RESULTS

12.3.1   Crop Biomass

Crop biomass at Horseshoe Bend has been measured intermittently since 1984. The majority of variation in aboveground crop biomass is explained by the species of crop being grown. Biomass was greatest for corn, then kenaf, followed by sorghum and cotton ($F_{3,36}$= 101.40, $p < 0.001$; Figure 12.1). There are also some modest effects of cover crop, tillage, and year on crop biomass. For example, a clover cover crop provides a substantial boost to corn production, whereas its effects on other crops are negligible ($F_{3,36}$ = 8.06, $p = 0.0003$; Figure 12.2). Likewise, tillage has its greatest impact on corn and sorghum production. In general, corn biomass was higher under CT than under NT, with the exception of the year 1996. In contrast, sorghum biomass was higher under NT than CT ($F_{11,184}$ = 3.64, $p < 0.0001$; Figure 12.3).

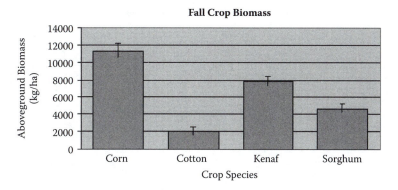

FIGURE 12.1   Aboveground biomass of crops measured during the fall at Horseshoe Bend.

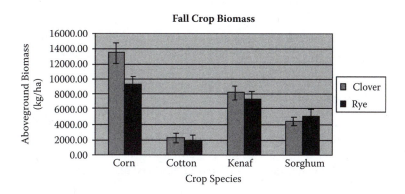

FIGURE 12.2   Effects of winter cover crop on crop biomass at Horseshoe Bend.

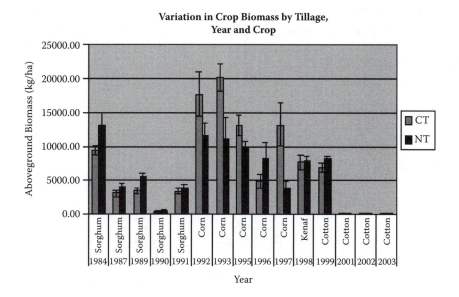

FIGURE 12.3   Yearly variation in the effects of tillage on crop biomass at Horseshoe Bend.

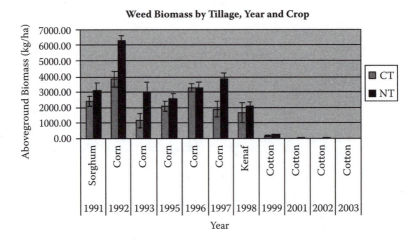

FIGURE 12.4 Yearly variation in the effects of tillage on weed biomass at Horseshoe Bend.

Overall, these data suggest that tillage and cover crop have relatively minor effects on crop biomass with the clear exception of corn. Corn biomass was favored by a cover crop of clover and CT.

12.3.2 WEED BIOMASS

The aboveground biomass of weeds in the fall was estimated, starting in 1991. Again, the species of crop planted explained the greatest variation in the fall biomass of weeds ($F_{3,36}$ = 133.78, p < 0.0001) with very low weed production under cotton (Figure 12.4). Weed biomass was generally higher under NT than under CT, with the greatest differences observed under corn production ($F_{7,131}$ = 2.74, p = 0.0109; Figure 12.4).

Overall, weed biomass under cotton was very low, although this was likely driven in part by significant drought during the cotton years. As expected, weed biomass was higher under NT.

12.3.3 LITTER BIOMASS

During 1991 (sorghum) and 1992 (corn), estimates were made of aboveground litter biomass during the fall sampling period. Litter biomass was greater under NT than under CT, with a larger effect of tillage on litter biomass under corn production ($F_{1,12}$ = 32.10, p = 0.0001; Figure 12.5).

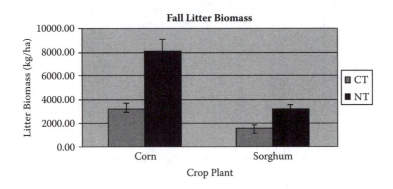

FIGURE 12.5 The effects of tillage on litter biomass at Horseshoe Bend.

12.3.4 Changes in Water-Stable Aggregates

Our hypothesis was that there will be more accumulation of carbon in aggregates in NT and hence more rapid loss and turnover of C in the CT treatments.

The distribution of water-stable aggregates was influenced significantly by tillage management in both depths of soils (Figure 12.6a and b). Macroaggregates (>2000 μm) comprised the larges percentage of the total soil, and they were 1.6 times greater ($p < 0.001$) in NT than in CT. For aggregates of 250 to 2000, 53 to 250, and <53 μm, the aggregates in CT plots were 1.3, 2.9, and 2.2 times greater than NT, respectively ($p < 0.05$). At 5 to 15 cm depth, the distribution of macroaggregates from NT was 1.2 times greater than in CT. There was no significant difference in the smaller water-stable aggregates (WSA) size classes.

Total C and N were significantly different by tillage, aggregate size, and their interaction ($p < 0.01$; Figure 12.7). Total sand-free C and N in 0 to 5 cm soils were significantly higher for all

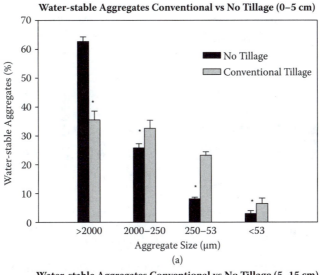

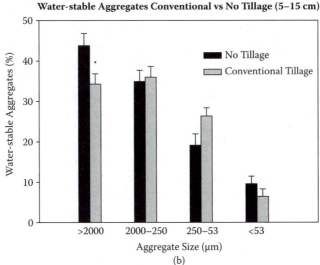

FIGURE 12.6 Distribution of water-stable aggregates from conventional and NT soils at 0–5 (a) and 5–15 (b) depth. Bars are means ± S.E. Asterisks indicate significant differences ($P < 0.05$, Tukey's HSD) between tillage treatments within size class; $n = 32$.

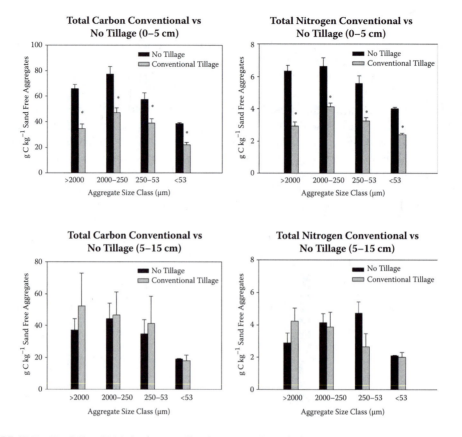

FIGURE 12.7 Total C and N (g kg^{-1}; normalized to a sand-free basis) in water-stable aggregates from conventional and NT soils at 0–5 and 5–15 depth. Bars are means ± S.E. Asterisks indicate significant differences ($P < 0.05$, Tukey's HSD) between tillage treatments within size class; $n = 32$.

aggregate size classes in NT than CT. In the 0 to 5 cm layer, sand-free C and N were highest in the 250 to 2000 µm aggregates, with total C and N 1.6 and 1.5 times greater. In contrast, there were no significant differences in aggregate C and N in the 5 to 15 cm depths.

The differences in POM-associated C and N were significantly affected by tillage and aggregate size class for all aggregate size classes in surface soils (N) and in aggregates >2000 µm and 250 to 2000 µm (C) (Figure 12.8). In general, the C/N ratios were much greater between size classes for the POM-associated organic matter than for the mineral-associated POM.

We compared total C within the same tillage practices for the three different sampling dates 1997, 2000, and 2005 (Figure 12.9). Total C in NT surface soils showed a significant loss for aggregate size classes > 53 µm, and a constant concentration of C in the aggregates < 53 µm. The net loss of C from 1997 to 2005 was reduced an average of 1.6 times (40 g C kg^{-1}) of C concentration in the NT treatment. In contrast, surface soils of CT showed no significant differences in time for the >53 µm size classes, but significantly different total C concentration in <53 µm aggregates.

12.3.5   CHANGES IN δ^{13}C IN AGGREGATES

There were significant differences in ^{13}C ratios in time within tillage method (Figure 12.10). Ratios for all aggregate size classes were shifting towards the ^{13}C ratios of C$_3$ plants. The values for NT in the surface soils were generally lower (more negative) compared to CT. In both NT and CT, the aggregates > 53 µm had the youngest age (most negative ^{13}C ratio) and the aggregate < 53 µm

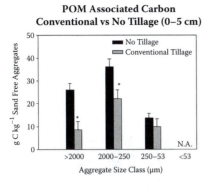

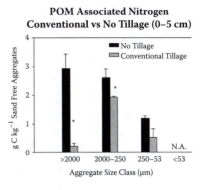

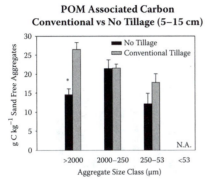

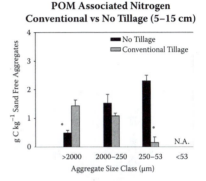

FIGURE 12.8 POM-associated C and N (g kg⁻¹; normalized to a sand-free basis) in water-stable aggregates from conventional and NT soils at 0–5 and 5–15 depth. Bars are means ± S.E. Asterisks indicate significant differences ($P < 0.05$, Tukey's HSD) between tillage treatments within size class; $n = 32$.

showed the oldest signature, being older in CT than NT. Both the 0 to 5 and 5 to 15 cm depths in CT were significantly more negative in the year 2000 than in 2005, and the 5 to 15 cm depth in NT was not significantly different in 1997 and 2000.

For surface soils there were similar values of new C in CT and NT (Table 12.1); in lower depths, the larger aggregates in CT had more new C incorporated than NT. The proportion of new C did not show a significant variation for both depths within the NT practice, with a similar behavior in CT except for the deeper soil profile.

Table 12.2 shows the average rate constant (k) for loss and turnover time ($1/k$) of the mixture of C_4 and C_3-C and average net input rate of C_3-C in aggregate size fractions as determined by ^{13}C natural abundance. The k was lower for microaggregates than for macroaggregates in both tillage treatments and the turnover was faster in CT than in NT for both surface and deep soil in most aggregate size fractions. The turnover of aggregate-associated old C in microaggregates was 27.08 years for NT and 26.12 for CT (0 to 5 cm) and for the deeper soil profile the difference was greater (34.5 and 12 years, respectively). In the 5 to 15 cm depth the aggregates < 53 μm for both NT and CT had very little net input rate of new C. Furthermore, within tillage there was a difference of turnover within aggregate size fractions.

12.4 DISCUSSION

12.4.1 Aggregate Distribution

Macroaggregates (>2000 μm) from surface soils of NT were more abundant than those of CT soils, and the situation was reversed for smaller aggregates in the surface horizon. Several researchers found

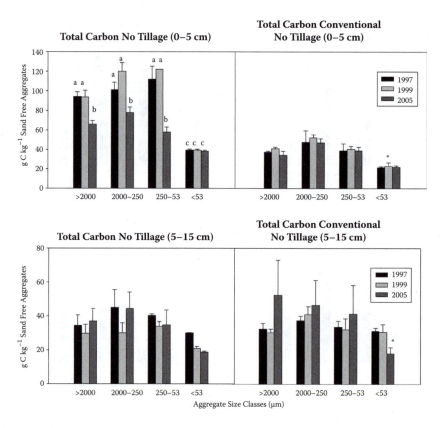

FIGURE 12.9   Total C concentration in time for NT and CT in different depths (0–5 and 5–15 cm) and aggregate size fractions.

more macroaggregates in NT versus CT soils (Beare et al. 1994a; Six et al. 2000; Bossuyt et al. 2002). Our results are similar to those obtained by Beare et al. (1994a), Hendrix et al. (1998), and Collins et al. (2001) in the same field site. Tillage management changes soil conditions (aeration, temperature, moisture) and decomposition rates of litter (Cambardella and Elliott 1992). In NT, residue accumulates at the surface where litter decomposition rate is slowed due to drier conditions and reduced contact between soil microorganisms and litter (Salinas-Garcia et al. 1997). A greater proportion of the microbial biomass is composed of fungi, which contribute to macroaggregate formation and stabilization (Tisdall and Oades 1982; Beare et al. 1992). In CT, subsurface soil is brought to the surface and exposed to wet–dry, freeze–thaw cycles and raindrop impact (Beare et al. 1994; Paustian et al. 1997).

Higher microbial activity (e.g., in CT soils) depletes SOM, which eventually leads to decreased microbial biomass and activity and consequently a lower production of microbial-derived binding agents and a loss of aggregation (Jastrow et al. 1996; Six et al. 1998). Six et al. (1999) developed a conceptual model of aggregate turnover that shows the faster macroaggregate turnover in CT than in NT results in fewer macroaggregates being maintained, and more free microaggregates being present in CT than in NT soils. Our results (Figure 12.1) support this model, with more macroaggregates existing in NT than in CT and more microaggregates in CT than in NT (0 to 5 cm).

12.4.2   CARBON AND NITROGEN CONCENTRATIONS

The total sand-free concentrations of aggregate C and N were up to 1.6 and 1.5 times greater in the surface layers of NT than in CT (Figure 12.7). With aggregate disruption, more organic substrates are made available for microbial attack, with ensuing increased SOM decomposition, and hence a

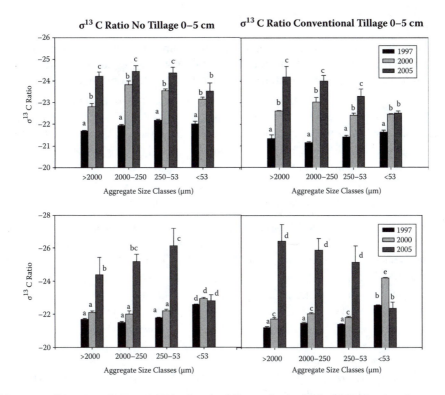

FIGURE 12.10 ^{13}C ratios of NT and CT in time in different depths (0–5 and 5–15 cm) and aggregate size fractions.

TABLE 12.1
Fraction of New Carbon in 2000 and 2005 Compared to 1997 at Two Depths, 0–5 and 5–15 cm, at Horseshoe Bend, Athens, GA

Aggregate Size, μm	No Tillage (0–5 cm)		Conventional Tillage (0–5 cm)	
	2000	2005	2000	2005
>2000	0.18	0.40	0.18	0.43
2000–250	0.31	0.41	0.27	0.41
250–53	0.23	0.38	0.15	0.28
<53	0.19	0.25	0.10	0.13

Aggregate Size, μm	No Tillage (5–15 cm)		Conventional Tillage (5–15 cm)	
	2000	2005	2000	2005
>2000	0.07	0.43	0.07	0.77
2000–250	0.08	0.57	0.09	0.67
250–53	0.07	0.70	0.06	0.56
<53	0.07	0.04	0.30	0.001

TABLE 12.2

Average Rate Constant (k) for Loss and Turnover Time ($1/k$) of the Mixture of C_4 and C_3 Carbon and Average Net Input Rate of C_3-C in Aggregate Size Fractions at 0–5 and 5–15 cm, Depths as Determined by ^{13}C Natural Abundance

	No Tillage (0–5 cm)			Conventional Tillage (0–5 cm)		
Aggregate Size, μm	k, yr^{-1}	$1/k$, yr	Net Input Rate, g kg^{-1} Fraction yr^{-1}	k, yr^{-1}	$1/k$, yr	Net Input Rate, g kg^{-1} Fraction yr^{-1}
>2000	0.08	12.51	3.30	0.12	8.58	1.86
2000–250	0.10	10.25	3.98	0.10	10.43	2.42
250–53	0.12	8.66	2.74	0.10	10.49	1.36
<53	0.04	27.08	1.21	0.04	26.12	0.36

	No Tillage (5–15 cm)			Conventional Tillage (5–15 cm)		
Aggregate Size, μm	k, yr^{-1}	$1/k$, yr	Net Input Rate, g kg^{-1} Fraction yr^{-1}	k, yr^{-1}	$1/k$, yr	Net Input Rate, g kg^{-1} Fraction yr^{-1}
>2000	0.06	15.89	2.00	0.13	7.49	5.03
2000–250	0.08	13.17	3.16	0.10	9.69	3.89
250–53	0.15	6.58	3.03	0.08	12.18	2.89
<53	0.03	34.50	0.10	0.08	12.01	0.001

decrease in C content. Beare et al. (1994a), Dick et al. (1997), and Six et al. (1999) also found higher C concentrations in surface samples of NT compared to CT soils.

POM-associated C and N values were greater in NT than in CT in all aggregate size classes in surface soils, and they were greater for larger than smaller aggregates (Figure 12.8). Thus, aggregate formation and stabilization processes are directly related to the decomposition of root residue and the dynamics of POM C in the soil (Gale et al. 2000). Beare et al. (1994a) noted that the differences in distributions of POM between depths in NT and CT may be a function of biological activity near the soil surface, including fungi (Doran 1980), roots (Cheng et al. 1990), and soil fauna (Parmelee et al. 1990), that assist in incorporating POM within macroaggregates and to increase their stability. Examining surface soils in CT (Figure 12.9), we found there were no significant differences in time for all of the larger size classes, but there was a significantly different concentration of total C for aggregates < 53 μm, indicating that for CT and NT of this size class the C in this fraction is stabilized by intimate associations with mineral particles. Gregorich et al. (1995) also found that the most stable organic matter is associated with this small size fraction. For both depths the old C associated with microaggregates may be physically protected, as was observed also by Christensen (1992).

12.4.3 Carbon-13 Concentrations and Carbon Turnover

Carbon turnover for NT and CT soils was calculated over time, showing δ^{13}C values from 1997 onward (Figure 12.10). We observed significant differences in time within tillage method. Increased tillage intensity enhanced turnover of SOM and decreased soil aggregation (Six et al. 1998). Because the average annual inputs of aboveground crop plus weed residues to NT and CT are very similar in Horseshoe Bend (HSB) (Beare et al. 1994a; this chapter), we attribute the differences in SOM content to differences in assimilation and decomposition of SOM under both tillage treatments. However, in this study, total C was reduced in NT plots for 2005 compared to previous years (Figure 12.9), which was not the case with CT. Perhaps there was an unusually large biomass of cotton that remained to decompose after being cut. Therefore, the woody tissue was incorporated into the soil faster in CT than in NT.

The fraction of new C in the soil is a direct expression of C turnover (Table 12.1). The similar values of new C in CT and NT suggest that the increase of C was proportional in both tillage practices over time. Similar results were noted by Six et al. (1998). In lower depths (5 to 15 cm), more new C was incorporated into new C for the larger aggregates of CT versus NT, again suggesting that new organic matter movement through the soil profile in CT. More details on this study of $\delta^{13}C$ change in soil macro- and microaggregates are given in Arce-Flores and Coleman (2007).

Our research shows considerably increased mean residence time of SOM under NT versus CT managements, similar to results measured by Paustian et al. (1997). Most of the aggregate size fractions experienced faster turnover in CT than in NT in both surface and deeper depths (Table 12.2). The turnover of aggregate-associated old C in microaggregates was 27.08 years for NT and 26.12 years for CT (0 to 5 cm) and was greater for the 5 to 15 cm depth (34.5 and 12 years) for NT and CT, respectively. In the deeper profile the microaggregates < 53 um in both NT and CT had very little net input rate of new C. The measured turnover times for micro- and macroaggregates are in accord with the aggregate hierarchy concept of Oades and Waters (1991), and further demonstrate the mechanisms involved in the binding of micro- versus macroaggregates (Tisdall and Oades 1982).

The net input rate of C_3 pathway organic matter to aggregates increased from micro- to macroaggregates, which supports the concept that larger aggregates are bound together initially by root exudates and exfoliates and mycorrhizal hyphae. As they senesce and begin to undergo comminution, they are then incorporated into the intraaggregate POM of larger macroaggregates first (Jastrow et al. 1996). The roles of mycorrhizal mycelia and their products are probably important in microaggregate formation, but little research has yet been carried out on this phenomenon (Rillig and Mummey 2006). In earlier studies at Horseshoe Bend, Bossuyt et al. (2002) measured more young C accumulated in the subsurface soil of CT than in NT, but it was not stabilized over the long term. They found, similar to our study, greater long-term stabilization of C in the surface layers of NT compared to CT.

Because the total C concentration in the three different sampling times between 1997 and 2005 remained relatively constant suggests that old C associated with microaggregates may be physically protected. As surface soils in CT are exposed to variable abiotic conditions, these factors contribute to the more frequent disruption of soil aggregates, releasing the aggregate protected SOM for mineralization (Beare et al. 1994b) and also to a lower production of aggregate-stabilizing agents (Angers et al. 1993).

12.5 CONCLUSIONS

Long-term studies of soil detrital food webs and the dynamics of soil organic matter at Horseshoe Bend have been illustrative of several basic principles of ecology. We feel that it has grown and matured with the influence of Ben Stinner and his colleagues who were so influential in establishing the studies nearly three decades ago.

REFERENCES

Angers, D. A. et al. 1993. Microbial and biochemical changes induced by rotation and tillage in a soil under barley production. *Canadian Journal of Soil Science*, 73, 39–50.

Arce-Flores, S. and Coleman, D. C. 2008. Comparing water-stable aggregate distributions, organic matter fractions and carbon turnover using ^{13}C natural abundance in conventional and no-tillage soils. *Soil Biology and Biochemistry*. (in revision).

Balesdent, J. and Balabane, M. 1992. Maize root-derived soil organic carbon estimated by natural ^{13}C abundance. *Soil Biology and Biochemistry*, 24, 97–101.

Balesdent, J. and Mariotti, A. 1996. Measurement of soil organic matter turnover using ^{13}C natural abundance. In *Mass Spectrometry of Soils,* Boutton, T. W. and Amasaki, S. I., Eds. Marcel Dekker, New York, 83–111.

Balesdent, J., Mariotti, A., and Guillet, B. 1987. Natural ^{13}C abundance as a tracer for soil organic matter dynamics studies. *Soil Biology and Biochemistry*, 19, 25–30.

Balesdent, J., Chenu, C., and Balabane, M. 2000. Relationship of soil organic matter dynamics to physical protection and tillage. *Soil and Tillage Research,* 53, 215–230.

Beare, M. H. and Bruce, R. R. 1993. A comparison of methods for measuring water-stable aggregates: Implications for determining environmental effects on soil structure. *Geoderma*, 56, 87–104.

Beare, M. H. et al. 1992. Microbial and faunal interactions and effects on litter nitrogen and decomposition in agroecosystems. *Ecological Monographs,* 62, 569–591.

Beare, M. H., Hendrix, P. F., and Coleman, D. C. 1994a. Water-stable aggregates and organic matter fractions in conventional and no-tillage soils. *Soil Science Society of America Journal,* 58, 777–786.

Beare, M. H. et al. 1994b. Aggregate-protected and unprotected pools of organic matter in conventional and no-tillage soils. *Soil Science Society of America Journal*, 58, 787–795.

Bossuyt, H., Six, J., and Hendrix, P. F. 2002. Aggregate-protected carbon in no-tillage and conventional tillage agroecosystems using carbon-14 labeled plant residue. *Soil Science Society of America Journal*, 66, 1965–1973.

Cambardella, C. A. and Elliott, E. T. 1992. Particulate organic matter across a grassland cultivation sequence. *Soil Science Society of America Journal*, 56, 777–783.

Cerri, C. et al. 1995. Application du traçage isotopique naturel en ^{13}C a l'etude de la dynamique de la matiére organique dans les sols. *Comptes Rendus de l'Académie des Sciences* (Paris), 300, 423–428.

Cheng, W., Coleman, D. E., and Box, Jr., J. E. 1990. Root dynamics, production, and distribution in agroecosystems on the Georgia Piedmont using minirhizotrons. *Journal of Applied Ecology,* 27, 592–604.

Christensen, B. T. 1992. Physical fractionation of soil and organic matter in primary particle size and density separates. *Advances in Soil Science*, 20, 1–90.

Coleman, D. C., Cole, C. V., and Elliott, E. T. 1984. Decomposition, organic matter turnover, and nutrient dynamics in agroecosystems. In *Agricultural Ecosystems—Unifying Concepts*, Lowrance, R., Stinner, B. R., and House, G. J., Eds. Wiley/Interscience, New York, 83–104.

Coleman, D. C. et al. 1994. The impacts of management and biota on nutrient dynamics and soil structure in sub-tropical agroecosystems: Impacts on detritus food webs. In *Soil Biota Management in Sustainable Farming Systems*, Pankhurst, C. E. et al., Eds. CSIRO, Melbourne, 133–143.

Collins, K., Bossuyt, H., and Hendrix, P. 2001. Tillage effects on carbon turnover in soil organic matter fractions measured using ^{13}C natural abundance. B.S. thesis, University of Georgia, Athens.

Denef, K. et al. 2004. Carbon sequestration in microaggregates of no-tillage soils with different clay mineralogy. *Soil Science Society of America Journal*, 68, 1935–1944.

Dick, W. A., Edwards, W. M., and McCoy, E. M. 1997. Continuous application of no-tillage to Ohio soils: Changes in crop yields and organic matter-related soil properties. In *Organic Matter in Temperate Agroecosystems: Long-Term Experiments in North America,* Paul, E. A. et al., Eds. CRC Press, Boca Raton, FL, 171–182.

Doran, J. W. 1980. Soil microbial and biochemical changes associated with reduced tillage. *Soil Science Society of America Journal*, 44, 765–771.

Gale, W. J., Cambardella, C. A., and Bailey, T. B. 2000. Root-derived carbon and the formation and stabilization of aggregates. *Soil Science Society of America Journal*, 64, 201–207.

Gregorich, E. G., Ellert, B. H., and Monreal, C. M. 1995. Turnover of soil organic matter and storage of corn residue carbon estimated from natural ^{13}C abundance. *Canadian Journal of Soil Science*, 75, 161–167.

Hassink, J., Whitmore, A. P., and Kubát, J. 1997. Size and density fractionation of soil organic matter and the physical capacity of soils to protect organic matter. *European Journal of Agronomy*, 7, 189–199.

Hendrix, P. F., Franzluebbers, A. J., and McCracken, D. V. 1998. Management effects of carbon accumulation and loss in soils on the southern Appalachian Piedmont of Georgia, USA. *Soil and Tillage Research*, 47, 245–251.

Hendrix, P. F. et al. 2001. Horseshoe Bend research: Old-field studies (1965–1975) and agroecosystem studies (1976–2000). In *Holistic Ecology: The Evolution of the Georgia Institute of Ecology (1940–2000)*, Barrett, G. W. and Barrett, T. L., Eds. Taylor & Francis, New York, 164–177.

Jastrow, J. D., Boutton, T. W., and Miller, R. M. 1996. Carbon dynamics of aggregate-associated organic matter estimated by carbon-13 natural abundance. *Soil Biology and Biochemistry*, 28, 665–676.

Oades, J. M. and Waters, A. G. 1991. Aggregate hierarchy in soils. *Australian Journal of Soil Research*, 29, 926–928.

Odum, E. P. et al. 1974. The effect of late winter litter burn on the composition, productivity, and diversity of a 4-year old fallow-field in Georgia. *Proceedings Annual Tall Timbers Fire Ecology Conference*, 13, 399–415.

Parmelee, R. W. et al. 1990. Earthworms and enchytraeids in conventional and no-tillage agroecosystems: A biocide approach to assess their role in organic matter breakdown. *Biology and Fertility of Soils*, 10, 1–10.

Paustian, K., Collins, H. P., and Paul, E. A. 1997. Management controls on soil carbon. In *Soil Organic Matter in Temperate Agroecosystems: Long-Term Experiments in North America*, Paul, E. A. et al., Eds. CRC Press, Boca Raton, FL, 15–49.

Paustian, K. et al. 1998. CO_2 mitigation by agriculture: An overview. *Climate Change*, 40, 135–162.

Rillig, M. C. and Mummey, D. L. 2006. Tansley lecture. Mycorrhizas and soil structure. *New Phytologist*, 171, 41–53.

Salinas-Garcia, J. R., Hons, F. M., and Matocha, J. E. 1997. Long-term effects of tillage and fertilization on soil organic matter dynamics. *Soil Science Society of America Journal*, 61, 152–159.

Schlesinger, W. H. 1997. *Biogeochemistry: An Analysis of Global Change,* 2nd ed. Academic Press, San Diego.

Six, J. et al. 1998. Aggregation and soil organic matter accumulation in cultivated and native grass-land soils. *Soil Science Society of America Journal*, 62, 1367–1377.

Six, J., Elliott, E. T., and Paustian, K. 1999. Aggregate and soil organic matter dynamics under conventional and no-tillage systems. *Soil Science Society of America Journal*, 63, 1350–1358.

Six, J., Elliott, E. T., and Paustian, K. 2000. Soil macroaggregate and soil microaggregate formation: a mechanism for C sequestration under no-tillage agriculture. *Soil Biology and Biochemistry*, 32, 2099–2103.

Skjemstad, J. O., LeFeuvre, R. P., and Prebble, R. E. 1990. Turnover of soil organic matter under pasture as determined by ^{13}C natural abundance. *Australian Journal of Soil Research,* 28, 267–276.

Tisdall, J. M. and Oades, J. M. 1982. Organic matter and water-stable aggregates in soils. *Journal of Soil Science*, 33, 141–163.

Yoder, R. E. 1936. A direct method of aggregate analysis of soils and a study of the physical nature of erosion losses. *Journal of American Society of Agronomy*, 28, 337–351.

Section IV

*Managing Acroecosystems and
Research to Support Multiple
Functions and Outcomes*

13 Challenges and Benefits of Developing Multifunctional Agroecosystems

John Westra and George Boody

CONTENTS

13.1 INTRODUCTION

In this chapter, we summarize previous works describing theoretical and applied aspects of multifunctional agriculture or agroecosystems, focusing primarily on agricultural policy and the economic literature. Then we concentrate our discussion on efforts to model and quantify some of the multiple benefits that the public can derive from agroecosystems. Specifically, we describe research in two watersheds that modeled multiple outputs from agricultural systems, including environmental benefits and ecosystem services. Results from a biophysical process simulation model were integrated into an economic policy model to evaluate potential environmental, agricultural, and economic outcomes from possible changes to current farming practices in these watershed study areas. In contrast to other modeling efforts of changes in land management, researchers in our project worked with residents of both watersheds to envision and develop potential future scenarios for their watersheds (alternative future trends in agricultural management). We next describe modeling results from those two watershed study areas. Findings from that research indicated environmental and economic benefits were attainable with changes in land management, without increased costs to the public. As one might expect, the extent of the benefits depended on the magnitude of changes to agricultural practices. Environmental benefits from multifunctional agroecosystems that were modeled included better water quality, better habitat and health of fish assemblages, decreased greenhouse gas emissions, and increased carbon sequestration. Economic benefits included greater farm profitability, reduced government transfer payments, and reduced cost of externalities from sedimentation and flooding.

We conclude our discussion with policy implications of extending our analysis to evaluate farming practices or systems in a more inclusive context. Given the current paradigm for agricultural programs and policies, the true potential of multifunctional agroecosystems is severely limited. Multiple benefits from agriculture can occur only if U.S. farm policy transitions from a policy that emphasizes mostly commodity-based food and fiber production to a policy that emphasizes more inclusive goals. One potential transition would be to redirect current farm payments that encourage commodity production to programs that reward farmers who produce these multiple benefits. The challenge of such a transition is transferring the $16 billion in direct government payments from farmers producing a few program commodities to farmers producing multiple benefits. However, by using alternative incentives we can encourage farmers to improve environmental conditions substantially, at little or no additional cost to the public.

13.2 MULTIFUNCTIONAL AGROECOSYSTEMS

For many, the function of agriculture is to produce private goods like food and fiber, and increasingly industrial products like bioenergy. However, agriculture can provide many public goods and services or externalities including land conservation, maintenance of landscape structure, reduction of soil erosion and runoff, biodiversity preservation, wildlife habit protection, nutrient recycling and loss reduction, greenhouse gas reduction and soil carbon sequestration; and contribution to the socioeconomic viability of rural areas (OECD 2001, 2003; Batie 2003; Abler 2004; Hartell, 2004; Lant et al. 2005; Farber et al. 2006). In all of this literature, researchers emphasize the critical need to incorporate human (socioeconomic) and ecological dimensions into any analysis of ecosystem functions and services in agricultural landscapes. This integration of economics and ecology yields better models and understanding of the reciprocal relationships between human behavior, ecosystem functions, and human welfare (Farber et al. 2006).

Within the last decade, the multiple functions of agriculture have become prominent fixtures in global trade negotiations (Romstad et al. 2000; Vatn 2002), especially as member countries of the European Union (EU) have embraced multifunctional agriculture as a means to support the production of traditional commodities and noncommodity outputs (NCOs) (OECD 2003). Harte and O'Connell (2003) argue that many of the EU agricultural policies are severely lacking when it comes to actually supporting multifunctionality; essentially these policies act as standard income support programs. They indicated that many policies of EU countries intervene when no intervention is needed; pay farmers for ecosystem services they already provide; and limit payments per farm or payment rates despite increasing participation by the farmer. Corroborating analysis by Bohman et al. (1999) argues that EU members promoted multifunctional agriculture for protectionist reasons, justifying price and income support programs and trade restrictions.

Japan, South Korea, and several European countries (including Norway and Switzerland) have argued that small to moderate-sized, independent farms can improve the economic, environmental, and social health of rural areas and preserve their cultural heritage (DeVries 2000; Romstad et al. 2000; Mann 2003). People in these countries, Romstad et al. (2000) and Mann (2003) argue, value the nonmarket goods and services (public benefits) agriculture can provide and therefore they encourage their governments to promote multifunctional agriculture through "green box payments," so called because they are minimally trade-distorting and are not direct price supports. Nonetheless, Peterson et al. (2002) indicate that policies designed to efficiently provide socially optimal levels of multiple benefits from agriculture may change commodity outputs so that nations that import commodities (like many EU nations) will favor NCO subsidies while commodity-exporting countries (like the United States) will oppose NCO payments.

Mitigation of greenhouse gases is an ecosystem service or potential environmental benefit from production agriculture, despite that agriculture is a major contributor to the greenhouse effect (Lal et al. 1999). In the United States, about 43 million metric tons carbon equivalent (MTCE) per year of CO_2 is released from agricultural energy use and soil carbon losses each year (USEPA 2000a, 2000b). Direct energy use accounts for 15 million MTCE, and indirect energy use (such as energy used to produce farm inputs like fertilizer) results in an additional 13 million MTCE. Tillage and conversion of land between uses (from wetland to cropland, for example) yield 15 million MTCE of CO_2 emissions (Faeth and Greenlaugh 2000). An alternative agricultural practice, pasture-raised animals, requires less fuel for operations and feed than confined animals and could lead to 27 to 33 percent less soil erosion and 23 to 26 percent less fuel use in crop production (Rayburn 1993). An additional benefit of such systems is that they could tie up 14 million to 21 million metric tons of CO_2 in the organic matter of pasture soils.

Emissions of N_2O from agriculture are about 88 million MTCE each year in the United States, including 49 million MTCE directly associated with inputs such as nitrogen fertilizer application and N fixation by crops (Faeth and Greenlaugh 2000). Pasture-raised livestock systems require fewer field crops for their feed than confined systems and could reduce N_2O emissions by 5.2 million to 7.8 million metric tons if adopted nationwide (Rayburn 1993).

Another benefit from changes in land management that researchers and policy makers are interested in exploring is the potential for agriculture to sequester carbon and offset other sources of carbon (industrial and transportation) through practices like no-till, cover crops, and planting perennials. In addition to tillage, soil carbon content in agricultural settings is affected by temperature, soil moisture, soil type, frost depth, animal activity, and biomass production. Robertson et al. (2000) estimated that reduced tillage cropping systems in the United States could sequester 30.0 g carbon per m^2 per year (0.3 metric tons per ha per year). Other studies show wide variability in carbon sequestration from no-till, with some suggesting relatively little if any sequestration from this practice (Baker et al. 2007). Even without conservation tillage, an increase of up to 0.1 metric tons per ha per year in soil organic carbon (SOC) may occur from conventional management of cropland (U.S. Department of State 2000). Perennial crops have the potential to capture and hold significant quantities of carbon as SOC, accumulating up to 0.9 metric tons carbon per ha per year in Minnesota (Paustian et al. 1997). Though agriculture has great potential to accumulate SOC across agricultural practices (Huggins et al. 1998; Post and Kwon, 2000; West and Marland 2002; Kucharik et al. 2003), it is unclear how long SOC will continue to accumulate with changes in management practices.

Despite increased funding and importance placed on conservation programs under recent farm bills, farmers, policy makers, environmentalists, and the public have become increasingly aware that U.S. farm policies can adversely affect farmers and the environment. At the same time, it has become increasingly clear that farmers can produce nonmarket "goods," such as environmental and social benefits, in addition to food and fiber (Cochrane 2003). Often what prevents farmers from producing these goods is that the "socially optimal land management" differs from the farmers' private solution, in part because farmers and the public place a different value on the ecosystem services of interest (Lankoski and Ollikainen 2003). To avoid potential social welfare losses and minimize trade distortions, it is critical to develop clear estimates of the quality, accessibility, and other characteristics of ecosystem services that agricultural can provide and that society values (Randall 2002). For example, using of spatially explicit modeling of three different scenarios Santlemann et al. (2004) found that a biodiversity scenario ranked higher than the current land management for biodiversity, water quality, farmer preference, and profitability and as well or better than the production and water quality scenarios on all three metrics except water quality for the water quality scenario. Developing a more comprehensive framework for this process is critical for understanding how U.S. farmers can be encouraged to produce more of these multiple goods, what value society places on those ecosystems services, and how much will it cost farmers and the public to produce more of these multiple benefits from agriculture.

Direct government payments in U.S. agriculture have a significant effect on what is produced and how it is produced. Though nearly one-half of farms in the United States receive some type of direct government payment in 2005, nearly all of the farmers who grow at least one of the five main commodity crops received direct government payments in 2005 (USDA 2007a). For this selected set of commodities (corn, wheat, soybeans, cotton, and rice) 90 to 99 percent of farms with these crops as the major source of income received payments averaging nearly $44,000 annually (USDA 2007a). These farms received 89 percent of the $91.2 billion in commodity payments from 1995 through 2002 with corn and soybean alone accounting for 56 percent of those payments (EWG 2003). Consequences of U.S. agricultural policy include environmental pollution, resource degradation, fewer agricultural producers, and depressed rural economies (Mitsch et al. 2001; Rabalais et al. 2001; Tilman et al. 2001; Cochrane 2003). Examination of agricultural policies in France corroborate that commodity-linked policy instruments (similar to those in the United States) do not result in environmental goods (NCOs) being produced, even when there is uncertainty of output prices and farmers are risk averse (Havlik et al. 2005).

One way that the conservation programs in the Farm Bill have attempted to mitigate environmental problems is by setting aside or retiring land that is fragile or marginal, reestablishing wetlands that had been converted into farmland, and providing technical assistance or financial assistance to producers on working lands. All these programs influence farming practices, with the expectation that resources are conserved and that ecosystem services and environmental amenities are improved. Over the past 10 years, nearly $2 billion has been spent annually on all conservation programs provided under the Farm Bill (USDA 2007a). Between 1985 and 2002, approximately 70 percent of conservation spending was for land retirement programs while the remaining 30 percent was used for working agricultural lands. The payments for working lands (approximately $600 million annually on average) were allocated to half the area of privately held land in the contiguous United States (Claassen et al. 2001). Thus, relatively little of the average $16 billion spent each year on all direct government farm payments went directly toward encouraging producers to create multifunctional agroecosystems.

To illustrate the potential scope and range of environmental benefits possible when farmers are encouraged to produce them, we present results from research in two Minnesota watersheds (Boody et al. 2005). The research team (17 scientists and farmers) worked with leaders and residents from these watersheds to develop alternative future agricultural scenarios that encouraged farmers to produce various environmental benefits. These scenarios ranged from increased adoption of minimum tillage to the reestablishment of perennial plants and wetlands. For each scenario, researchers estimated potential economic and environmental outcomes: specifically, (1) water quality, (2) fish health, (3) greenhouse gas emissions, (4) carbon sequestration, (5) net farm income (profitability), (6) certain environmental damage costs avoided, and (7) social capital formation. This case study provides a general watershed approach that has broad applicability to developing scenarios for multifunctional agroecosystems in other regions throughout the world.

13.3 CASE STUDY WATERSHEDS

For this analysis, we selected two study areas, Wells Creek (Figure 13.1) and Chippewa River (Figure 13.2), to reflect the variation in agricultural practices, native habitat and wildlife, fish assemblages, soil properties, and agroclimatic conditions in the Upper Midwest of the United States.

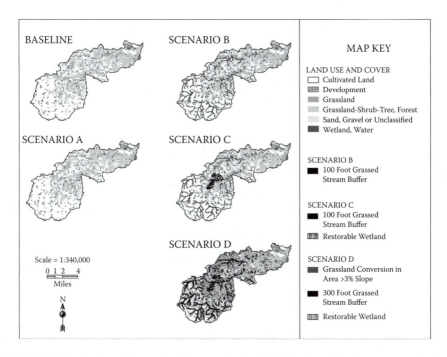

FIGURE 13.1 Wells Creek study area baseline conditions in 1999 and predicted outcomes of various scenarios for potential future land use.

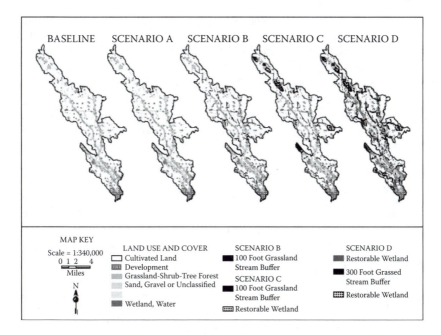

FIGURE 13.2 Chippewa River study area baseline conditions in 1999 and predicted outcomes of various scenarios for potential future land use.

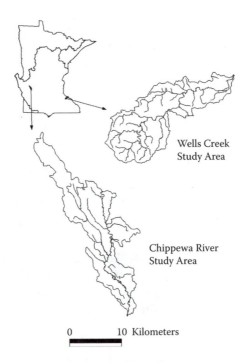

Wells Creek
Study Area

Chippewa River
Study Area

0   10 Kilometers

FIGURE 13.3 Location of the Wells Creek study area (Goodhue County in southeastern Minnesota) and the Chippewa River study area (Chippewa and Swift Counties in southwestern Minnesota).

13.3.1 Wells Creek

A tributary to the Mississippi River, Wells Creek is a 16,264-ha watershed located in Goodhue County, in southeastern Minnesota (Figure 13.1 and Figure 13.3). Farmers make up 54 percent of the study area's 1500 residents; an additional 30 percent of the population lives in rural areas (Boody et al. 2005). With an average slope of 6.5 percent, much of the land is cultivated (61 percent), primarily to corn and soybeans with some small grains and alfalfa hay. The remaining area is woodland (26 percent) or grassland and pasture (10 percent) (USDA 2004). In Wells Creek, 1500 farms operated in 1997, down 12 percent from 10 years previous; during the same period, land planted to corn increased by 22 percent and to soybean by 74 percent, while land devoted to dairy farming decreased. Wells Creek historically supported a cold-water fish assemblage, but nine species collected in 1999 were primarily fish that tolerate high temperatures (Zimmerman et al. 2003). Brown trout, intolerant of high temperature and sediment, were present in low numbers.

13.3.2 Chippewa River

The Chippewa River study area is a 17,994-ha subbasin of the Chippewa River basin, primarily located in Chippewa County in western Minnesota (Figure 13.2 and Figure 13.3). In the Chippewa River study area, about 89 percent of the 6357 residents live in the city of Montevideo (Boody et al. 2005). This relatively flat study area (slopes less than 2 percent) has extensive tile drainage, and 81 percent of the area is planted primarily to corn and soybeans, managed with both conventional and conservation tillage (USDA 2004; Boody et al. 2005). In Chippewa County the number of farms fell 25 percent in 10 years, to 618 farms by 1997. During that 10-year period, the area planted to corn increased by 72 percent and to soybeans increased by 37 percent, replacing small grain and hay. The Chippewa River is a warm-water river, with a fish assemblage of 19 species (Zimmerman et al. 2003), though game fish sought by anglers are present in low numbers.

TABLE 13.1
Land-Use Characteristics of the Wells Creek Study Area under the Current Baseline and Four Hypothetical Scenarios: Scenario A (Continuation of Current Practices), Scenario B (BMPs), Scenario C (High Diversity and Profitability), and Scenario D (Increased Vegetative Cover)

Land Use (ha)	Baseline	Scenario A	Scenario B	Scenario C	Scenario D
Agriculture	11,588	11,588	11,588	11,372	11,372
Grassland	1,656	1,656	1,656	1,656	3,319
Cultivated	9,932	9,932	9,932	9,716	8,053
Small grains–alfalfa (CT)	1,312	1,046	2,022	5,253	3,467
Small grains–alfalfa (CN)	834	665	0	0	0
Corn (CT)	1,111	0	1,348	1,318	870
Corn (CN)	320	0	0	0	0
Corn–soybeans (CT)	2,758	3,626	5,585	2,184	1,441
Corn–soybeans (CN)	3,172	4,171	0	0	0
Cover crops	0	0	0	0	424
Riparian buffer	0	0	553	537	1,851
Deciduous, wooded	4,223	4,223	4,223	4,223	4,223
Developed	372	372	372	372	372
Wetlands	21	21	21	238	238
Open water	32	32	32	32	32
Not classified	28	28	28	28	28

Note: CN, conventional tillage; CT, conservation tillage.

13.3.3 SCENARIO DEVELOPMENT

Four scenarios of possible future land use were developed (Table 13.1 and Table 13.2). Agricultural, socioeconomic, and environmental results from modeling of these four scenarios were compared to current conditions (1999) in the watersheds. Scenarios were citizen driven, based on historical materials created by basin residents and through focus groups and interviews. Focus groups of approximately 40 rural residents and producers from the study areas outlined their desires and expectations for future agricultural land use in each study area. The focus groups provided direction on broad goals or outcomes in production practices under several scenarios. Specific farming systems were not described in detail by focus group members, but sufficient information was obtained so that researchers could model a set of production activities that represented the range of possible practices outlined. Once production activities for each scenario were developed, focus groups were reconvened to solicit additional feedback. This ensured that the farming systems were representative of the range of reasonably practical production activities for either watershed. Scenarios varied slightly between study areas to account for local conditions and differences in agricultural systems.

Researchers developed more detailed descriptions of the scenarios to evaluate quantitative changes in environmental parameters (water quality, global warming potential, wildlife habitat, and fisheries health), agricultural production and associated affects on farmer profits (net farm income), commodity and conservation program payments, and potentially avoided damages (certain externality costs from sedimentation). Potential social impacts were described on the basis of information gathered through interviews, reviews of institutional mission statements, and other approaches. Last, researchers estimated citizen willingness to pay for the potential environmental benefits associated with these changes in land management.

TABLE 13.2
Land-Use Characteristics of the Chippewa River Study Area under the Current Baseline and Scenarios A (Continuation of Current Practices), Scenario B (BMPs), Scenario C (High Diversity and Profitability), and Scenario D (Increased Vegetative Cover)

Land Use (ha)	Baseline	Scenario A	Scenario B	Scenario C	Scenario D
Agriculture	16,013	16,013	16,013	15,515	15,515
Grassland	1,464	1,464	1,464	1,464	2,712
Cultivated	14,550	14,550	14,550	14,051	12,803
Small grains–alfalfa (CT)	229	185	527	7,413	4,962
Small grains–alfalfa (CN)	312	83	0	0	0
Corn–soybeans (CT)	3,796	6,232	11,082	4,607	3,082
Corn–soybeans (CN)	7,587	5,309	0	0	0
Corn–sugar beets (CN)	1,494	1,610	1,454	560	375
Cover crops	0	0	0	0	3,144
Riparian buffer	0	0	356	340	1,240
Deciduous, wooded	1,080	1,080	1,080	1,080	1,080
Developed	637	637	637	637	637
Wetlands	154	154	154	653	653
Open water	108	108	108	108	108
Not classified	2	2	2	2	2

Note: CN, conventional tillage; CT, conservation tillage.

The final scenarios were based on: a continuation of current trends (scenario A); best management practices, or BMPs (scenario B); maximizing diversity and profitability (scenario C); or increased vegetative cover (scenario D). Scenario A (based on current trends) projected an increase in farm-size (area per farm) and an increase in land planted to corn, soybeans, and sugar beets. Small, diversified farms were present, although at decreased levels. Scenario B (based on BMPs) involved conservation tillage on all land currently under conventional tillage; 30-m riparian buffers along both sides of all streams; and following research-based fertilizer application rate recommendations. Scenario C (based on high diversity and profitability) attempted to increase farm profitability and move beyond BMPs. In addition to all scenario B changes, scenario C included wetland restoration and increased crop diversity, with more land in 5-year crop rotations, perennial crops, and managed intensive rotational grazing (MIRG). The 5-year crop rotation included an increase in small grains and alfalfa and a reduction in area under corn–soybean and corn–sugar beet rotations. Scenario D (based on increased vegetative cover) extended scenario C by adding perennial cover; grasslands replaced cultivated lands on an additional 7 to 14 percent of the area, riparian buffers were widened to 90 m, and all row crops included cover crops. The increased grassland reflected conversion to MIRG, restored prairie, and other grasslands. Most of the grasslands and prairie were located on steeper lands (greater than 6 percent in Wells Creek and greater than 3 percent in the Chippewa River study area).

13.4   PROJECTED ENVIRONMENTAL AND ECOSYSTEM EFFECTS

13.4.1   WATER QUALITY

Boody et al. (2005) estimated sediment, nitrogen (N), and phosphorus (P) loadings for baseline land use in 1999 and for the four scenarios using ADAPT (agricultural drainage and pesticide transport), a field-scale biophysical process model designed for water table management research (Gowda et

TABLE 13.3
Percentage Change Sediment, N, and P Loading; Greenhouse Gas Emissions; SOC; Lethal Fish Events; Estimated Aggregate Production Costs; Net Farm Income; Commodity Payments; and CRP Payments in the Wells Creek Study Areas for Each Scenario in Relation to Baseline Conditions

	Baseline	Scenario A	Scenario B	Scenario C	Scenario D
Sediment (Mg/yr)	36	4	−31	−56	−84
Nitrogen (kg/yr)	1,364	−6	−37	−63	−74
Phosphorus (kg/yr)	3,430	−4	−54	−70	−71
Greenhouse gas (MTCE)	5,003	−2	−13	−19	54
SOC (metric tons/yr)	3,902	−3	31	41	86
Lethal fish events (no./yr)	6.7	10	−57	−72	−98
Production costs (M$/yr)	13.5	−1	−3	−8	45
Nitrogen fertilizer (kg/yr)	851,260	−7	−47	−73	−85
Net farm income (M$/yr)	2.1	−1	−1	12	105
Commodity payment (M$/yr)	1.6	−1	−6	−44	−63
CRP payment (M$/yr)	0.1	0	113	110	378
Commodity + CRP (M$/yr)	1.7	−1	3	−27	−24

Note: MTCE, metric tons carbon equivalent.

al. 1999). Detailed descriptions of data, the GIS supporting that research, and biophysical and economic modeling are in Zimmerman et al. (2003), Westra et al. (2004), and Boody et al. (2005). Modifications of ADAPT have been calibrated in several river basins in Minnesota for different soil types, crops, slopes, and other land characteristics similar to those found in the two study areas (Davis et al. 2000; Dalzell et al. 2001; Westra et al. 2002). These studies suggested that the output from ADAPT modeling is unbiased and precise. We present results as relative changes from the baseline for the four scenarios.

As one might expect, changing farming practices reduced the estimated delivery of sediment, N, and P to the mouth of the river in both study areas. On the other hand, researchers found little projected change in sediment or nutrient loading if current trends continued (scenario A). The greatest reductions in sediment and nutrient loading occurred under scenario D, with Wells Creek having reductions of more than 80 percent (Table 13.3). In Wells Creek under all scenarios except A, it would be possible to reduce N by 30 percent. This 30 percent reduction in N is mentioned as the level needed to achieve a goal set by USEPA (2001) for reducing the 5-year running average of the areal extent of the hypoxia zone in the Gulf of Mexico to less than 5000 km^2 by 2015 (Rabalais et al. 2002). In the relatively flat, more intensively farmed Chippewa River study area, implementing BMPs (scenario B) was insufficient to meet the goal (Table 13.4). More diverse farming systems with increased vegetation were needed (scenario C or D) to meet goals of the Hypoxia Action Plan. In both watersheds, all scenarios except continuing current trends (scenario A) met the State of Minnesota's goal of reducing P loading by 40 percent. These findings highlight the importance of using diversified systems to reduce nutrient loading locally and into the Gulf of Mexico as recent analysis indicates that agricultural sources contribute more than 70 percent of the delivered N and P (Alexander et al. 2008).

13.4.2 Fish Populations

Using sediment loading estimations from the ADAPT model, historic stream flow data, and stream bank erosion estimates, Zimmerman et al. (2003) calculated the daily suspended sediment

TABLE 13.4
Percentage Change Sediment, N, and P Loading; Greenhouse Gas Emissions; SOC; Lethal Fish Events; Estimated Aggregate Production Costs; Net Farm Income; Commodity Payments; and CRP Payments in the Chippewa River Study Areas for Each Scenario in Relation to Baseline Conditions

	Baseline	Scenario A	Scenario B	Scenario C	Scenario D
Sediment (Mg/yr)	1.8	−9	−25	−35	−49
Nitrogen (kg/yr)	6,348	1	−17	−51	−62
Phosphorus (kg/yr)	2,322	−5	−42	−70	−75
Greenhouse gas (MTCE)	2,065	0	−6	−39	−37
SOC (metric tons/yr)	4,792	17	37	59	112
Lethal fish events (no./yr)	11.2	2	0	0	−10
Production costs (M$/yr)	9.2	1	−3	−19	−38
Nitrogen fertilizer (kg/yr)	875,205	1	−8	−62	−90
Net farm income (M$/yr)	1.0	2	3	58	32
Commodity payment (M$/yr)	1.4	2	−3	−56	−70
CRP payment (M$/yr)	0.3	0	27	26	245
Commodity + CRP (M$/yr)	1.7	2	3	−41	−13

Note: MTCE, metric tons carbon equivalent.

concentrations in each stream over the 50-year simulation period. By calculating the total number of days each year that concentrations of suspended sediment reached lethal or sublethal thresholds for the respective fish assemblages, they estimated the magnitude of sublethal or lethal events due to suspended sediment for the resident fish in both study areas. Newcombe and Jensen's (1996) meta-analysis, which quantitatively related fish response to concentrations of suspended sediment and duration of exposure, was used to calculate the number of sublethal and lethal events. Sublethal effects were defined as moderate habitat degradation, impaired homing, physiological stress (such as coughing or increased respiration), and reduction in feeding rates or feeding success. Lethal effects included reduced growth rate, delayed hatching, reduced fish density, increased predation, severe habitat degradation, and mortality (Newcombe and Jensen 1996). In general, lethal and sublethal effects on fish increased as the suspended sediment concentrations and duration of exposure increased, although the nature of this relationship varied by fish assemblage. Thresholds corresponding to juvenile and adult salmonids were used to represent the Wells Creek assemblage, whereas the Chippewa River was represented by adult freshwater nonsalmonids, comprising mainly warm-water species.

Of the two study areas, the Chippewa River had more mean annual days with sublethal and lethal events under baseline conditions. Under the four scenarios, changes in sediment loading decreased lethal events up to 98 percent in Wells Creek (Table 13.3), but had only a minor effect in the Chippewa River (Table 13.4). In the Chippewa River, the number of days with sublethal and lethal events did not change significantly across scenarios. One reason for this may have been the relatively flat topography of that study area, which may have resulted in sediment concentrations that were often lower than in Wells Creek, but longer in duration. It was also likely that the fish species in the Chippewa River were more sensitive to extended exposure of suspended sediments than the fish in Wells Creek (Newcombe and Jensen 1996).

These findings suggest significant improvements in fish assemblages are possible if sediment concentrations are reduced, but the level of improvement may vary due to characteristics of the fish assemblage and watershed. Although difficult to predict, reductions in lethal and sublethal events from suspended sediment for fish in Wells Creek could be the catalyst for a change in the fish assemblage to one with more cool-water or cold-water species. If expanded riparian areas modeled under scenarios B through D provided shade for 50 percent of the stream surface, trout populations in Wells Creek would be expected to increase (Blann et al. 2002). However, policy makers and resource managers need to be mindful that if one of the underlying goals is to increase fish population, targeting land management efforts based on physical criteria such as total maximum daily load (TMDL) regulations may not yield the desired results or be the most economically efficient approach (Watanabe et al. 2006; Hascic and Wu 2006).

13.4.3 GREENHOUSE GASES

Among other factors, estimated changes in greenhouse gas emissions for each scenario were calculated from changes in N fertilizer use based on farmer surveys. These estimates include projected changes in land use, changes in the number and type of livestock (to determine the contribution of ruminants and manure to projected emissions), potential reductions in N fertilizer use when farmers take credit for N in legumes, and animal manure fertilizer. Emissions of N_2O from altered fertilizer use were calculated using guidelines established by the Intergovernmental Panel on Climate Change (IPCC 2001). For all livestock except ruminants, animal numbers and management were held constant across scenarios. Ruminants were differentiated among dairy, beef heifers, cows, and steers and were classified into those housed in confinement, conventionally grazed, and grazed using MIRG. Emissions of CH_4 and N_2O, with respective global warming potentials of 21 times and 310 times that of CO_2 (IPCC 2001), were calculated and converted to carbon equivalents using guidelines of the IPCC. For MIRG, we assumed that animals grazed in paddocks for up to 24 hours, eight times per season, to allow for the recovery and continued growth of pasture plants both above and in the soil.

In the Chippewa River watershed, converting land to pasture, increasing rotations, and widening riparian buffers reduced atmospheric N losses from fields. These land use changes were predicted to reduce losses of N_2O by 83 percent, from 17,562 to 2,958 kg per year (Table 13.4). Under scenario D, the area of pastured grasslands grazed using MIRG doubled in each study area. Therefore, the number of ruminant animals in scenario D increased by 6785 dairy animals (+125 percent) and 1710 beef animals (+125 percent) in Wells Creek, and by 640 dairy animals (+252 percent) and 515 beef animals (+90 percent) in the Chippewa. A reduction in greenhouse gases of as much as 37 percent is predicted in the Chippewa River study area if scenario D is adopted (Table 13.3). In the Wells Creek study area, the predicted reductions are smaller for scenarios B and C, because dairy animals generate more CH_4 than beef cattle. Under scenario D, if dairy animals were increased by an additional 125 percent in the Wells Creek study area, greenhouse gas emissions would increase by 54 percent (Table 13.3). Thus, certain changes in farming practices have the potential for exacerbating specific environmental problems instead of mitigating them. However, if the full range of benefits is understood a practice such as increasing grass-based dairying has significant benefits.

13.4.4 CARBON SEQUESTRATION

In these two watersheds, the potential increase in SOC (in metric tons per hectare per year) for land uses projected under our four scenarios were calculated using a value of 0.10 for conventional cropping (U.S. Department of State 2000), 0.30 for conservation tillage (Robertson et al. 2000), 0.90 for pastures (Follett et al. 2001), and 2.0 for wetlands (Lal et al. 1999). Kahn et al. (2007) found that long-term N fertilization, especially at high levels, may deplete soil carbon. Boody et al. (2005) estimated that in the Wells Creek study area, SOC had the potential to increase by 86 percent, from 3902 to 7245 metric tons per year, under scenario D due to the conversion of row crop land to perennials and the use of cover crops in rotation with annual crops. This increase in SOC had the potential to offset 22 percent of the greenhouse gas emissions (CH_4) associated with increasing the number of cattle in Wells Creek (Table 13.3). In the Chippewa River study area under scenario D, SOC was estimated to increase from 4,792 to 10,147 metric tons per year (an increase of 112 percent) and total net carbon storage (SOC minus total greenhouse gas production) *increased by 328 percent* (Table 13.4). There have been few studies measuring carbon sequestration under MIRG systems after conversion from row crops in northern temperate climate zones such as Minnesota (Contant et al. 2003). As a result, Boody et al. (2005) estimated SOC using information from studies converting row cropland to perennial cover such as in the Conservation Reserve Program (CRP) (Follett et al. 2001).

13.5 ECONOMIC EFFECTS

13.5.1 Farm Production Costs

Under any of the scenarios analyzed in both watersheds, changes in production practices changed the set of inputs used, the intensity of equipment use, and the costs of production for commodities. For example, under baseline conditions, 851,260 kg of N fertilizer per year was applied in the Wells Creek study area (Table 13.3), and 875,205 kg per year was applied to cropland in the Chippewa River (Table 13.4). Because many producers are risk averse, nutrients are often applied at rates that exceed agronomic recommendations, resulting in higher than needed production costs and oftentimes affecting water quality (e.g., reducing dissolved oxygen; Berka et al. 2001). For all scenarios other than the baseline, producers followed University of Minnesota Extension Service recommendations and took credit for the N content of legumes in rotations or in manure applied as fertilizer. As a result, N and P application rates and production costs were reduced.

Aggregate production costs, calculated as an area-weighted summation of production costs for each system, changed under all scenarios. For the current-trend scenario (scenario A), there was a negligible change in production costs. Cost savings of 3 percent with BMPs (scenario B) resulted from reductions in N use of 47 percent in Wells Creek (Table 13.3) and 8 percent in the Chippewa River (Table 13.4), reduced tillage costs, and reduced production costs from land converted into buffers. For conservation tillage systems, all producers surveyed had the necessary equipment to switch from conventional tillage to conservation tillage without purchasing additional equipment. As a result, variable production costs were reduced for a conservation tillage system, relative to the conventional tillage system, as fuel and equipment repair costs were reduced (Westra et al. 2004). Production costs declined further under scenario C, because costs associated with small grains and alfalfa are lower than those for corn and soybeans. Because most producers surveyed in this project had the necessary equipment for the transition to small grains or hay in a crop rotation (scenarios C and D), no additional transition costs were assumed (Westra et al. 2004). Perennial cover (scenario D) had higher costs in Wells Creek, but lower costs in the Chippewa, compared with the baseline; this was not surprising, considering the increased dairy production and associated expenses under MIRG in Wells Creek.

13.5.2 Net Farm Income

Net farm income, a measure of returns to management effort, was a function of (1) the output produced by farmers under the various systems modeled, and (2) a 5-year weighted average of real output prices (2000) for crop and livestock products in Minnesota, and production costs for each system. Output prices were assumed to remain unchanged in all scenarios, because the quantity supplied or produced was small relative to the market for all commodities and livestock products. The output for each system represented the surveyed producer's estimated average production, adjusted to reflect differences in soil quality in each watershed (Westra et al. 2004).

Net farm income was projected to fall slightly under scenarios A and B. In Wells Creek, under scenario A, farm income declined because revenues declined more than production costs, even though slightly more land was planted to corn and soybeans (Table 13.3). For scenario B, farm income declined because of reductions in yield for conservation tillage and land converted to buffers. By contrast, net farm income increases under scenarios C and D in both study areas (Table 13.3). These results demonstrate that it is possible for producers to create multiple benefits while increasing their farm income (at least that portion that is derived from the market; not government commodity programs).

Government commodity program payments were separated from farm income estimates, to demonstrate how both income and commodity program payments would be affected by changing farming practices. Commodity payments were estimated by adjusting mean commodity payments per farm enrolled in the area Farm Business Management Association by the percentage of the total farm program payments in Chippewa and Goodhue Counties for each crop (EWG 2003). Because Wells Creek had dairy payments, 9 percent of the Goodhue County dairy subsidies were applied to the total government program payments in that study area across each scenario. That percentage is based on the size of the study area in relation to the county.

Under baseline conditions, commodity program payments approached (Wells Creek) or exceeded (Chippewa) the level of farm income derived from the market. The size of these payments relative to net farm income demonstrates how much the commodity programs motivate farmer behavior in terms of what crops to plant and how intensively to grow them. Results for scenario A indicated that commodity payments encouraged production that lowered farm income, increased government costs, and increased environmental costs relative to what could potentially be achieved under the other scenarios (B, C, and D). Commodity payments were projected to decrease significantly under scenarios C and D (Table 13.3 and Table 13.4) because land planted to corn and soybeans decreased (it was planted to perennial cover (buffers) or to more diverse rotations).

The researchers assumed that land converted to riparian buffers in scenarios B, C, and D was eligible for enrollment in the Conservation Reserve Program (CRP). The CRP payments, based on the average CRP rental rates in each county for 2000, could partially offset lower commodity program payments (Table 13.3 and Table 13.4), especially under scenario B. There could also be significant one-time payments, totaling from $525,000 in Wells Creek to $1,285,000 in the Chippewa if new wetlands (scenario C and scenario D) were enrolled in a program like the Wetland Reserve Program. In economic terms, the marginal cost to taxpayers for environmental changes projected under scenarios B, C, and D was likely zero because these increased conservation payments would be offset by reduced commodity program payments.

13.5.3 EXTERNALITY COSTS

13.5.3.1 Reduced Sedimentation

One of the only estimated costs of externalities associated with sedimentation in freshwater systems was developed by Ribaudo (1989). Economic damages from sedimentation by Ribaudo were estimated for physical costs associated with sedimentation, including dredging stream channels. Deleterious impacts on wildlife and fish were not included in these damage estimates. Using an inflation-adjusted cost of $5.38 per metric ton (real 2000 dollars) for damages caused by waterborne sediment, the economic damages avoided through reductions in sediment loading under each scenario were estimated. For the baseline, economic damages associated with sedimentation were $213,131 per year for Wells Creek and $10,525 per year in the Chippewa. The cost per unit damage was assumed constant throughout so that cost reductions across scenarios were identical to the percentage reductions in sediment load (Table 13.3 and Table 13.4). As with sediment, much greater reductions (absolute and relative) in economic damages were achieved in Wells Creek than in the Chippewa study area. Thus, the potential benefits of similar changes in land management practices differ across watersheds. This finding highlights the importance of efficiency and equity gains from targeting practices or areas with resource concern, especially when budgets of agencies addressing these resource concerns are limited.

13.5.3.2   Reduced Flooding

Flooding often results in short-run and long-term economic losses. Flood magnitudes have increased in the Mississippi River Valley over the past several decades, in part because of extensive land use change and greater channel confinement (Miller and Nudds 1996). In these case studies, increased soil infiltration capacity from increased conservation tillage and more perennial cover reduced runoff and could reduce flooding in both study areas (Zimmerman et al. 2003). Riparian buffers reduce overland runoff into streams (Smith 1992; Daniels and Gilliam 1996), and wetland restoration can reduce flood flow volumes (Demissie and Khan 1993; Schultz and Leitch 2003). Modeling has shown that reducing runoff by 10 percent in a watershed may reduce the flood peaks with a 2- to 5-year return period by 25 to 50 percent, and might reduce a 100-year flood by as much as 10 percent (USACE 1995). In Wells Creek, an increase in wetland area from 21 to 238 ha reduced peak flow and flood flow volumes approximately 10.4 percent, while in the Chippewa, an increase in wetland area from 154 to 653 ha reduced flows by 5.8 percent.

13.5.4   Economic Benefits

If changes in land management practices create environmental benefits, what are citizens willing to pay for those environmental benefits? Many changes in environmental quality have no market mechanism by which people can reveal their willingness to pay for benefits. Economists theorize that because people have no mechanism for directly revealing their willingness to pay for these benefits, a less than optimal amount of these benefits is created. Economist in this study used contingent valuation techniques to estimate the economic value associated with the environmental benefits created by changes in agricultural practices in the two study areas (Bishop and Heberlein 1990). Welle (2001) used focus groups that covered valuation of environmental changes to identify questions for a mail survey and personal interviews. Based on these focus groups, the contingent valuation (CV) survey centered on a 50 percent reduction in soil erosion and agricultural nutrient runoff, a 25 percent reduction in small to moderate flooding from agricultural lands, a 10 to 20 percent reduction in greenhouse gases from agriculture, and a 50 percent increase in bird and wildlife habitat on Minnesota farmland—levels consistent with scenarios C and D.

Using a statewide mail survey of randomly selected households in Minnesota, respondents indicated that they were willing to pay $201 annually per household to reduce environmental impacts from agriculture. Personal interviews conducted in the study areas indicated a higher willingness to pay for these benefits ($394 annually per household). The higher value in study area could be due to the personal nature of the interviews compared with the mail survey. On the other hand, residents in the study watersheds may place higher values on environmental benefits, because they perceive these as more localized and tangible than to statewide respondents (Welle 2001; Boody et al. 2005). These findings were in line with results presented in a meta-analysis of valuation of water quality improvements in the United States by van Houtven et al. (2007).

The approach and findings by Boody et al. (2005) conform to results of others who determined the public's willingness and level of support for farmer's providing environmental and social benefits beyond market commodities. Moran et al. (2007) used two different approaches, analytical hierarchy process (AHP) and choice experiments (CE), with focus groups in Scotland to determine the public's preference and willingness to pay farmers subsidies (financed through increased taxes) for enhanced water quality, improved wetlands, locally produced food, and other landscape and environmental benefits (increased habitat and ecosystem services). Similar support by the public for multifunctional agriculture was found in Spain using the CE methodology (Kallas et al. 2007) and the CV and AHP approaches (Kallas et al. 2007). Last, Bennet et al. (2004) used the CE approach and found that urban dwellers were willing to support farmers who provided environmental benefits and maintained viability of rural communities in Australia. Though the level of support varied across valuation techniques and locations, results from all approaches demonstrated the existence of significant demand for ecosystem services and other social benefits provided by farmers in a variety of locations throughout the world.

13.6 CONSEQUENCES OF DIVERSIFYING AGRICULTURE

The findings from these two case studies indicate that diversifying agriculture on working lands could provide environmental, social, and economic benefits, and that citizens would be willing to pay for these benefits. These results corroborate findings by Tilman et al. (2001) and Wackernagel et al. (2002) in that if present land use patterns continue, environmental, social, and economic problems will be exacerbated. To prevent that future scenario, U.S. farm policy will need to be modified so that farmers are rewarded for producing the environmental and economic benefits of a diversified agriculture. Additionally, the deleterious environmental and economic effects of the current commodity programs in U.S. farm policy need to be eliminated or diminished. Unfortunately, recent changes to agricultural and energy policy (RFA 2008), ostensibly designed to increase energy independence through biofuels, have increased the area planted to corn and have exacerbated the environmental problems associated with row crops (Marshall 2007).

In the two case studies, changes in land management under scenarios B, C, and D would lead to changes in nutrient and sediment losses and to reductions in production expenses associated with a decline in N fertilizer use and increased conservation tillage. These scenarios also would increase perennial cover and diversify rotations with more small grains rather than corn and soybeans. The increased net farm income associated with scenarios C and D, coupled with the increased conservation payments from CRP and WRP, suggests that the environmental benefits from these two scenarios are achievable at similar or lower government expenditures than are currently occurring with commodity program payments. Nonmarket environmental benefits realized under scenarios B, C, and D included reduced nutrient and sediment losses, improved water quality and fish health, increased carbon sequestration, decreased greenhouse gas emissions, and reduced runoff and flooding potential. These significant environmental changes could be attained through a combination of land use changes, ranging from individual practices like BMPs to more comprehensive systemic changes such as the establishment of perennial plant systems and wetlands (Mitsch et al. 2001). Thus, in either watershed, under scenarios B, C, and D net social benefits or welfare would increase.

These findings highlight the fact that different types of geography, climate, soil type, and even social infrastructure may require a variety of strategies to attain environmental benefits in different watersheds. For example, the adoption of agricultural BMPs alone may not be sufficient to reduce the nutrient load causing the hypoxic zone in the Gulf of Mexico (a goal of the Hypoxia Action Plan). In relatively flat landscapes, such as the Chippewa River study area, meeting such goals would require diverse farming systems that include perennial plant systems. In steeper landscapes, such as Wells Creek, BMPs using recommended fertilizer rates might suffice, but additional reductions could be achieved with a diversified landscape. One critical impediment to achieving these goals is the structure of the current U.S. farm programs. If we are to produce the multiple benefits U.S. farmers are capable of producing, then U.S. farm policy must be designed to create incentives for farmers to use practices appropriate to local conditions that provide environmental benefits. Such a policy would be more economically efficient than the current conservation programs at producing environmental benefits (Abler 2004).

Focus group participants in these case studies indicated that present commodity programs discouraged diversified agriculture and conservation efforts. Farmers in both study areas agreed and suggested that innovations in farming, including more diversified agricultural systems, are more likely to occur if local institutions are willing to change with farmers (Salomonsson et al., Chapter 16). Innovations in the ways farmers gather and share information are necessary because many current governmental programs focus on a few crops and reinforce production of traditional commodities, such as corn and soybeans in the Midwest, cotton and rice in the South, and sorghum and wheat in the Plains.

13.7   POLICY IMPLICATIONS

Present U.S. conservation programs operate within a system of income- and commodity-support programs that encourage producers to maximize production. Over the 10 most recent calendar years (1996 through 2005), conservation program payments were $2 billion annually, on average (USDA 2007a). This accounted for 12 percent of total direct government payments to producers in the United States during that period. Between 1985 and 2002, about 70 percent of conservation program payments were for land retirement programs. Although it may be appropriate for marginal or ecologically sensitive lands to be retired from production agriculture, results from the case studies above demonstrate that it is possible for agricultural lands to produce environmental benefits while simultaneously producing agricultural commodities. We suggest a reorientation of U.S. farm policy to foster the creation of such multiple benefits. Rather than supporting commodity production, government policy should support agricultural diversification to enhance nonmarket ecosystem services. Future farm programs should reward farmers for producing environmental benefits, especially if by providing these social benefits their production of conventional market products decreases. Policies that help create options for farmers, help support farm income via safety nets for all farmers, and offer incentives for pilot and demonstration projects will help restore vibrancy and diversity to working agricultural landscapes. By initially implementing policy changes on a pilot basis in a variety of select watersheds across the country, we may identify advantages and limitations of rewarding producers who provide ecosystem services. These policies need to be integrated across environmental goals, because such an integrated, holistic approach will likely be more efficient than the current set of programs.

To create additional environmental benefits from pasture and hay production, U.S. farm policy needs to be modified to allow for more ruminant production on grass (as under scenarios C and D in the case studies). A coordinated change in U.S. policy could promote grass-finished beef by altering U.S. Department of Agriculture (USDA) meat-grading standards and by educating the public about its lack of use of antibiotics and its reduced risk of contamination by bovine spongiform encephalopathy (BSE) and microbes such as *Escherichia coli* O157 (Diez-Gonzalez et al. 1998). These case studies demonstrated that policy encouraging crop rotations and MIRG on working farmland, through programs like the Conservation Security Program (CSP) or other green payment programs, could support farm income, encourage land use changes from row crops, and provide significant environmental benefits to society (Westra et al. 2004; Dobbs and Pretty 2004). For instance, the CSP could be used instead of the CRP to enroll the expanded buffers found in scenario D. This would allow farmers to harvest perennial crops for energy or hay, or allow livestock grazing, while continuing to provide many of the same environmental benefits from this land under CRP.

Agricultural policy can be used more effectively to promote strategic preservation and restoration of wetlands. As these case studies demonstrated, the total area of working lands restored to wetlands might be much less than the area currently in land retirement programs. Such changes would be less expensive to taxpayers in the long run. The value and benefits of investing in wetlands restoration stem from the joint production of several "environmental services" (nitrogen abatement, buffering and storage capacity, and biodiversity) and the net natural growth implies increased future supply of these benefits—increasing the returns to scale of environmental benefits (Gren 1994). However, the means by which farmers are compensated for wetlands restoration and the location of that restoration must be carefully determined (Gren 2004). To increase the potential benefits, areas for wetland restoration and preservation might be targeted to areas historically in wetlands, and efforts should be made to locate wetlands to maximize hydrological connection with upland areas (Shultz and Leitch 2003).

Much research of multifunctional agriculture has focused primarily on food production, environmental quality, and continuing agriculture in a changing landscape. However, if the full potential of multifunctional agriculture in the United States is to be realized, additional thought and energy should be expended to broaden the focus of farm policy to include biomass energy production under perennial systems, recreation, education, and other activities besides food production that bring income and economic development to farmers and the rural areas in which they live.

In 2007, the U.S. government set a goal of producing 36 billion gallons of ethanol by 2022. This includes 15 billion gallons of corn-based ethanol and 21 billion gallons from alternatives that focus on cellulosic sources (RFA 2008). Marshall (2007) predicted that to produce 11 billion gallons of ethanol annually would require an increase in area harvested for corn to 92.5 million acres in the United States. Such a situation would be associated with a 3.7 percent increased loss of N to water, a 2.5 percent increased loss of P to water, and a 2.8 percent increase in sediment loss. Actual corn acreage planted in 2007 was 92.9 million acres (USDA 2007b). In contrast, scenario D from Boody et al. (2005), which simulated a cellulosic energy and grazing scenario doubling grassland acres in Wells Creek and increasing grassland by 85 percent in Chippewa, showed steep declines in pollutant losses to water and air. More recently, Boody et al. (2007) predicted losses to water for the Rock Creek, a small tributary of Lake Erie in Ohio. In that study, they found that a 25 percent increase in corn acreage over the baseline production was associated with N loss that increased by 6 percent, sediment loss increased by 2 percent, and total P loss increased by 1 percent. Scenarios with increasing perennials for cellulosic energy (such as 500-ft grass buffers) or pastures for grazing that converted 2800 acres from soybeans dramatically decreased those nutrient and sediment losses by 18 percent to more than 40 percent.

Advancing multifunctional agriculture will require diligence and foresight into designing policies that buy ecosystem services in a cost-effective manner (Claassen 2007). The U.S. policy environment recently has become more complex with agricultural, environmental, and energy policy all playing roles in prompting significant, large-scale changes in agricultural land use related to energy production. We need holistic analyses to understand the impacts on farms, food, energy availability, and ecosystems so that policy making takes into account projected impacts on our natural resources, farmers of all sizes, and rural communities.

Critics of proposals to reorient agricultural policy assert that these changes will reduce production, thereby exacerbating worldwide food shortages. However, on a per-unit area basis, diversified farming systems can produce more food and ecosystem services than more simplified, conventional systems. It is anticipated that the policy changes described above and simulated in the case studies would reduce the total production of corn and soybean commodities; slightly under scenario B, more so under scenarios C and D. However, meat production probably would show no net change, dairy production would probably increase, and production of crops that are directly consumed by humans might even increase. From a broader perspective, current U.S. agricultural policy has produced flooding in the north-central United States, hypoxia in the Gulf of Mexico, and increasing production of greenhouse gases. All these unintended products of U.S. agriculture have real economic costs to society. Furthermore, it is neither feasible nor sustainable in the long run to exchange future agricultural productivity and environmental quality for short-term increases in commodity production. Dealing with current and future global food shortages requires an agriculture that produces rather than consumes ecosystem services. If we can muster the political will to encourage our farmers to produce ecosystem services, in addition to food and fiber, U.S. agriculture will once again become a sustainable model for the rest of our world.

REFERENCES

Abler, D. 2004. Multifunctionality, agricultural policy, and environmental policy. *Agricultural and Resource Economics Review*, 33, 9–17.

Alexander, R. et al. 2008. Differences in phosphorus and nitrogen delivery to the Gulf of Mexico from the Mississippi River Basin. *Environmental Science and Technology*, 42, 822–830.

Baker, J. et al. 2007. Tillage and soil carbon sequestration—What do we really know? *Agriculture, Ecosystems, and Environment,* 118, 1–5.

Batie, S. 2003. The multifunctional attributes of Northeastern agriculture: A research agenda. *Agricultural and Resource Economics Review*, 32, 1–8.

Bennet, J., van Bueren, M., and Whitten, S. 2004. Estimating society's willingness to pay to maintain viable rural communities. *Australian Journal of Agricultural and Resource Economics*, 48, 487–512.

Berka, C., Schreier, H., and Hall, K. 2001. Linking water quality with agricultural intensification in a rural watershed. *Water, Air, and Soil Pollution*, 127, 389–401.

Best, L. et al. 1995. A review and synthesis of habitat use by breeding birds in agricultural landscapes of Iowa. *American Midland Naturalist*, 134, 1–29.

Bishop, R. and Heberlein, T. 1990. The contingent valuation method. In *Economic Valuation of Natural Resources: Issues, Theory, and Application*, Johnson, R., and Johnson, G., Eds. Westview Press, Boulder, CO, 81–104.

Blann, K., Nerbonne, J., and Vondracek, B. 2002. Relationship of riparian buffer type to physical habitat and stream temperature. *North American Journal of Fisheries Management*, 22, 441–451.

Bohman, M. et al. 1999. The use and abuse of multifunctionality. Briefing paper, USDA, Economic Research Service, Washington, D.C. http://www.ers.usda.gov/Briefing/WTO/PDF/multifunc1119.pdf (accessed June 24, 2008).

Boody, G. et al. 2005. Multifunctional agriculture in the US. *BioScience*, 55, 27–48.

Boody, G. et al. 2007. Results and policy implications of modeling diversified farming systems in watershed in OH and MN. Multiple Benefits of Agriculture Initiative briefing paper, Land Stewardship Project. http://www.landstewardshipproject.org/programs (accessed February 20, 2008).

Claassen, R. 2007. Buying environmental services: Effective use of economic tools. In *Managing Agricultural Landscapes for Environmental Quality*, Schnepf, M., and Cox, C., Eds. Soil and Water Conservation Society, Ankeny, IA, 92–103.

Claassen, R. et al. 2001. Agri-environmental policy at the crossroads: Guideposts on a changing landscape. Agricultural Economic Report no. 79. USDA, Economic Research Service, Washington D.C. http://www.ers.usda.gov/publications/aer794/aer794.pdf (accessed June 24, 2008).

Cochrane, W. 2003. *Curse of American Agricultural Abundance: A Sustainable Solution*. University of Nebraska Press, Lincoln.

Contant, R., Six, J., and Paustian, K. 2003. Land use effects on soil carbon fractions in the southeastern United States; management-intensive versus extensive grazing. *Biology and Fertility of Soils*, 38, 386–392.

Dalzell, B., Mulla, D., and Gowda, P. 2001. Modeling and evaluation of alternative agricultural management practices in Sand Creek watershed. In *Proceedings International. Symposium on Soil Erosion Research for the 21st*, Ascough II, J., and Flanagan, D., Eds. American Society of Agricultural Engineers, St. Joseph, 637–640.

Daniels, R. and Gilliam, J. 1996. Sediment and chemical load reduction by grass and riparian filters. *Soil Science Society of America Journal*, 60, 246–251.

Davis, D. et al. 2000. Modeling nitrate nitrogen leaching in response to nitrogen fertilizer rate and tile drain depth or spacing for southern Minnesota, USA. *Journal of Environmental Quality*, 29, 1568–1581.

Demissie, M. and Kahn, A. 1993. Influence of wetlands on streamflow in Illinois. Contract Report no. 561. Illinois State Water Survey Hydrology Division, Champaign.

DeVries, B. 2000. Multifunctional agriculture in the international context: a review. Working paper, Land Stewardship Project. http://www.landstewardshipproject.org/mba/MFAReview.pdf (accessed June 24, 2008).

Diez-Gonzalez, F. et al. 1998. Grain feeding and the dissemination of acid-resistant Escherichia coli from cattle. *Science*, 281, 1666–1668.

Dobbs, T. and Pretty, J. 2004. Agri-environmental stewardship schemes and "multifunctionality." *Review of Agricultural Economics*, 26, 220–237.

Environmental Working Group (EWG). 2003. Farm Subsidy Database. http://www.ewg.org/farm/ (accessed June 24, 2008).

Faeth, P. and Greenlaugh, S. 2000. *A Climate Change Strategy for US Agriculture*. World Resources Institute, Washington, D.C.

Farber, S. et al. 2006. Linking ecology and economics for ecosystem management. *BioScience*, 56, 121–133.

Follett, R. et al. 2001. Carbon sequestration under the Conservation Reserve Program in the historic grassland soils of the United States of America. In *Soil Carbon Sequestration and the Greenhouse Effect*, Lal, R., Ed. Special Publication 57 Soil Science Society of America, Madison, WI, 27–40.

Gowda, P. et al. 1999. The sensitivity of ADAPT model predictions of streamflows to parameters used to define hydrologic response units. *Transactions of the American Society of Agricultural Engineers*, 42, 381–389.

Gren, I. 1994. The value of investing in wetlands for nitrogen abatement. *European Review of Agricultural Economics*, 22, 157–172.

Gren, I. 2004. Uniform or discriminating payments for environmental production on arable land under asymmetric information. *European Review of Agricultural Economics*, 31, 61–76.

Harte, L. and O'Connell, J. 2003. How well do agri-environmental payments conform to multifunctionality? *EuroChoices*, 2, 36–41.

Hartell, J. 2004. Pricing benefit externalities of soil carbon sequestration in multifunctional agriculture. *Journal of Agricultural and Applied Economics*, 36, 491–505.

Hascic, I. and Wu, J. 2006. Land use and watershed health in the United States. *Land Economics*, 82, 214–5239.

Havlik, P. et al. 2005. Joint production n under uncertainty and multifunctionality of agriculture: Policy considerations and applied analysis. *European Review of Agricultural Economics*, 32, 489–515.

Huggins, D. et al. 1998. Soil organic C in the tallgrass prairie-derived region of the corn belt: Effects of long-term crop management. *Soil and Tillage Research*, 47, 219–234.

Intergovernmental Panel on Climate Change (IPCC). 2001. Good practice guidance and uncertainty management in national greenhouse gas inventories. National Greenhouse Gas Inventories Programme, Hyama, Kanagawa Japan. http://www.ipcc-nggip.iges.or.jp/public/gp/english/ (accessed June 24, 2008).

Kahn, S. et al. 2007. The myth of nitrogen fertilization for soil carbon sequestration. *Journal of Environmental Quality*, 36, 1821–1832.

Kallas, Z., Gomez-Limon, J., and Arriaza, M. 2007a. Are citizens willing to pay for agriculture multifunctionality? *Agricultural Economics*, 36, 405–419.

Kallas, Z., Gomez-Limon, J., and Hurle, J. 2007b. Decomposing the value of agricultural multifunctionality: Combining contingent valuation and analytical hierarchy process. *Journal of Agricultural Economics*, 58, 218–241.

Kucharik, C., Roth, J., and Nabielski, R. 2003. Statistical assessment of a paired-site approach for verification of carbon and nitrogen sequestration on Wisconsin Conservation Reserve Program land. *Journal of Soil and Water Conservation*, 58, 58–67.

Lal, R. et al. 1999. Managing cropland to sequester carbon in soil. *Journal of Soil and Water Conservation*, 5, 374–381.

Lankoski, J. and Ollikainen, M. 2003. Agri-environmental externalities: A framework for designing targeted policies. *European Review of Agricultural Economics*, 30, 51–75.

Lant, C. et al. 2005. Using GIS-based ecological-economic modeling to evaluate policies affecting agricultural watersheds. *Ecological Economics*, 55, 467–484.

Mann, S. 2003. Doing it the Swiss way. *EuroChoices*, 2, 32–35.

Marshall, L. 2007. *Thirst for Corn: What 2007 Plantings Could Mean for the Environment*. WRI policy note 2. World Resources Institute, Washington, D.C.

Miller, M. and Nudds, T. 1996. Prairie landscape change and flooding in the Mississippi River Valley. *Conservation Biology*, 10, 847–853.

Mitsch, W. et al. 2001. Reducing nitrogen loading to the Gulf of Mexico from the Mississippi River Basin: Strategies to counter a persistent ecological problem. *BioScience*, 51, 373–388.

Moran, D. et al. 2007. Quantifying public preferences for agri-environmental policy in Scotland: A comparison of methods. *Ecological Economics*, 63, 42–53.

Newcombe, C. and Jensen, J. 1996. Channel suspended sediment and fisheries: a synthesis for quantitative assessment of risk and impact. *North American Journal of Fisheries Management*, 16, 693–727.

Organization for Economic Co-operation and Development (OECD). 2001a. *Multifunctionality: Towards an Analytical Framework*. OECD, Paris.

Organization for Economic Co-operation and Development (OECD). 2001b. *Multifunctionality: The Policy Implications*. OECD, Paris.

Paustian, K. et al. 1997. Agricultural soils as a sink to mitigate CO_2 emissions. *Soil Use and Management*, 13, 230–244.

Peterson, J., Boisvert, R., and de Gorter, H. 2002. Environmental policies for a multifunctional agricultural sector in open economies. *European Review of Agricultural Economics*, 29, 423–443.

Post, W. and Kwan, K. 2000. Soil carbon sequestration and land-use change: processes and potential. *Global Change Biology*, 6, 317–327.

Rabalais, N., Turner, R., and Wiseman, W. 2001. Hypoxia in the Gulf of Mexico. *Journal of Environmental Quality*, 30, 320–329.

Rabalais, N., Turner, R., and Scavia, D. 2002. Beyond science into policy: Gulf of Mexico hypoxia and the Mississippi River. *BioScience*, 52, 129–142.

Randall, A. 2002. Valuing the output of multifunctional agriculture. *European Review of Agricultural Economics*, 29, 289–307.

Rayburn, E. 1993. Potential ecological and environmental effects of pasture and BGH technology. In *The Dairy Debate: Consequences of Bovine Growth Hormone and Rotational Grazing Technologies*, Liebhardt, W., Ed. University of California Printing Services, Davis, 247–276.

Renewable Fuels Association (RFA). 2008. Renewable fuel standard. http://www.ethanolrfa.org/resource/standard (accessed June 24, 2008).

Ribaudo, M. 1989. Water quality benefits from the Conservation Reserve Program. Agricultural Economic Report 606. USDA, Economic Research Service, Washington, D.C.

Robertson, G., Paul, E., and Harwood, R. 2000. Greenhouse gases in intensive agriculture: contributions of individual gases to the radiative forcing of the atmosphere. *Science*, 289, 1922–1925.

Romstad, E. et al. 2000. Multifunctional Agriculture: Implications for Policy Design. Report 21. Agricultural University of Norway, Ås.

Santlemann, M. et al. 2004. Assessing alternative futures for agriculture in Iowa, USA. *Landscape Ecology*, 19, 357–374.

Shultz, S. and Leitch, J. 2003. The feasibility of restoring previously drained wetlands to reduce flood damage. *Journal of Soil and Water Conservation*, 58, 21–29.

Smith, V. 1992. Arbitrary values, good causes, and premature verdicts. *Journal of Environmental Economics and Management*, 22, 71–89.

Tilman, D. et al. 2001. Forecasting agriculturally driven global environmental change. *Science*, 292, 281–284.

U.S. Army Corps of Engineers (USACE). 1995. *Floodplain Management Assessment of the Upper Mississippi River and Lower Missouri Rivers and Tributaries*, Vol. 6. USACE, Washington, D.C.

U.S. Department of Agriculture (USDA). 2004. 2002 Census of agriculture. Vol. 1, Geographic area series. Part 23, Minnesota State and county data. USDA, National Agricultural Statistics Services, Washington, D.C. http://www.nass.usda.gov/census/census02/volume1/mn/MNVolume104.pdf (Accessed June 24, 2008).

U.S. Department of Agriculture (USDA). 2007. Farm income and costs: farms receiving government payments; which farms receive government payments? Briefing room data USDA, Economic Research Service. http://www.ers.usda.gov/Briefing/FarmIncome/govtpaybyfarmtype.htm (accessed June 24. 2008).

U.S. Department of State. 2000. US Submission on land use, land use change, and forestry to conference parties. 6th Meeting of the Conference of the Parties to the Climate Convention, The Hague, Netherlands. http://www.state.gov/www/global/global_issues/climate/000801_unfccc1_subm.pdf (accessed June 24, 2008).

U.S. Environmental Protection Agency (USEPA). 2000a. Minnesota greenhouse gas emissions and sinks inventory: Summary. http://yosemite.epa.gov/oar%5Cglobalwarming.nsf/UniqueKeyLookup/JSIN5DQSZ8/$file/MNSummary.PDF.

U.S. Environmental Protection Agency (USEPA). 2000b. Recent trends in US greenhouse gas emissions. http://www.ece.umr.edu/links/power/Energy_Course/energy/Environment/Global%20Warming/trends.html.

U.S. Environmental Protection Agency (USEPA). 2001. Action plan for reducing, mitigating, and controlling hypoxia in the northern Gulf of Mexico. Mississippi River/Gulf of Mexico watershed nutrient task force. http://www.epa.gov/msbasin/actionplan.htm.

Van Houtven, G., Powers, J., and Pattanayak, S. 2007. Valuing water quality improvements in the United States using meta-analysis: Is the glass half-full or half-empty for national policy analysis? *Resource and Energy Economics*, 29, 206–228.

Vatn, A. 2002. Multifunctional agriculture: Some consequences for international trade regimes. *European Review of Agricultural Economics*, 29, 309–327.

Wackernagel, M. et al. 2002. Tracking the ecological overshoot of the human economy. *Proceedings of the National Academy of Sciences*, 99, 9266–9271.

Watanabe, M., Adams, R., and Wu, J. 2006. Economics of environmental management in a spatially heterogeneous river basin. *American Journal of Agricultural Economics*, 88, 617–631.

West, T. and Marland, G. 2002. Net carbon flux from agriculture: Carbon emissions, carbon sequestration, crop yield, and land-use change. *Biogeochemistry*, 63, 73–83.

Westra, J., Easter, K., and Olson, K. 2002. Targeting nonpoint source pollution control: Phosphorus in the Minnesota River Basin. *Journal of the American Water Resources Association*, 38, 493–505.

Westra, J., Zimmerman, J., and Vondracek, B. 2004. Do conservation practices and programs benefit the intended resource concern? *Agricultural and Resource Economics Review*, 33, 105–120.

Zimmerman, J., Vondracek, B., and Westra, J. 2003. Agricultural land use effects on sediment loading and fish assemblages in two Minnesota basins. *Environmental Management*, 32, 93–105.

14 Conceptual Model for Integrating Ecological and Economic Sustainability in Agroecosystems
An Example from Subtropical Grazing Lands

Patrick J. Bohlen and Hilary M. Swain

CONTENTS

14.1 INTRODUCTION AND OVERVIEW

There has been an upsurge in interest in recent years among ecologists and social scientists in the concept of coupled human and natural systems (Turner et al. 2003; Pickett et al. 2005; McPeak et al. 2006; Farber et al. 2006; Liu et al. 2007). Much recent literature emphasizes the trade-offs in managed ecological systems and the need to integrate our understanding of the ecological, physical, economic, and social effects of management, oftentimes with an emphasis on sustainability (Gunderson and Holling 2002; Tallis and Kareiva 2006; Bennett and Balvanera 2007; Kareiva et al 2007). Although it is tempting to view the current emphasis on integrating the natural and social science as progressive and new, such ideas actually have a long history in ecology (Hanson 1939; Kingsland 2005). Now subsumed under a larger effort among ecologists to understand coupled human–environment systems, agroecology was one of the first fields within ecology to explicitly focus on human-dominated systems and the application of ecological principles to management of these systems (Altieri 1995; Gliessman 2007). More than 20 years ago, Ben Stinner and his colleagues grappled with the idea of viewing agroecosystems primarily as natural or social systems (Lowrance et al. 1984; Stinner, Chapter 2). This conundrum remains largely unresolved, but reflects that agroecosystems cannot be understood or managed ecologically without including perspectives from both the ecological and social sciences (Rickerl and Francis 2004; Boody et al. 2005).

Agroecology strives to provide the knowledge base and methodology for developing an agriculture that is environmentally sound, highly productive, and economically profitable. It is based on the concept of the *agroecosystem* as a framework of analysis, and emphasizes integration of ecological, social, and economic perspectives (Lowrance et al. 1984; Altieri 1995). From its very beginnings, agroecology has had a mixed emphasis on understanding ecological processes in agroecosystems, and being a force for change toward more sustainable food systems (Gliessman 2007).

The call for a more ecological approach to agriculture appeals to concepts of stability, internal cycling, and local controls, but agroecosystems, and the people who manage them, are subject to forces operating at vastly different scales. Rapid technological change, global economic and environmental factors, agricultural and other natural resource policies, and different human aspirations operating at local, regional, and national levels complicate any analysis of change or sustainability in agroecosystems (Turner et al. 2003; Robertson and Swinton 2005). There is a substantial disjunction between academic concepts of sustainable ecosystem management and the daily decisions and trade-offs made at the local level by people who own and manage the land. While academics develop frameworks and models for analysis of sustainability at the regional and global scale, farmers and ranchers make daily decisions within their own contexts that affect the ecological or economic sustainability of their individual operations. Although both of these viewpoints contribute toward outcomes affecting sustainability (Polasky et al. 2005), it is the aggregation of local decisions, driven by a myriad of endogenous and exogenous forces, that determine the persistence of agricultural systems through time (Bland and Bell, Chapter 8). Recent developments in a new class of social models, known an agent-based models, is beginning to characterize how ground-level decisions aggregate up to these higher levels of organization (Liu et al. 2007).

In this chapter, we use our vantage point as professional ecologists who oversee the management and finances of a 3000-head working cattle ranch in south Florida to explore the trade-off decisions made by local landowners or managers, and the factors that influence those decisions. Although we focus on a particular agricultural operation, in this case a beef cattle ranch in south Florida, our ideas have broad relevance for agricultural sustainability because they link general ecological concepts with the practical reality of managing an economically viable agricultural operation. We recognize that we address only how local decisions aggregate into larger wholes affecting sustainability at the regional level; we do not weigh up regional sustainability in a global perspective (how do our local and regional decisions compare with gains and losses in other global ecosystems), although we recognize the importance of issues at that scale. Our goal in this chapter is to inform perspectives from the bottom up, drawing from our experience managing a working ranch, to discuss challenges of implementing ecological management, explore gaps between concepts and reality, and, we hope, shine some new light on an age-old problem.

14.2   CONCEPTUAL MODEL FOR A LANDSCAPE OF SUSTAINABILITY

14.2.1   Feasible Ecological and Economic Domains in Agriculture

Achieving both ecological and economic sustainability on managed agricultural lands is a balancing act. Both are necessary to meet the goal of an enduring agricultural landscape. Ecological durability is sought by the public, and is often considered in terms of the ecosystem services provided by private lands (Millennium Ecosystem Assessment 2005). Agricultural lands are essential to providing many of the ecosystem services desired by society, beyond merely food and fiber (Westra and Boody, Chapter 13; Clay 2004), and appropriate management on these lands offers the potential to increase ecosystem services (Robertson and Swinton 2005; Havstad et al. 2007). Ecological durability is also a goal shared by many farmers and ranchers, whose families often have an intergenerational long view. Balanced against public demand and the private desire for ecological integrity is the requirement for economic endurance. As Aldo Leopold (1949) recognized: "It of course goes without saying that economic feasibility limits the tether of what can or cannot be done for land. It always has and it always will." Financial return is measured directly by the farmer or rancher, who needs, at least, to remain solvent or, in the case of large agribusiness, to satisfy stockholders. Financial pressures in agriculture often promote production of food and fiber at the expense of biodiversity, water quality, and soil conservation (e.g., Bennett and Balvanera 2007; Kareiva et al. 2007). Linking financial security of the farmer and rancher to enhanced environmental sustainability is ultimately important to the public because the alternatives are abandonment or, more typically, to a narrowing of objectives toward homogenized intensive production systems, or conversion to suburban and urban land uses, which correspond to greatly diminished ecosystem services.

Farmers or ranchers make the proximate and ultimate management decisions on agricultural landscapes. They work within a "feasible domain" defined by, for example: endogenous constraints of physical conditions such as precipitation, temperature, and soil type; availability of suitable crops or livestock varieties; development of products such as pesticides or vaccines; and regulatory barriers to the use of products or practices. A myriad of exogenous forces beyond the manager's control affect economic feasibility including market prices for crops and animal products, distances to market, and global trade conditions. Individual farmers and ranchers may hold very different views of such internal constraints and external forces. They must grapple with the changes and timeframe necessary to reduce environmental impacts, their willingness to rely on intensive inputs to maintain productivity, their lifetime experience of the consequences of weather and other physical extremes, and how family demography and status affects planning for the future. Although Leopold (1949) recognized the importance of economic considerations, he also noted that the "bulk of all land relations ... is determined by the land-user's tastes and predilections, rather than by his purse." Predilections and general philosophy drive decisions about products, practices, and alter both (1) the external boundaries of the feasible domain within which appropriate decisions are made, and (2) the type and pace of decisions they make within the feasible domain (Bland and Bell, Chapter 8). Changing products, practices, or philosophy can be framed by, but do not alter, most external drivers, but have direct effects on internal ecological and economic sustainability. Understanding the complex web of choices farmers and ranchers face, and how they weigh questions of sustainability, is vital to informing public policy.

14.2.2   CONCEPTUALIZING SUSTAINABILITY AS A FITNESS LANDSCAPE

We propose here to illustrate the integration or the balance between ecological and economic concerns as an overall conceptual index of sustainability, akin to fitness landscapes originally used in evolutionary biology. This notion for a conceptual landscape of sustainability was inspired by the original models of evolutionary fitness and the stability landscapes proposed for social–ecological systems (e.g., Walker et al. 2004; McGhee 2007). Adding economic viability is an extension of the established definitions of sustainability to the case of privately owned and managed lands that function as landscapes in ways favorable toward environmental values, and where public ownership is neither feasible nor affordable. Displayed graphically, sustainability on private lands is portrayed as a "landscape" with surface topographic highs forming under conditions of highest ecological integrity and greatest economic returns, which represent areas with the highest values of sustainability (Figure 14.1). The ability to move from one position to another on the landscape of sustainability is akin to rolling a ball: steep uphill gradients and deep valleys are harder to traverse; plateaus, ridges, and shoulders allow easier passage. Equilibrium peaks on the landscape are relatively stable because traversing to different although higher peaks incurs costs. Some transitions may result in very steep fall-offs, moving rapidly downhill to lower sustainability. The ability to traverse the landscape to new positions of higher sustainability is constrained by exogenous forces and driven by endogenous decisions. Movements are a function of the financial, regulatory, and social costs of transformation. No one agricultural operation has exactly the same sustainability landscape as another, since all have unique physical and financial constraints under which they operate. However,

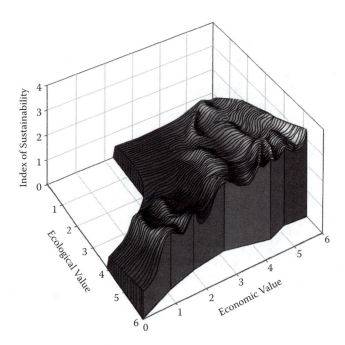

FIGURE 14.1   Conceptual diagram of a sustainability landscape. x axis represents a measure of ecological integrity, y axis is a measure of the economic returns to the landowner, and z axis, sum of x and y, as an overall index of sustainability. Topographic highs in this conceptual landscape represent higher returns both ecologically and economically, and are adaptive positions representing local stable equilibria, the resonant configurations to which most farms and ranches migrate. Topographic lows represent maladaptive or poorly functioning positions that have poor chance of persisting through time.

one might envisage those sharing a common agricultural operation, in the same geographic region, may occupy a broadly similar equilibrium position.

Location on the landscape is an indication of how efficiently coupled or decoupled a farm or ranch is as a human–natural system. There may be multiple equilibrium points on the landscape, as different combinations of ecological values and economic returns may have similar index values. To persist through time as prices, policies, and environmental conditions change, farms and ranches have to migrate to new stable peaks in the landscape. Under certain conditions a farm or ranch could fall "off the edge" of the landscape of sustainability because it is either no longer able to function economically or, alternatively, it is so degraded that it can no longer be considered ecologically viable. In some cases this fall could be a transfer to public ownership with continued or enhanced environmental services, although without remaining an economically viable operation. For low- or medium-intensity ranch operations falling "off the edge" could represent regime shifts (Carpenter and Gunderson 2001; Scheffer et al. 2002; Scheffer and Carpenter 2003) away from the environmental services provided by ranches to highly intensive agriculture or suburban–urban land uses.

14.3   THE SUSTAINABILITY LANDSCAPE FOR FLORIDA CATTLE RANCHES

14.3.1   The Ecological and Economic Context of South Florida's Ranches

Cattle ranches in south-central Florida are an example of agricultural operations on private lands striving to balance ecological and economic sustainability. These ranches form a vast subtropical landscape, comprising much of the Northern Everglades region, including the Lake Okeechobee watershed. Ranches in this region overlap with unique environmentally sensitive lands, and support

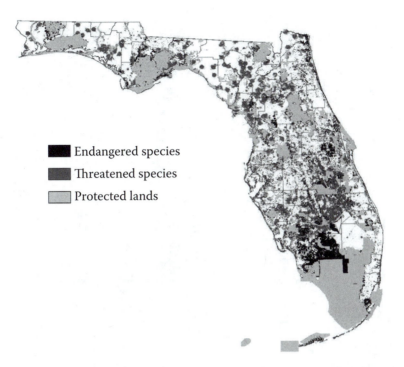

FIGURE 14.2 Known locations of state and federal threatened and endangered species in Florida outside of existing protected lands (federal, state, local, private, and other managed areas) in Florida. (From Florida Natural Areas Inventory, 2006.)

large numbers of rare, threatened, and endangered species (Figure 14.2). Florida is one of the leading beef cattle regions in the United States, second or third in beef production east of the Mississippi River. More than 1 million beef cow-calf units graze over 5 million acres, mostly in a 10-county region in south-central Florida (Figure 14.3). A high proportion of the landscape is privately owned cow-calf operations ranging from 500 to 10,000-ha, with a few over 50,000-ha in size.

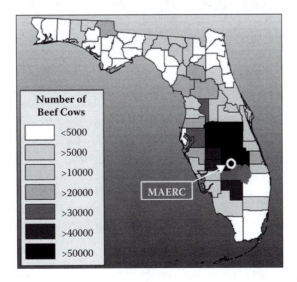

FIGURE 14.3 Distribution of beef cattle production in Florida based mainly on number of head of beef cows per county. (From USDA, 2006.)

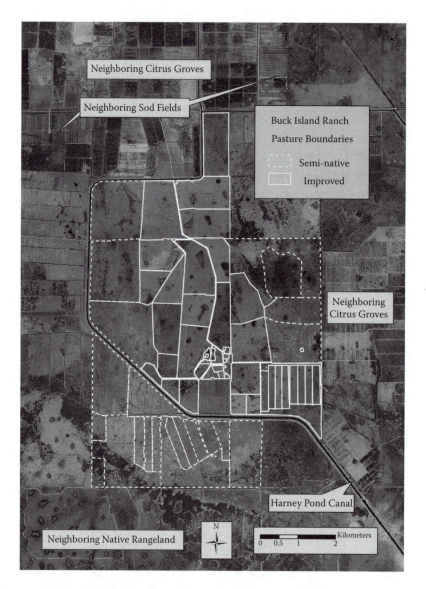

FIGURE 14.4   Aerial image of Buck Island Ranch (N 27°09′12″; W 81°11′51″) showing property boundary, agriculturally improved and seminative pastures, native communities (wetlands and hammocks), and agricultural land uses on neighboring properties. (2004 imagery, R. Pickert, Archbold Biological Station GIS Laboratory, 2008.)

In this chapter we explore issues related to the ecological and economic factors using our experience with data from Buck Island Ranch, the site of the MacArthur Agro-ecology Research Center, which is both an agroecology research site and also a full-scale 3000-head, 4252-ha working cattle ranch, representative of the region (Figure 14.4). Ranches in this region, as is true for much of the southeastern United States, are cow-calf operations in which breeding cows are maintained year-round on grass pastures producing calves which are shipped at 8 to 10 months of age, mostly to central United States and Texas, where they are grown to slaughter weight in feedlots, or with a combination of stocker grazing and feedlot.

Florida ranches are classed as temperate/subtropical humid grasslands (40.5 percent of the global terrestrial area is grassland; south-central Florida is included under the 8.3 percent that is nonwoody grassland; GLCCD 1998). Embedded within these cattle ranches are extensive native habitats (including marshes and wet prairies, pine flatwoods, cypress domes and strands, oak and palm hammocks, scrub and dry prairie) and also diversified citrus, sod, and row crop operations. The Florida ranch landscape currently provides an array of complex ecosystem services under the four broad categories of the Millennium Ecosystem Assessment (2005): provisioning, regulatory, supporting, and cultural. Ranches provide beef and other agricultural products which are consumed by the public, and provide financial security for the rancher or landowner (Table 14.1 a to e, columns 1, 3). Ranches maintain regulatory ecosystem services, such as fire regimes and abating floods, that are in the public interest (Table 14.1 f to l, column 1) as well as supporting services including soil formation, nutrient cycling, and photosynthesis (Table 14.1 m, column 1). They also preserve cultural values, not only the cherished way of life, but also, for example, extensive habitat for publicly valued threatened and endangered species (Table 14.1 n to t, column 1). Many rare species are dependent on the open spaces of these agricultural enterprises for their continued existence in Florida (see Figure 14.2). Benefits from regulatory, supporting, and cultural ecosystem services accrue to the public, but the rancher also sees some direct return from these same services (Table 14.1 f to t, column 3), in addition to their general philosophy of land stewardship. For example, wetland retention is used locally to limit freeze damage for adjacent citrus and other row crops.

The degree to which grazing lands provide environmental services may be counterbalanced by the extent to which the ecological processes on which ecosystem services depend have been modified and compromised by agricultural operations (Bohlen and Swain 2008). Many grazing land management practices were implemented without anticipating their environmental costs, resulting in a concomitant decline in regulatory and cultural services compared to those derived from pristine natural ecosystems (Table 14.1, column 2, a to m, o, p). Resource costs of production have included large-scale manipulation of nutrient dynamics, intensive drainage, introduction of nonnative forages, and a widespread modification of the spatial structure of the landscape. One example is that prior applications of phosphorus (P) fertilizer in pastures continued to contribute to excess P in downstream Florida watersheds, even many years after P applications ceased (Capece et al. 2007; Swain et al. 2007).

Although ecosystem services may be higher from native habitats, at present such habitats offer very limited annual economic return. To remain on the Florida landscape ranches must be economically viable, deriving income from selling agricultural products. Calf prices in recent years have held relatively steady. The total annual agricultural revenue for Buck Island Ranch, although increasing over time (Figure 14.5), has been offset by increasingly expensive feed, labor, veterinary care, fertilizer, and fuel costs such that net revenues for the ranch, as is true of other a cow-calf operators in this region, are slim (Figure 14.6). Current product diversification at Buck Island Ranch does not adequately spread financial risk (Figure 14.7), although many ranchers regionally are diversified through more intensive production of sod, citrus, or row crops to achieve diversification (Gornak and Zhang 1999).

Completely dwarfing these concerns is that regional land values have tripled, or even quintupled from 2002 to 2007 (Figure 14.8), although they leveled off in 2008. Ranching on private lands provides a low return on asset value, primarily land, which means opportunity costs of remaining in agriculture are high and exacerbated by federal estate (death) taxes based on gains in assets. In contrast, ranchers grazing on public lands do not incur full cost accounting for land values. Florida ranchers face tough choices in terms of economic sustainability and are faced with the temptation to sell for development, especially ranches bordering cities or residential sprawl. Opposing this overwhelming force will require public–private compacts that recognize, reward, and enhance the ecosystem services ranches can provide. Agribusiness market forces alone are unable to support this mutual benefit; they will always seek cheaper product or production costs elsewhere globally (Roberts 2008). There is the concern that remaining cattle ranches could become a social anachronism, with inadequate support from feed suppliers, equipment operators, large animal veterinarians, livestock markets, or university extension and research. The Florida ranchlands agroecosystem is poised on the edge of sustainability; understanding how land use decisions and management practices will affect economic viability and ecological sustainability has never been more pressing.

TABLE 14.1

Comparing Public and Private Costs and Benefits of Maintaining Ecosystem Services on Florida Cattle Ranches

Benefits to Public	Costs to Public (Financial or Impacts on Resources)	Benefits to Private Landowner	Costs to Private Landowner
Provisioning Services			
a. **Food: beef.** Minimal use of pesticides and herbicides; *supports local rural economies and sustains rural communities.*	Habitat loss and fragmentation; nutrient runoff; introduced exotic plants; methane emissions from cows; soil degradation; hydrologic modifications	Income; employment of family and others; sense of purpose and way of life	High input costs; low net revenue; high financial risks and liabilities; high opportunity cost of forgone land use alternatives
b. **Turf sod** (for landscaping) and landscape plants **Food products:** *Citrus, sugar, row crops*	Habitat loss and fragmentation; increased fertilizer use; surface and groundwater use; increased herbicide and pesticide use; soil loss and depletion	Employment; relatively high net revenue per acre; diversification of income stream	Financial risks high; opportunity costs: sale for urban/suburban land use is more profitable
c. **Raw materials** *like timber, bio-fuels*, other renewable bioproducts	Extraction can be unsustainable; biofuels and silviculture replace native habitats and can cause habitat loss/ fragmentation	Employment; net revenue; diversification of income stream	Opportunity costs for conversion to more profitable intensive agriculture or urban/ suburban
d. **Native products:** *Cabbage palms, alligators, alligator eggs;* helps retain native habitats.	Potential overexploitation; disturbance and habitat fragmentation from cabbage palm harvesting	Employment; net revenue, although marginal	Opportunity costs for conversion to more profitable intensive agriculture or urban/ suburban
e. **Hunt leases.** Hunting opportunities; retains native habitat and ecosystem services	Support limited number of lessees; disproportionate emphasis on game species in land management decisions	Employment; net revenue from lease can be high; personal enjoyment	Opportunity costs for conversion to more profitable intensive agriculture or urban/ suburban
Regulatory Services			
f. **Fire management:** Critical for maintaining native habitats and regional fire regimes.	Prescribed fire on ranches is often dormant season rather than preferred growing season burns	Prescribed burns provide forage and control brush	High liability risk; permits, training, and certification requirements
g. **Flood control:** Retained wetlands and pastures *reduce downstream flooding;* less flashy hydroperiods; water storage and recharge	Floods trigger disaster payments via federal funds; new programs offer public payments for ecosystem services, and flowage or conservation easements	Livestock watering; game species habitat; enhanced productivity; reduced impact of winter freezes	Water management costs; opportunity costs for conversion to more profitable intensive agriculture and urban/ suburban

TABLE 14.1 (*Continued*)
Comparing Public and Private Costs and Benefits of Maintaining Ecosystem Services on Florida Cattle Ranches

Benefits to Public	Costs to Public (Financial or Impacts on Resources)	Benefits to Private Landowner	Costs to Private Landowner
h. **Carbon sequestration** in pastures, range, or woodlands: Large C stocks in native habitats, grassland, wetlands	May be future credit payments (public or private dollars) for C sequestration on ranches; reduced C stocks due to drainage, soil loss, or wetland conversion	Reduced soil erosion; maintenance of soil fertility; possible income source	Opportunity costs for conversion to more profitable intensive agriculture and urban/suburban
i. **Nutrient storage:** Leads to higher productivity, also feeds into carbon sequestration	Nutrient loadings in surface runoff, with negative impacts on downstream freshwater and estuarine ecosystems	Nutrient cycling drives forage biomass directly benefiting cattle operation	Reducing nutrient load can be costly and raises regulatory and permitting issues
j. **Groundwater and aquifer recharge:** Large rain-fed systems	Consumptive water use reduces water for public use and for native ecosystems	Irrigation and livestock watering	Permitting and regulatory issues
k. **Maintaining regional weather:** Evapotranspiration (ET) feeds summer rains; remaining wetlands reduce winter freezes	Past drainage on ranch landscapes contributed to decrease in ET below historic levels regionally and increased likelihood of winter freezes	Reduces freeze impacts Contributes toward maintaining rainfall regionally	Opportunity costs for conversion to more profitable intensive agriculture and urban/suburban
l. **Invasive species control** *both plant* (e.g., Brazilian pepper) and *animals* (e.g., wild hogs)	Some invasive control measures are supported by public dollars; ranching has exacerbated some invasive plants	Improves productivity and maintains desired native habitats and species	Expensive to control in time and dollars
Supporting Services			
m. **Maintain natural processes** *such as* soil formation, nutrient cycling, and photosynthesis	Uncertain; may be reductions in some services such as hydrologic function and soil fertility, and wetland services	Maintains ecosystem functions on which ranch depends	Opportunity costs for conversion to more profitable intensive agriculture and urban/suburban
Biodiversity and Cultural Services			
n. **Cultural heritage of ranching** is fundamental to the history, arts, and literature of Florida	None	Cultural heritage, family heritage, intergenerational values; sense of place	Fighting negative public perception of ranching, e.g., environmental concerns

Continued

TABLE 14.1 (*Continued*)
Comparing Public and Private Costs and Benefits of Maintaining Ecosystem Services on Florida Cattle Ranches

Benefits to Public	Costs to Public (Financial or Impacts on Resources)	Benefits to Private Landowner	Costs to Private Landowner
o. **Habitat for common native species:** At low public cost	Public dollars for conservation programs like easements, management, although low for services provided regionally	Access to dollars for limited cost share conservation programs	Opportunity costs for conversion to more profitable intensive agriculture and urban/suburban
p. **Habitat for state and federal threatened and endangered species**	Public dollars for managing endangered species; small sums in relation to services provided	Some access to cost share programs for conservation	High costs of regulatory restrictions; hassle factor
q. **Open space** in large parcels for *animals with large-area needs*	None	Cost share and other conservation programs	Opportunity costs for conversion to more profitable land use
r. **Recreational opportunities**, *aesthetic* values, education	Limited access for public	May be some revenue from recreational access	High liability costs and risks with public access
s. **Buffers for public conservation lands**	None	May reciprocally enhance wildlife on ranch	Neighboring public land may increase public oversight
t. **Security for wildlife:** Restricted access	Public dollars for enforcement on private lands	Limits poaching	Costs and risks of security

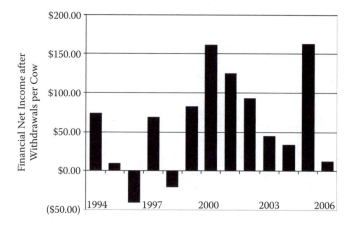

FIGURE 14.5   Financial net income per grazing cow at Buck Island Ranch from 1994 to 2006 as derived from Standardized Performance Analysis, a financial and production accounting program developed for beef cow-calf operations. (Data provided by Gene Lollis, Ranch Manager, Buck Island Ranch.)

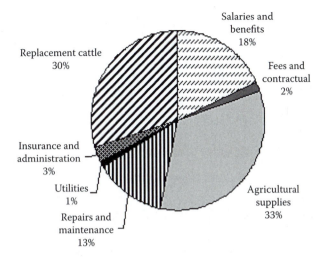

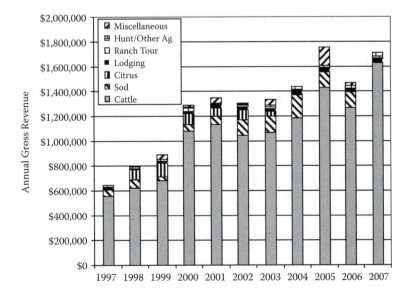

FIGURE 14.6 Annual operational expense categories and percent (annual average of $1,139,968 for 1997 to 2007) for Buck Island Ranch. (Data from Eric Stein, Archbold, and Gene Lollis, Buck Island Ranch Manager.)

FIGURE 14.7 Annual total agricultural revenue, and categories of sources of revenue at Buck Island Ranch from 1997 to 2007.

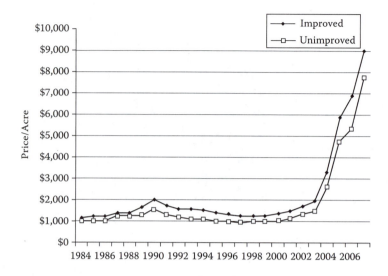

FIGURE 14.8   Changes in land values for pasture in south Florida from 1984 to 2007 based on annual surveys by J. E. Reynolds, Florida Land Value Survey, conducted by the Food and Resource Economics Department at the University of Florida; most recent survey http://edis.ifas.ufl.edu/fe625710 (accessed July 26, 2008).

14.3.2   REAL-WORLD SUSTAINABILITY SCENARIOS FOR FLORIDA RANCHLANDS

To illustrate our need for integration, we return to the conceptual landscape of sustainability, but this time consider a series of points, marked A to P, each of which represents different scenarios and decisions facing a typical Florida cattle ranch (Figure 14.9). The y-axis is a simplistic economic return to the rancher. The x-axis is a conceptual amalgam of all the ecosystem values of a ranch: the greater the retention of natural communities; the more the numbers and diversity of native species and the presence of rare, threatened, and endangered species; the higher the maintenance of ecosystem processes such as fire, hydrological regimes, and carbon sequestration; the broader the extent of acreage providing landscape linkages and limiting habitat fragmentation; the less the nutrient loading from the operation; the farther to the right on the axis of environmental sustainability. The interactive sum of the x and y values contribute to the z values which is a conceptual index of sustainability. A range of incentives are available to increase both environmental and ecological values—that is, to move to higher overall sustainability peaks. These incentives include "equity incentives," in which the ranchers capitalize on the value of the land, or "operational incentives," in which ranchers receive revenue or cost offsets for implementing practices or providing environmental services. The public cannot afford to rely solely on equity payments on such extensive areas of land. Ideally operational incentives could cumulatively generate enough annual revenue so that landowners would not have to seek equity payments to stay on the land. There are also events and decisions that would result in either environmental or financial collapse, falling off the edge of the sustainability landscape as a working cattle ranch.

As mosaics of improved pastures, some native range and woodlands, with citrus, sod, and row crops interspersed, Buck Island Ranch and most other south-central Florida ranches occupy a position on the landscape of sustainability somewhere around point A, depending on their predilection toward intensification and the efficiency of their management and finances. How can they stay or become more profitable without sacrificing ecological values?

A. Ecotourism is often promoted to increase ecological and economic sustainability. In an attempt to capitalize on their natural resources some Florida ranches offer public *ecotours*. In general, these have had minimal economic and ecological impact, and do not move

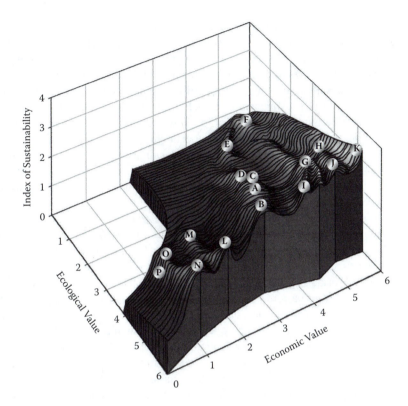

FIGURE 14.9 Sustainability landscape illustrating different optima discussed in text.

a ranch off sustainability point A. At Buck Island Ranch we charge $15 for our "Indian Prairie Safari" tour, but have grossed up to $15,000 annually, with little net revenue. Even the Babcock Ranch tour (http://www.babcockwilderness.com/eco.htm) near Fort Myers, with up to 50,000 visitors a year, barely breaks even (Arnie Sarlo, Ranch manager, pers. comm.). Hunt leases can generate substantial income. Perhaps a ranch totally dedicated to ecotourism and hunting could improve financial gains; for example, in the 1980s after decades of cattle ranching in southeastern Zimbabwe, the ecological and economic capacity had become so untenable that neighboring ranches converted to large wildlife "conservancies" without livestock or fences, managed collectively for tourism and hunting (Cumming 2005). This would represent an absolute transformation in the Florida landscape; whether or not this could compete with soaring land values is debatable.

B. State and federal cost share *conservation programs,* such as the U.S. Department of Agriculture Environmental Quality Improvement Program and Wildlife Habitat Improvement Program, the U.S. Fish and Wildlife Service Private Stewardship Grant Program, and the Florida Fish and Wildlife Conservation Commission FWC Landowner Incentive Program, are designed to enhance environmental values on private lands. But they have limited funds available to individual ranchers so they move operations only a short distance along the axis of ecological sustainability, to point B. Furthermore, because they often require on 50 percent cost share, such programs can affect economic sustainability negatively. Generally, ranchers enroll in these programs not for financial returns but because they are philosophically committed (GAO 2006), or want to reduce the risk of regulatory costs by, for example, in this region, implementing practices to reduce P loadings in surface waters.

C. More intensive pasture management could result in a decline in environmental values to point C. Most ranchers already harvest *Bahia* grass (*Paspalum notatum*) sod (low quality turf sod) intermittently from improved pastures to generate additional income, while maintaining the pasture for grazing between harvesting cycles. This produces substantial revenue in favorable years. However, increasing the frequency of lifting sod may reduce environmental values down to point C by requiring more fertilizer and incurring the long-term risk of losing soil organic matter and fertility. Planting intensive forages—for example, Florona stargrass (*Cynodon nlemfuensis*), Floralta limpograss (*Hemarthria altissima*), or Jiggs grass, a bermudagrass hybrid (*Cynodon dactylon*)—in existing improved pastures may increase productivity and improve economic returns, but requires more fertilizer. Only total cost accounting will clarify whether improved forages result in an overall decline in either ecological or economic sustainability to point C, in comparison to maintaining a larger area of less productive grass, especially because higher quality forages reduce dependency on supplementary feed.

D. No ranch in Florida can run an economically viable cow-calf operation on native range alone; all require a significant proportion of agriculturally improved pasture (ranging from ~40 to 70 percent of the landscape). However, further loss or fragmentation of remaining natural habitat, such as conversion from pine flatwoods, hardwood hammocks, or native range—largely Florida dry prairie (a threatened ecosystem) and seasonal wetlands—to agriculturally improved pasture has extremely negative ecological consequences, moving downwards to point D, with impacts on native species, large release of carbon stocks, and alteration in hydrological and nutrient cycles. At present, maintaining areas as native communities provides low economic returns other than hunting. Ranchers weigh the opportunity costs of maintaining these natural areas relative to converting to other land uses.

E. Increasing fuel and energy costs is generating massive interest in *biomass and biofuel production* in Florida, as elsewhere, even though the environmental sustainability of bio-fuels from a greenhouse gas perspective is debatable (e.g., Adler et al. 2007; Searchinger et al. 2008). Cellulosic biofuels grown efficiently in Florida may contribute positively by reducing global carbon emissions and slowing climate change. This has to be balanced against whether biofuel and biomass crops will reduce regional ecosystem services from grazing lands, down to point E because they (1) cause habitat loss or fragmentation, either directly or indirectly by displacing other agricultural crops which in turn result in conversion of native habitats (Fargione et al. 2008); (2) increase demand for water; (3) increase fertilizer inputs or nutrient loadings; or (4) have the potential to be invasive, such as *Arundo donax* or *Jatropha curcas* (Raghu et al. 2006; Barney and Ditomaso 2008; Quinn and Holt 2008).

F. The degree to which agricultural intensification leads to habitat loss and further decline in ecological sustainability beyond point D depends on the nature and extent of conversion. Large areas of former pasture and range were converted to high quality turf sod production and citrus over the past couple of decades (Gornak and Zhang 1999; McCarty 2006) on ranches north of Lake Okeechobee, with concomitant increases in consumptive water use, fertilizers, and pesticides, and loss of habitat for native species. At present, row crops and sugarcane are located largely on more organic soils south of Lake Okeechobee. Although such crops may improve revenue and agricultural diversification, they occupy a lower position in the overall landscape of sustainability. To remain fiscally viable it is critical for a ranch to have diversified production, and include some intensive operations, but when practices become highly intensive, moving below point F and representing limited ecological value, the ranch (or farm) may "fall off" the landscape of agroecosystem sustainability.

If, overall, we assume it is in the public interest to continue the provision of ecosystem services from private Florida ranches, how can we improve economic status and increase financial incentives without causing further losses in ecological value? Hoctor et al. (2008) have provided a thorough catalog of the wide range of incentives currently available to private landowners in Florida. Federal government crop subsidy commodity payments, even though they do not target environmental values, are not available to cow-calf producers to enhance economic status. U.S. Department of Agriculture (USDA) commodity payments from 1995 to 2005 in Florida totaled $314 million, ranking 36th nationally with only 0.2 percent of national payments, in comparison with top-ranked Iowa with $12.5 billion in crop subsidies, or 9.6 percent of total crop subsidy support (www.ewg. org). Commodity grain payments do support Florida ranchers indirectly by affecting costs in the feedlots, and current trade protection for sugar affects the price of molasses, a typical supplementary feed used by Florida ranchers. Nationally there has been widespread enhancement of farmland biodiversity with the USDA Conservation Reserve Program (CRP) (e.g., Van Buskirk and Willi 2004). CRP has had negligible impact in Florida because it is only available to a few north Florida peanut and cotton producers.

Incentives with the greatest financial impact on landowners, and greatest long-term conservation security, are those that provide payment for the equity value of the land—easements (see point K, below) and various land use planning options which improve the financial status of ranches significantly while retaining environmental values.

G. Transfer of development rights (TDRs) or other planning tools are designed to "transfer" residential densities already allocated to agricultural lands (typically 1 unit per 5 acres throughout much of rural Florida) to increased densities in urban or suburban settings, with associated easements placed on the lands remaining in agriculture. Despite the potential to enhance both economic and environmental sustainability for agricultural lands and some exemplary programs nationally, TDRs have been little used in Florida to date. Most county commissions seem willing to rezone for increased density without requiring offsetting decreases in density elsewhere. TDRs can be a clumsy tool if they give the same weight to transferring residential density from lands of high ecological value as transfers from highly degraded land.

H. The passage of the Rural Lands Stewardship Act RLS potentially ushered in a new era in planning for rural lands in Florida. For the rancher RLS offers the potential for realizing a large economic return on their land, while retaining considerable ecological value and maintaining an agricultural operation. The concept of RLS (Section 163.3177(11) (d), Florida Statutes, the Rural Land Stewardship Area Program) is that landowner(s) with parcels of at least 10,000 acres can apply to rezone for intensive integrated small town, village, and hamlet development on a portion of their land. The unique aspect of RLS is that this high density development (the receiving area) is facilitated by generating "credits" from the rest of the property, with an emphasis on areas with higher ecological value (the sending areas: e.g., rare habitats, important landscape linkages, floodplains, rivers, habitat for listed species). The higher the ecological value of the sending area credits, the higher the density that can be applied toward development in disturbed, intensively farmed, or otherwise degraded land. In theory, this focuses most development on the least environmentally sensitive land and protects, via permanent easement, valuable ecological and agricultural lands. To date, there are two RLS projects that have been completed and others have been under consideration, representing millions of acres. However, new proposed state rules for the implementation of RLS have put a hold on many RLS planning efforts. Optional Sector Planning, Regional General Permits, and Ecosystem Management Agreements are other Florida large area planning and regulatory programs that can be used to protect high-priority natural areas—without public land acquisition. For conservation RLS, sector plans and these other agreements are all something of a Faustian deal—maybe around point H—where environmental sustainability increases with set-aside of large areas of important conservation lands, but is compromised by the accelerated addition of new "towns, villages, and hamlets" outside previously urban and suburban areas. There are questions about the conservation science behind RLS, sector plans, and other credit-based transfer systems, because the process is not transparent and effectiveness remains unproven. However, the alternative—continued, unfettered approval of sprawling 1 unit per 5- or 10-acre ranchette and subdivision—will be an ecological disaster for Florida, realizing the alarming projections for state growth by 2060 (Zwick and Carr 2006).

Market-based and marketlike incentives have potential to improve the annual revenues and operational payments for Florida ranchers, but current markets are very immature and the logistics required to implement these programs will require an array of collaborating social, business, environmental and policy entrepreneurs (Bohlen et al. 2009) beyond anything we have seen in agriculture to date.

I. If beef consumers could generate sufficient market demand and were willing to pay financial premiums for cattle produced from "environmentally sustainable" or "green certified" grazing lands, or from grass-fed beef, this could increase both economic and ecosystem sustainability, maybe up to point I. At Buck Island Ranch we are also working on entering into a partnership with the feedlot to share in premiums derived from selling calves to the European market without growth hormone implants, but it is difficult for the cow-calf producer to get upfront premiums in these sorts of arrangements. These payments are related to consumer health preferences rather than environmental preferences. Although we use few pesticides other than fly treatments and worming, and rarely apply herbicides except occasionally to control invasive plants, we receive no market premiums for these environmentally sound approaches. Grass-fed beef is another niche market possibility, but the challenge of finding a reliable slaughterhouse and market, providing a year-round supply, as well as the lack of sufficient grass regionally to grow to market weight precludes grass-fed beef as a fiscally viable alternative for Buck Island Ranch at this time. Marketing environmentally sustainable beef at typical production levels found in Florida (e.g., Buck Island Ranch produces 2000 to 2600 calves annually) is difficult for individual producers with huge barriers to farm-based marketing. Early discussions with the feedlot industry have revealed many challenges for reconciling environmentally sustainable labeling for our calf crop through the entire production chain, which includes the feedlot and meat packing industry.

J. Paying ranchers directly to provide ecosystem services on private lands at a cost to society equal to or lower than the cost of the same ecosystem services from alternate sources such as large public works projects would help sustain ranches economically, as well as increasing the ecological values of the land. Increasing availability of payments for ecosystem services—in the near term potentially both water and carbon credits—may move ranchers to point J on the landscape. A exciting pilot project, Florida's Ranchland and Environmental Services Program, funded by the USDA, the South Florida Water Management District, and the Florida Department of Agriculture, is currently evaluating a pay-for-performance environmental services program, with public funding paying for water storage and water treatment (to reduce P loading) on eight Florida ranches (WWF 2008; Bohlen et al. 2009). Carbon credits hold potential for Florida where carbon stocks in natural systems are relatively high due primarily to the abundance of wetlands in the state, although the question of methane emissions from wetlands remains unresolved.

K. Sale of conservation easements, on land with high conservation value, has been an option nationally (e.g., Rissman et al. 2007) and for some Florida ranchers to realize land (equity) values while maintaining private ownership. The Florida Chapter of The Nature Conservancy reports a huge increase in interest in conservation easements among large landholding ranchers in 2007–2008; most of this is for sale of easements, although bargain sales and easement donations to help offset estate taxes are often a component of easement negotiations (Keith Fountain TNC pers. comm.). Clearly, the security of perpetuity conservation easements greatly enhances ecological values up to point K, representing a high point for both ecological and economic security. Easement payments are substantial, often more than 50 percent of fair market value, thus significantly improving economic sustainability, although the sale of rights limits future options for the landowner. State of Florida and Florida Water Management District acquisition programs have increasingly turned to the purchase of large conservation easements on ranches to achieve conservation goals, for example, Brighthour (5,137 ha), Lykes (17,010 ha), Smoak (3,416 ha), and Lightsey (5,103 ha). The state recently acquired the 29,970-ha Babcock Ranch by the state as an outright fee purchase that is to be maintained as a publicly owned, but privately managed working cattle ranch. The acres enrolled in federal conservation easement programs hovers around 1 percent of total farm land in Florida (ERS 2008). Ranchers prefer federal easement programs that allow continued agricultural use, like the USDA Grassland Reserve Program and the Farm and Ranch Protection Program, but there has been more funding available for the USDA Wetland Reserve Program, which restricts agricultural use. At a federal level, sound priority setting for conservation programs needs to be grounded in integrated information about what programs are achieving (Feather et al. 1999; SWCS 2001). An evolving aspect are private agreements in which a few Florida ranchers are offering easements and management agreements as part of mitigation or mitigation banking to meet offsite wetland and endangered species permit requirements.

We recognize the sustainability landscape here does not portray combinations of multiple incentives which may arrive at higher equilibrium positions of sustainability. In an ideal scenario, combinations of operational incentives would be high enough that a landowner need not seek equity payments in order to stay in business. There has been little discussion of potential trade-off costs to exercising equity incentives (like easements) if these legal agreements limit opportunities of adaptive management for operational incentives like payment for ecosystem services.

Although sufficient financial incentives may keep ranchers in business in Florida and increase ecological values, we recognize there are threshold conditions that will rapidly push the tipping point in the other direction, and drive the combination of economic and ecological values of private agricultural operations to precipitously low levels (L, M, N, O, P), or completely off the edge of agroecosystem sustainability.

L. Environmental "disasters," which include stochastic events such as floods, tornadoes, hurricanes, and droughts, can have an immediate economic impact but, depending on the severity, federal disaster-relief payments may buffer this downturn, and in general ranch landscapes are fairly resilient to this type of stochasticity, such as recovery from the five hurricanes in 2004–2005. Long-term climate change could be a looming disaster in Florida but climate predictions for the state have so much uncertainty it is hard to anticipate what will happen to agriculture, except that it may be squeezed out by new housing developments stemming from a "flight from the coast" under rising sea levels.

M. In comparison, a market-product, health-related, or biosecurity disaster tied to Florida could be instantly debilitating. A major incident of bovine spongiform encephalitis (mad cow disease), hoof (foot) and mouth disease, or other health-related diseases could be the final straw putting ranchers out of business. The U.K. experience suggests that restoring consumer confidence and the direct costs of destroying animals and reestablishing herds is very tough, even with disaster relief funding.

N. Increasing stringency of current regulatory provisions, for example, compliance with nutrient standards, wetland regulations, or threatened and endangered species covered by the U.S. Endangered Species Act, such as Florida caracara *Caracara cheriway* Florida scrub-jay *Aphelocoma coerulescens*, red-cockaded woodpecker *Picoides borealis*, Florida grasshopper sparrow *Ammodramus savannarum floridanus,* snail kite *Rostrhamus sociabilis*, Florida panther *Puma concolor coryi*, should increase environmental benefits but, if they result in financial insolvency and a decision to sell for development, ecological values will deteriorate.

O. Cyclical market prices have been typical for the cattle industry for decades, often associated with drought cycles. However, increasing costs for transport and feed, exacerbated by the diversion of corn to ethanol and increasing world food demand, foretell a decline in prices for beef producers, and portend little relief in sight for coming years. This may be the final push for many ranchers toward selling for development.

P. Paying federal estate (death duty) taxes often pushes a ranch over the edge, off the landscape of sustainability. Soaring land values have overwhelmed prior estate tax planning, forcing some families to sell. Family demography comes into play with members who have moved off the ranch often precipitating the decision to sell. Inevitably at least some land is sold, often to a more intensive producer or for development, to use the well-hackneyed phrase "concrete is the last crop" (Small 2002). Two neighbors' properties next to Buck Island Ranch were sold recently in response to death in the family, and converted immediately to intensive turf sod production.

Most public discussions about policies to support environmentally sustainable agriculture have focused to date on incentives to increase revenue (Hoctor 2008). Florida ranchers have, however, like the rest of agriculture, become dependent on cheap energy, fertilizer, transportation, water, and feed, much of which has been subsidized directly or indirectly by public policy. In recent years these "cheap" products have become increasingly costly, in many cases more environmentally damaging, and progressively more in demand for competing public uses. At Buck Island Ranch we have established 12 solar wells to reduce diesel pumping costs and our reliance on public canals for surface water, and have increased our rates of liming to reduce fertilization needs. Assuming it is in the public interest to help Florida ranchers remain in business, then market and policy incentives could help reduce their reliance on cheap energy and water. These challenges are global in nature and require coordinated international approaches; the need for integrated research and development for on-farm/ranch low-energy technologies or energy production schemes, and total cost accounting is essential.

A total of 28 percent of the State of Florida is already managed as state (15.2 percent), federal (11.6 percent), local (1.2 percent), or private (0.5 percent) conservation/protected/managed lands (FNAI 2008), but this is not enough to meet the need for Florida's long-term environmental sustainability. The public simply cannot afford to purchase and manage all the land needed to maintain sustainability over peninsular Florida, with its multitudes of rare species, natural communities, and vital ecosystem processes. Public funding for state acquisition of conservation lands in Florida has been extraordinary; since 1989 Florida Forever and its predecessor program Preservation 2000, expended $5.57 billion to protect 2.4 million acres (FDEP 2008). Reauthorization of Florida Forever in 2008 shows strong support for continuing to protect Florida's highest value conservation lands with public funding. But it is obvious that public acquisition, either fee simple or conservation easements, could never afford to acquire enough private land to meet the state's environmental sustainability goals. Nor is this desirable from a practical management or management cost perspective. Florida needs private landowners to continue to play a major role in the goals of sustainability and maintaining a vibrant agricultural industry is vital to meet this need. If we do not have alternatives and incentives in place soon, much of the remaining rural Florida landscape may convert to housing or intensive agricultural land uses with potentially huge costs to the public in terms of ecosystem services lost forever.

The Century Commission in Florida, established in 2005, has just completed a GIS analysis (CLIP), which has identified the considerable proportion of the state needed to protect Florida's critical green infrastructure. The Florida Fish and Wildlife Conservation Commission is heading up the Comprehensive Cooperative Conservation Blueprint (CCB) initiative to reach out to stakeholders, especially landowners, to outline incentives for private landowners to secure this critical green infrastructure into the future. Our scenario-based model provides one framework for evaluating incentives in terms of the sustainability of agricultural producers.

14.4   BROADER IMPLICATIONS FOR SUSTAINABLE AGROECOSYSTEM MANAGEMENT

Balancing long-term ecological and economic sustainability of Florida's grazing lands, like decisions for other agroecosystems, is complicated by the large number of agents involved; farmers and ranchers, government policy makers and regulators, industry representatives, and environmental groups. Landowner decisions rely on environmental cues, financial and production status, public perception, and the regulatory environment (Pahelke 1999). Landowner actions, which affect economic and ecological sustainability, are dependent on prior, current, and predicted future conditions, complex economic relationships at multiple levels, and the rate at which these relationships change over time. Our conceptual model presents alternative scenarios illustrating the balance between ecological and economic forces, using our knowledge of the particulars of Buck Island Ranch, a representative cow-calf enterprise in this region. This conceptual model illustrates regional circumstances, but does not provide general explanations by which solutions to other types of challenges can be addressed. Current tools available to farmers and ranchers to integrate economic and ecological decision making are typically simple and crop oriented, and do not address the complex dynamics facing these landscapes.

New approaches and tools using agent-based models (ABMs) might provide a better framework to examine interactions of socioeconomic and agroecosystem processes, and how these interactions will affect landowner agricultural choices, ecosystem services, economic status, and land use change over time (An et al. 2005; Bolte et al. 2006; Brown and Xie 2006). Such models are appropriate for examining these interactions in agroecosystems because of the complexity of in agent circumstances, interdependencies, and feedbacks among multiple levels, which are beyond traditional modeling (Berger 2001; Parker et al. 2003). ABMs can reveal feedbacks, nonlinearities, and ecologically or economically significant thresholds in response to changes in policy or agent decisions (Bousquet and Le Page 2004). They can also be coupled to spatially explicit factors (Grimm et al. 2005; Weiner et al. 2005), allowing them to link policy and agent decisions, the resultant responses in agroecosystems, and land use change at the landscape level (Bockstael 1996; Parker et al. 2003; Berger 2006). These models incorporate social factors through surveys of ranchers and other agents to characterize their predilections for different agricultural options and outcomes (Redman 2002; Pocewicz et al. 2008).

New tools to guide public policy are critical to predict which changes can cause threshold effects, moving us toward more, or less, sustainable equilibria. Gaining a basic understanding of how to pull back, when poised on the edge of sustainability, toward an agroecosystem stabilized around desirable equilibria, will be of tremendous value for designing public policy and incentive programs (Liu et al. 2007). However, it is as yet unclear whether new modeling tools, despite the exciting academic intellectual frontiers they represent, will actually be effective at influencing policies, decision making, or outcomes of agroecosystem design or management.

Resilience and adaptability are crucial to sustainable agroecosystem management, as they are to all sustainable social-ecological systems (Walker et al. 2004). In this chapter we have used the focal point of a working ranch to examine exogenous and endogenous forces that influence sustainability outcomes including factors operating at different social or ecological scales. Resilience is the capacity of the system to absorb disturbance while maintaining essentially the same function which, in the case of a Florida cattle ranch, is to reorganize in response to changing circumstances while maintaining a viable economic entity based largely on cattle production. Crucial aspects of resilience for this or any other agricultural operation include: the latitude of the system, which is the amount of change it can withstand before switching over to another state from which it cannot recover its former function; resistance, which is the ease or difficulty making changes; and precariousness, which gauges how close the current system is to a switching threshold (Walker et al. 2004). The alternative scenarios we describe for Florida ranches touch on all of these aspects of resilience and could be applied, with context-specific details, to any agroecosystem.

Adaptability, in the social sense, refers to the ability of humans to manage resilience. In the case of agroecosystem management, it is the ability of ranchers and farmers to manage their responses to changing conditions (e.g., weather, markets, competition, regulations, agricultural and environmental policies) in ways that allow them to continue to provide the basic functions of food production as well as other desired ecosystem services. Multiple examples of degraded agricultural systems throughout history suggest a failure of many social-ecological systems to retain adaptability. In developed countries like the United States, agricultural productivity has been maintained, but often at the expense of the environment and other ecological services, as well as at the expense of family farmers and rural communities. It has been argued that these "externalities" are inevitable consequences of advancing technologies and increased production efficiencies, and that these negative consequences can be remedied by subsidies or other offsets, or tolerated as acceptable consequences of "success." The problem with the tendency to seek such seemingly successful equilibria is that they often involve pursuing increased resource exploitation based on perceived efficiencies and specialization that can lead to an "efficiency trap." This locked position can reduce the ability to seek alternative options that might actually represent better positions for responding to changes and achieving the broader goals of sustainability (Scheffer and Westley 2007). Maintaining flexibility and resisting locked-in patterns of efficiency that limit future adaptability is a central challenge to sustainable agroecosystem management. Unfortunately, agricultural policies often encourage these inflexible and unsustainable patterns.

Forecasting global environmental change is critical for maintaining adaptability (Tilman et al. 2001; Pielke et al. 2007), but how to keep options open in relation to, for example, climate model projections, is not yet linked to aggregated decision making by individual farmers/ranchers. A rancher's response to, and understanding of, temporal trends is constrained by experiential knowledge, not informed by predictions for a changing world, especially one with significant climate change (Liu et al. 2007). Predicting the effects of agricultural decisions in relation to long-term climate change is unclear to both ecologists and ranchers in Florida, because changes are masked by huge interannual variability in precipitation, like ENSO or the 20- to 30-year cycles of the Atlantic Multi-decadal Oscillation (AMO), and the associated lag responses in, for example, water availability, fire frequencies, and production. Cow-calf operations are hard to manage strategically from a climate perspective because large herd size means ranchers cannot respond rapidly to changes in resource availability.

Sustainability in agroecosystem management depends on complex factors ranging from individual preferences to international agreements and global change, but it boils down fundamentally to the combined ecologic–economic realities of particular regions and systems. Understanding what influences the transition from one state to another is critical to developing incentives that can help achieve desired outcomes. In this chapter we have provided a framework within which to organize potential incentives for Florida ranches. Without including the options of converting agricultural lands to development, these range in our best estimation, from highest to lowest fiscal impact to landowners, as follows:

- Incentives for return on equity—State, federal, or private conservation and agriculture easement programs to protect lands most valuable for conservation and sustainability.
- Direct conservation payments for grazing lands set-asides (as in the USDA Conservation Reserve Program).
- Relief on estate tax payments if linked to conservation (e.g., conservation easements).

- Payment for ecosystem services like water services or carbon credits (market like).
- Premiums for agricultural products that are certified as environmentally friendly (market-based).
- Technical support and extension services to help reduce costs for energy, fertilizer, and feed.
- Cost-share conservation programs.

There are no simple solutions to the complex challenge of developing sustainable agroecosystems that support multiple ecosystem services, but every solution involves creating or responding to appropriate incentives. It may be possible to develop tools that help determine where there will be the highest public ecosystem benefits for incentive programs, and for private entrepreneurship, and which combinations of these incentives might be modified to suit different circumstances. Incentives cannot all be publicly funded, nor can it be expected that every rancher or farmer should to be compensated for every type of public benefit they provide. Unfortunately, many agricultural policies both in the United States and across the globe create perverse incentives that undermine ecological sustainability in pursuit of narrow economic interests (Flora 2004; Roberts 2008). Sustainable agroecosystem management requires polices and incentives that give ecological, economic, and social considerations equal footing.

REFERENCES

Adler, P. R., Del Grosso, J., and Parton, W. J. 2007. Life-cycle assessment of net greenhouse gas flux for bio-energy cropping systems. *Ecological Applications,* 17, 675–691.

Altieri, M. 1995. *Agroecology: The Scientific Basis for Sustainable Agriculture.* Westview Press, Boulder, CO.

An, L. et al. 2005. Exploring complexity in a human environment system: An agent-based spatial model for multidisciplinary and multiscale integration. *Annals of the Association of American Geographers,* 95, 54–79.

Barney J. N. and Ditomaso, J. M. 2008. Nonnative species and bioenergy: Are we cultivating the next invader? *BioScience.* 58, 64–70.

Bennett, E. M. and Balvanera, P. 2007. The future of production systems in a globalized world. *Frontiers in Ecology and the Environment,* 5, 191–198.

Berger, P. A. 2006. Generating agricultural landscapes for alternative futures analysis: A multiple attribute decision-making model. *Transactions in GIS,* 10, 103–120.

Berger, T. 2001. Agent-based spatial models applied to agriculture: A simulation tool for technology diffusion, resource use changes, and policy analysis. *Agricultural Economics,* 25, 245–260.

Bockstael, N. E. 1996. Modeling economics and ecology: The importance of a spatial perspective. *American Journal of Agricultural Economics,* 78, 1168–1180.

Bohlen, P. J. and Swain, H. 2008. Ranch for environmental services: Public benefits from private land. In *Proceedings of the 3rd National Meeting of the Grazing Lands Conservation Initiative,* Dec. 10–13, St. Louis, MO.

Bolte, J. P. et al. 2006. Modeling biocomplexity—Actors, landscapes, and alternative futures. *Environmental Modelling and Software,* 22, 570–579.

Boody, G. et al. 2005. Multifunctional agriculture in the United States. *Bioscience,* 55, 27–38.

Bousquet, F. and Le Page, C. 2004. Multi-agent simulations and ecosystem management: a review. *Ecological Modelling,* 176, 313–332.

Brown, D. G. and Xie, Y. 2006. Spatial agent-based modeling, Guest Editorial, *International Journal of Geographical Information Science,* 20, 941–943.

Capece, J. C. et al. 2007. Soil phosphorus, cattle stocking rates, and water quality in subtropical pastures in Florida. *Rangeland Ecology and Management,* 60, 19–30.

Carpenter, S. R. and Gunderson, L. H. 2001. Coping with collapse: Ecological and social dynamics of ecosystem management. *BioScience,* 51, 451–457.

Clay, J. 2004. *World Agriculture and the Environment: A Commodity-by-Commodity Guide to Impacts and Practices.* Island Press, Washington, D.C.

Cumming, D. H. M. 2005. Wildlife, livestock, and food security in the South East Lowveld of Zimbabwe. In *Conservation and Development Interventions at the Wildlife/Livestock Interface: Implications for Wildlife, Livestock, and Human Health,* Osofsky, S. A. et al., Eds. IUCN, Gland, Switzerland, 41–46.

ERS. 2008. State fact sheets: Florida. U.S. Department of Agriculture Economics Research Service, Washington, D.C. http://www.ers.usda.gov/StateFacts/FL.htm (accessed July 25, 2008).

Farber, S. et al. 2006. Linking ecology and economics for ecosystem management. *BioScience,* 56, 121–133.

Fargione, J. et al. 2008. Land clearing and the biofuel carbon debt. *Science,* 319, 1235–1238.

FDEP. 2008. Land conservation: Land acquisition as of March 31, 2008. Florida Department of Environmental Protection, Tallahassee. http://www.dep.state.fl.us/secretary/stats/land.htm (accessed June 25, 2008).

Feather, P., Hellerstein, D., and Hansen, L. 1999. Economic Valuation of Environmental Benefits and the Targeting of Conservation Programs. U.S. Department of Agriculture, Economic Research Service, Agricultural Economic Report 778. Washington, D.C.

Flora, C. B. 2004. Social aspects of moving to permaculture: Need for regime change. Paper presented at Agroecosystem Symposium, Un-Plowing the Land: Restoring Agroecosystem Health, Ecological Society of America Annual Meeting, Portland, OR. http://www.ncrcrd.iastate.edu/pubs/ESApresentation.pdf (accessed July 15, 2008).

FNAI Florida Natural Areas Inventory. 2008. Summary of Florida Conservation Lands. March 2008. http://www.fnai.org/PDF/maacres_200803_fcl.pdf (accessed July 21, 2008).

GAO. 2006. Report to the chairman, committee on environment and public works, U.S. Senate USDA Conservation Programs; stakeholder views on participation and coordination to benefit threatened and endangered species and their habitats. GAO-07-35, Washington, D.C., 1–69.

Gliessman, S. R. 2007. *Agroecology: The Ecology of Sustainable Food Systems,* 2nd ed. Taylor & Francis/CRC Press, Boca Raton, FL.

Gornak, S. and Zhang, J. 1999. A summary of landowner surveys and water quality data from improved pasture sites in the northern Lake Okeechobee watershed. *Applied Engineering in Agriculture,* 15, 121–127.

Grimm, V. et al. 2005. Pattern-oriented modeling of agent-based complex systems: lessons from ecology. *Science,* 310, 987–991.

Gunderson, L. H. and Holling, C. S., Eds. 2002. *Panarchy: Understanding transformations in human and natural systems.* Island Press, Washington, D.C., 103–120.

Hanson, H. C. 1939. Ecology in agriculture. *Ecology,* 20, 111–117.

Havstad, K. M. et al. 2007. Ecological services to and from rangelands of the United States. *Ecological Economics,* 64, 261–268.

Hoctor, T. S., Dubé, E. H., and Beyeler, S. 2008. Conservation incentives and programs for protecting critical lands and water. Century Commission for a Sustainable Florida. https://www.communicationsmgr.com/projects/1349/docs/ConsInc2008.pdf (accessed on June 30, 2008).

Kareiva, P. et al. 2007. Domesticated nature: Shaping landscapes and ecosystems for human welfare. *Science,* 316, 1866–1869.

Kingsland, S. E. 2005. The evolution of American ecology, 1890–2000. Johns Hopkins University Press, Baltimore, MD.

Leopold, A. 1949. *A Sand County Almanac, and Sketches Here and There.* Oxford University Press, New York.

Liu, J. G. et al. 2007. Complexity of coupled human and natural systems. *Science,* 317, 1513–1516.

Lowrance, R., Stinner, B. R. and House, G. J. 1984. *Agricultural Ecosystems: Unifying Concepts.* John Wiley & Sons, New York.

McCarty, L. B. 2006. Sod Production in Florida. BUL260 of the Environmental Horticulture Department, Florida Cooperative Extension Service, Institute of Food and Agricultural Sciences, University of Florida. Original publication date May 1, 1991. Reviewed September 2006. Available from http://edis.ifas.ufl.edu (accessed July 21, 2008).

McGhee, G. R. 2007. *The Geometry of Evolution: Adaptive Landscapes and Theoretical Morphospaces.* Cambridge University Press, Cambridge.

McPeak, J. G., Lee, D. R., and Barrett, C. B. 2006. Introduction: The dynamics of coupled human and natural systems. *Environment and Development Economics,* 11, 9–13.

Millennium Ecosystem Assessment. 2005. Ecosystems and their services. In *Ecosystems and Human Well-Being,* Island Press, Washington, D.C., chap. 2.

———. 2005a. *Ecosystems and Human Well-Being: Scenarios.* Island Press, Washington, D.C.

———. 2005b. *Ecosystems and Human Well-Being: Current State and Trends.* Island Press, Washington, D.C.

———. 2005c. *Ecosystems and Human Well-Being: Policy Responses.* Island Press, Washington, D.C.

Pahelke, R. 1999. Towards defining, measuring and achieve sustainability: Tools and strategies for environmental valuation. In *Sustainability and the Social Sciences,* Becker, E. and Jahn, T., Eds. Zed, London.

Parker, D. C. et al. 2003. Multi-agent systems for the simulation of land-use and land-cover change: A review. *Annals of the Association of American Geographers,* 93, 316–340.

Pickett, S. T. A., Cadenasso, M. L., and Grove, J. M. 2005. Biocomplexity in coupled natural human systems: A multidimensional framework. *Ecosystems, 8,* 225–232.

Pielke, Sr., R. A. et al. 2007. A new paradigm for assessing the role of agriculture in the climate system and in climate change. *Agricultural and Forest Meteorology, 142,* 234–254.

Pocewicz, A. et al. 2008. Predicting land use change: Comparison of models based on landowner surveys and historical land cover trends. *Landscape Ecology, 23,* 195–210.

Polasky, S. et al. 2005. Conserving species in a working landscape: land use with biological and economic objectives. *Ecological Applications* 15, 1387–1401.

Quinn, L. D. and Holt, J. S. 2008. Ecological correlates of invasion by *Arundo donax* in three southern California riparian habitats. *Biological Invasions,* 10, 591–601.

Raghu, S. et al. 2006. Adding biofuels to the invasive species fire? *Science,* 313, 1742.

Redman, C. L. 2002. Agrarian landscapes in transition: A cross-scale approach. NSF BE/CNH: Project 216560. http://sustainability.asu.edu/agtrans/ (accessed July 21, 2008).

Rickerl, D. and Francis, C. 2004. Multidimensional thinking: a prerequisite to agroecology. In *Agroecosystems Analysis,* Rikerl, D. and Francis, C., Eds. American Society of Agronomy, No. 43 in the Agronomy Series, Madison, WI, 1–18.

Rissman, A. et al. 2007. Conservation easements: Biodiversity protection and private use. *Conservation Biology,* 21, 709–718.

Roberts, P. 2008. *The End of Food.* Houghton Mifflin, New York.

Robertson, G. P. and Swinton, S. M. 2005. Reconciling agricultural productivity and environmental integrity: A grand challenge for agriculture. *Frontiers in Ecology and the Environment,* 3, 38–46.

Scheffer, M. and Carpenter, S. R. 2003. Catastrophic regime shifts in ecosystems: linking theory to observation. *Trends in Ecology and Evolution,* 12, 648–656.

Scheffer, M. and Westley, F. R. 2007. The evolutionary basis of rigidity: Lock in cells, minds, and society. *Ecology and Society,* 12(2), 36 [online]. www.ecologyandsociety.org/vol12/iss2/art36/ (accessed 10 June 2008).

Scheffer, M. et al. 2002. Dynamic interaction of societies and ecosystems; linking theories from ecology, economy, and sociology. In *Panarchy: Understanding Transformations in Human and Natural Systems,* Gunderson, L.H. and Holling, C. S., Eds. Island Press, Washington, D.C., 195–240.

Searchinger, T. et al. 2008. Use of U.S. croplands for biofuels increases greenhouse gases through emissions from land-use change. *Science,* 319, 1238–1240.

Small, S. J. 2002. Preserving Family Lands; Book III: New Tax Rules and Strategies and a Checklist. Law Office of Stephen J. Small, Esq., Boston.

Swain H. M. et al. 2007. Integrated ecological and economic analysis of ranch management systems; an example from south central Florida. *Rangeland Ecology and Management,* 60, 1–11.

Tallis, H. and Kareiva, P. 2006. Shaping global environmental decisions using socio-ecological models. *Trends in Ecology and Evolution,* 21, 562–568.

Tilman, D., Farigione, J., and Wolff, B. 2001. Forecasting agriculturally driven global environmental change. *Science,* 292, 281–284.

Turner, B. L. II et al. 2003. Illustrating the coupled human environment system for vulnerability analysis: Three case studies. *Proceedings of the National Academy of Sciences,* 100, 8080–8085.

Van Buskirk, J. and Willi, Y. 2004. Enhancement of farmland biodiversity within set-aside land. *Conservation Biology,* 18, 987–994.

Walker, B. et al. 2004. Resilience, adaptability, and transformability in social–ecological systems. *Ecology and Society,* 9(2), 5. http://www.ecologyandsociety.org/vol9/iss2/art5 (accessed 10 June 2008).

Weiner, J., Wiegand, T., and DeAngelis, D. L. 2005. Pattern-oriented modeling of agent-based complex systems: Lessons from ecology. *Science,* 310, 987–991.

WWF. 2008. Florida ranchlands environmental services project. World Wildlife Fund, http://www.worldwildlife.org/what/globalmarkets/agriculture/FRESP.html (accessed July 26, 2008).

Zwick, P. D., and Carr, M. H. 2006. Florida 2060: A population distribution scenario for the state of Florida. A research project prepared for the 1000 Friends of Florida. Geoplan Center, University of Florida, Gainesville. http://www.1000friendsofflorida.org/PUBS/2060/Florida-2060-Report-Final.pdf (accessed July 24, 2008).

15 Principles of Dynamic Integrated Agricultural Systems
Lessons Learned from an Examination of Southeast Production Systems

Gretchen F. Sassenrath, Jon D. Hanson, John R. Hendrickson, David W. Archer, John F. Halloran, and Jeffrey J. Steiner

CONTENTS

15.1 SUMMARY

In the past, American agriculture was focused solely on its ability to produce sufficient food, fuel, and fiber to meet local, national, and global demands. While productivity will continue to be a major factor in food production systems, increased societal demands for environmentally sound management, the need for rural community viability, and a rapidly changing global marketplace have resulted in challenges for the current agricultural system. New production systems developed around principles of integrated agricultural systems may assist in addressing some of these challenges. However, when helping to design and manage these systems, researchers need to be aware of how external influences may affect these systems. A framework for agricultural management systems is being developed that increases the use of renewable resources, decreases the reliance of agricultural production on fossil-derived fuels and fertilizers, and enhances producer flexibility to meet individual and societal goals. The four main categories that influence agricultural systems include (1) social/political factors, (2) economic factors, (3) technological factors, and (4) environmental factors. A case study from the southeastern United States is used to demonstrate the evolution of the current production system from these four factors. From these case studies, we examine sustainability issues for future agriculture, and potential changes needed to attain economic and environmental sustainability.

15.2 INTRODUCTION

Agricultural production capacity is currently sufficient to meet the food and fiber needs of most of the world's population. World population has grown from 2.5 billion in 1950 to an estimated 6.7 billion today (U.S. Census Bureau, 2008). Global average per capita food availability has risen from <2400 calories to >2700 calories (Ruttan 1999), in part because of increasing cereal yields, while land area in farms has remained relatively stable since 1950 (Trewavas 2002). Today's agriculture uses only 0.2 ha of land per person (Trewavas 2002). However, agricultural producers are operating in an increasingly complex and rapidly changing environment. Agricultural production systems are characterized by a high level of rapid and continuous change in response to external drivers (Hendrickson et al. 2008b). A striking example of the turbulence in U.S. agriculture is the current rapid shift in production in Southeast agriculture to corn from more traditional cotton systems in response to rising corn prices from new expansion in the biofuels industry (Smith 2007). The impact of these production shifts on rural communities, the U.S. economy, and global markets will potentially be substantial. In addition to traditional problems of production, agricultural producers today also face social and political changes, changes in consumer expectations, market fluctuations, technological advances, and a rapidly evolving U.S. farm policy. New agricultural production systems developed around the concept of integrated agriculture may address current and future challenges.

The Integrated Agricultural Workgroup was developed to explore current agricultural production systems and develop principles that underlie integrated systems. In 2005, a workshop was held at Auburn, Alabama, and producers from the Southeast were interviewed to discuss their observations regarding production agriculture. These producers were passionate about their chosen livelihoods. One producer said, "I want to stay on the farm ... keep growing...." Another said, "Farmers ... they want to stay on their land, like to grow livestock, be out in the woods ... to make a living...." Yet, they still had major concerns with how their industry functions. A poultry producer made the statement, "The Company owns the feed, the chicken houses, the processing plant, and the chickens.... All I own are the dead chickens." He felt he was limited to being an indentured servant because of the type of written contract that has emerged within the industry. With these thoughts in mind, we would like to spend some time examining three fundamental questions. First, how did our agricultural system get to where we are today? Second, what are the most pressing sustainability issues today? Finally, what will future agricultural systems need to look like to truly incorporate economic and environmental sustainability?

15.3 FARM MAKEUP

The first meeting of the Integrated Agricultural Workgroup, held in Mandan, North Dakota, in 2004, identified four key drivers to agriculture: social/political, economic, environmental, and technological. These drivers are presented in detail in a series of manuscripts in the journal *Renewable Agriculture and Food Systems* (introduced in Hendrickson et al. 2008a). Following the initial meeting, the Integrated Agricultural Workgroup hosted a meeting in Auburn, Alabama, in the fall of 2005. Five producers were invited to present their production systems and discuss production practices and concerns with the group. We selected this small group of successful farmers to better examine production systems in detail. Presentations were made by two chicken/hay/cattle producers, one catfish/cattle/row crop farmer, one row crop/hay/cattle producer, and one row crop producer. Following the 1 hour producer presentations, participants discussed principles and characteristics common to the production systems. Reports from the discussion groups were compiled into lists of drivers, characteristics, and potential principles. Here, we present summaries of the producer presentations in light of the drivers of production systems studied in our first set of manuscripts (Hendrickson et al. 2008a).

The farmers invited to the workshop did not represent "average" farmers for Alabama, or the Southeast. Most notably, they were progressive farmers who regularly work closely with scientists and extension, and were willing to give up valuable time to commit to our process of examination. The average farm size in excess of 2000 acres for each farmer easily exceeded the state average of 198 acres per farm* (NASS 2007). The farmers represented a younger, more educated group than the average Alabama farmer, and averaged more than four products per farm, which exceeds the national average of just over one product per farm (Dimitri et al. 2005). All participants relied on farming as their primary source of income. Although farm size exceeded national and regional averages, all the farms were family farms and none considered its operations agribusinesses. We selected these farmers to examine how primary agriculture has changed in the Southeast United States, the drivers responsible for that change, and the potential future challenges of agricultural production.

15.4 THE PATH OF AGRICULTURE

15.4.1 SOCIAL CHANGES

Changes in the farm structure have resulted in fewer people being directly involved in production agriculture. The percentage of people who considered farming their primary occupation decreased about 63 percent from 1940 to 2000. Over the same period, the average farm size increased about 67 percent. Currently, less than 2 percent of the U.S. population is directly involved in farming. Farmers are increasingly dependent on off-farm employment and a higher degree of farm specialization (a movement from having a diversity of farm enterprises to having only one or two enterprises) (Dimitri et al. 2005).

Archer et al. (2008) examined social and political factors influencing agriculture. They defined internal social factors as those factors arising from within the farm system, and directing the decision-making process of the farmer. In our group of farmers interviewed, the strongest factor dictating the farming operation was the internal social driver to continue farming as a way of life. Each farmer on the panel described their desire to continue the farming tradition, and pass that tradition on to their children. The farmers saw heritage, love of the land, family, and the farming lifestyle as primary factors in their decision to farm, and in the production choices they made.

A second internal social driver was the desire for independence within the farming system. The poultry producers in particular were frustrated at the lack of input they had in the production process. The increased vertical integration of the poultry industry has led to the development of stringent contracts that remove management decisions from producers' control. One farmer put it this way, "Right now, I can go out there and that cow will listen to me. Talking to an integrator (poultry company representative) is like talking to a wall." Similar frustrations were described in working through government programs, such as conservation programs, that gave producers some funds to implement conservation practices. The end result is that, while the growers saw some value in the conservation programs, they did not want the government directing their farming operations.

This is in direct contrast to the primary internal social drivers identified by an external group of expert panelists, most likely because the panelists viewed internal social drivers as those (external) social factors having an influence at the farm level (Archer et al. 2008). The panelists identified globalization and low margins that required increased scale and efficiency as the two most significant internal social drivers. Although the farmers interviewed here were sensitive to those concerns, other factors took precedence.

A common external social factor influencing all production systems was the lack of labor and, in particular, the lack of skilled labor. This reflects the decline in people directly involved in agriculture.

---

* Note that this state average includes all farms, including recreational and retirement farms, whose primary function is not the production of agricultural products.

Most of the farming operations were handled by the farmer himself, with assistance from family members, both older (parents) and younger (children). Again, quality of life was a consideration in whether to expand the farming operation, though lack of skilled labor was noted as a factor limiting further expansion. As one farmer put it, "the current labor pool … can barely pour corn out of a bucket," let alone function under the more complex environment in which the producers operate. Technological advancements, together with a declining rural population and, in particular, fewer people with farming experience, contribute to this lack of skilled labor.

The market consolidation/concentration identified by Archer et al. (2008) was another major external social factor influencing the production systems, especially for the poultry producers. The poultry producers' loss of control of the production system through aggressive company contracts and demands was seen as a major negative factor influencing the farm system. According to the farmers, legislation and litigation have not been successful in protecting producers' interests. Poultry producers feel company policies have resulted in the loss of control of the production operations on their own farms, and the very limited avenues available to them for input into the operation or the contracts. The producers said: "They (the company) have the pencil, and they can put you anywhere on the chart they want to," meaning the company controls how much feed is purchased and the live weight of chickens after harvest, which directly determine the producer's profit. As Archer et al. (2008) point out, the vertical integration has resulted in an imbalance in distribution of wealth, and pressure from that imbalance may be bringing changes to the industry, albeit slowly.

Farmers in other areas have formed cooperatives or associations to give them more bargaining power. In catfish production, for example, the formation of a catfish marketing association has stabilized supply and demand by coordinating between producers and processors to make informed judgments on production levels needed. Now, producers are able to sell fish above the cost of production, and processing plants have sufficient, but not excessive, fish to fill demand.

15.4.2   Farm Policy Changes

Changes in farm policy over the last 100 years have resulted in a shift in production to commodity crops, with a concomitant decrease in diversity and shift to monoculture production (Hendrickson et al. 2008b). Changes in farm policy have also led to a greater reliance on price and income supports, and supply control. Archer et al. (2008) identified Farm Bill commodity programs and environmental regulations as the two largest political factors influencing farming in the United States. These factors were also noted by the producers as major factors affecting their production decisions. While dependence on farm supports is recognized by the producers, as one farmer put it, "farming the supplements" is viewed as a necessary evil. As another farmer noted, he tried to pay attention mostly to agronomics, but recognized that he had to "play the game" to stay in business. Overall, the farmers would much rather be able to generate their income from sales of their products than to rely on farm supports. Through numerous farm bills, the price and income supports have evolved from commodity-specific prices, to flexible prices, to target prices and deficiency payments, loan rates, and commodity loan programs (Halloran and Archer 2008). With the 1996 Farm Bill, a movement toward decoupled agricultural payments started in an effort to allow farmers to respond to market signals instead of program benefits. In 2003, about 39 percent of all farms received some type of government payment with between 71 and 84 percent of medium-sales small family farms and large-scale family farms receiving government payments (Hoppe and Banker 2006).

Although the farm commodities programs have sometimes been noted as decreasing the diversification of individual farms and increasing the dependence on single enterprise systems, the group of farmers in this discussion group each had several different products. The least diversified were the poultry producers, with three products, who used hay and pasture fields to dispose of chicken litter and waste. The hay and pasture were used to graze cattle. The youngest producer had the most enterprises at seven (peanuts, cotton, small grains, corn, cows, Bahia grass, and timber). This philosophy of diversification came from his father: "If you can't put a crop on it, you can put a cow on it, and if you can't put a cow on it, then you put a tree on it." The underlying philosophy is to make every piece of ground productive in some way. This young farmer felt he had an advantage because he did not grow up under the old farm program; therefore it did not limit his expectations. The greater diversification of this group of farmers than the national average may also arise from their self-identified role as innovators.

15.4.3   ECONOMICS AND MARKET CHANGES

Money is a key driver of the farm production system. Volatility of external (e.g., fuel prices, farm policy, markets) and internal (e.g., management decisions) factors to agriculture make short-term economics concerns paramount for farmers. Production decisions are made based on minimizing risk and maximizing return. Farmers also noted the importance of reinvesting to the farm. Often, however, a lack of funds kept them tied to their current production practices even when they saw benefit in alternative or newer technologies. Debt load also limited expansion of operations. As one farmer noted, the first deed on the land was made in 1831, and they still owed the bank.

One approach that was common to all producers in the group was to create income from several sources. Additional enterprises expand the diversity of the individual farms and provide alternative sources of income. The return on investment for each of the enterprises changed with year. In any given year, different enterprises would contribute positively to the overall farm budget. By relying on several enterprises, the farmers were able to spread risk and continue in business. To this end, the row crop farmer used his combine and cotton picker to do custom harvesting. Alternative enterprises associated with the current production systems that allowed additional income included a cotton gin, small grain seed sales directly to hunters, and part ownership in catfish feed and processing plants. To be considered, any ancillary enterprise needed to contribute to and fit into the current production system. Everything produced was sold or used elsewhere, such as for cattle feed.

One source of frustration for the producers was the marketing of products. All the farmers recognized the globalization of markets, and as one noted, they "feel like we're in an unfair trading market." Formation of marketing associations gave producers a stronger voice in marketing decisions. A good example of this was with the catfish producer. Because catfish is a relatively new crop, the marketing structure is somewhat different from that for other crops. Catfish production is unsubsidized, and fish are sold on the open market. Recent decreases in profitability led producers to form a marketing association, which allows closer coordination between processors and farmers, which in turn provides a steady supply of fish while holding profits at acceptable levels. Although formation of the marketing association has not led to a large price increase, it has stabilized the supply and demand. Because of uncertainties in the market, and the needs of the processing plants to ensure adequate supplies, growers are increasingly selling fish on contract to specific processing plants. These contracts are more analogous to standard marketing contracts made in traditional row crop production than those used in poultry production. Moreover, the catfish producer retains full control of his or her production facilities. Coordination of marketing and promotion of catfish has led to increased market shares for what was formerly a niche market. Aggressive technological developments and close attention to quality has led to development of a high-quality product.

The same forces that influenced U.S. agriculture, namely, specialization, technological innovation, and the commensurate increase in scale, have also influenced the markets for agricultural products. The contribution of agriculture to the gross domestic product (defined as the market value of all final goods and services produced within a country in a given period of time) has decreased sharply from just under 8 percent in 1930 to less than 1 percent in 2002 (Dimitri et al. 2005). Other forces such as increased demand for variety, convenience, packaging, quality, and recently, how agricultural commodities are produced, have also shaped the food marketing system. With changing demographics (e.g., women in the workforce, growth in ethnic markets), the demand for convenience and variety has grown (Halloran and Archer 2008). Vertical integration of the meat industry has largely evolved because of increasing consumer demand for convenient, readily available, and consistent products (Martinez and Stewart 2003). According to U.S. Department of Agriculture (USDA) analysis, over 40 percent of the consumer's food dollar is now spent on meals away from home (Harris et al. 2002).

Poultry production is highly vertically integrated (Martinez 2000) and concentrated in the southeastern United States (Drabenstott 2000). Processing companies establish contracts with individual growers, and dictate most aspects of the production cycle. Contracts specify the batch of chicks that producers start with, the feed for the chicks, the construction, maintenance, and upgrading of the chicken houses, and the environmental conditions within the houses during chicken growth and development. The processing company owns the chickens and the feed. Farmers are paid for the weight of the birds at harvest, less cost of feed. Rigid requirements for growth and development of the chickens result in consistent, high-quality birds and ensure adequate supplies for the processing plants (Drabenstott 2000). Initial contracts may be for several years, which help farmers get funding to cover start-up costs. After an initial period, farmers may be put on batch-to-batch contracts. A primary constraint of some vertical integrators is the contracts that limit farmers' control over production practices and management. Binding arbitration clauses in more recent contracts give farmers little recourse to correct problems. Other processing companies have gone to paying farmers by the square foot for production, and one producer indicated that growers are happier with that arrangement.

Economic constraints greatly limit profitability of chicken production systems. Companies are interested in improving the size of the birds and the feed conversion rates, for greater return and a more consistent product. Improvements in technologies of confinement building construction and animal care require farmers to continue to invest in expensive updates. Chicken houses are expensive, and the high debt load encumbered in establishing the business and making required updates prevents the growers from moving to potentially more profitable ventures. Debt also keeps growers in line, further reducing bargaining power in contract development.

Catfish production stands in sharp contrast to the highly vertically integrated chicken production system. Catfish is a relatively new crop, which began in the 1960s as a niche market and has grown substantially, especially in the last few years. In contrast to the lack of management control in chicken production systems, catfish producers are quite involved in the decisions associated with animal production. In addition to on-farm fish production, producers have also developed ancillary enterprises which contribute to catfish production and postharvest processing. For example, producers in Alabama realized a need for catfish feed, and so established their own feed mill. In addition to getting a break on the cost of feed, which accounts for nearly 50 percent of the production expense, the feed cooperative, working in collaboration with researchers, has developed an optimal feed for their area. Some producers are also involved in cooperatives that process and market the fish after harvest.

A major concern for the catfish industry has been the importation of cheaper fish from other countries. One serious issue for fish production is food safety because of unsanitary production practices in other countries and the use of banned antibiotics. There are concerns that imported fish, grown in less-than-desirable conditions, may undercut the quality of U.S. catfish. Through catfish associations, the catfish industry has worked to control the importation of competing fish, particularly basa and tra, through country of origin labeling and information on use of banned antibiotics by importers. Aggressive lobbying by states and production groups has led to bans on importation, and introduced requirements for country-of-origin labeling. By forming associations, catfish producers have kept greater control of their industry.

15.4.4 TECHNOLOGICAL CHANGES

The primary goal of technology is to provide a benefit to society by solving a problem or circumventing a functional constraint (Sassenrath et al. 2008). Agricultural technologies include both engineering and biological inventions, such as variable-rate applicators and insect-resistant crops, as well as knowledge itself. Farmers today must determine how best to use new technologies such as precision agriculture technologies, and decision support and management systems. Mechanization has reduced the labor required to operate modern farms. The advent of cheap chemicals and the "perfection" of modern equipment have also led to increased yields and overall productivity. New technologies continue to emerge, including innovations in genomics, animal production systems, conservation systems, and information systems (Sassenrath et al. 2008).

All the farmers in our study group saw interaction with scientists and extension as key components benefiting their production systems. This reflects the early to mid-innovator status of the participants. While they were not willing to take undue risk in implementing a new procedure or practice, they had no hesitancy to call a scientist or extension specialist to help them with a problem. And they were more than willing to implement a new technology that had been proved, if only in research tests. Key limiting factors to implementation of proven technologies were time and money. As one producer said, he has a mold-board plow, so that's what he uses. For him, investing in conservation systems was just not possible at the moment, as the money (for new equipment) was simply not there.

The producers in the discussion group had implemented a variety of new technologies, including Global Positional System (GPS) and precision technologies, ginning technology, conservation systems, knowledge systems and decision aids, and biological and engineering tools. Producers also took advantage of marketing tools for better crop sales. Farmers were very cognizant of the need to keep crop value high to successfully compete on the world market.

One factor that has not been addressed in previous discussions is the importance of farmer's knowledge to choices of production practices. Several farmers noted that they were willing to get involved in enterprises that were entirely new to them because someone in their family had technical expertise to help them. While it is not clear how much lack of knowledge may impede adoption of alternative practices, it is apparent that the more closely the farmer has access to experts or resources, the more amenable the farmer is to trying new practices. This is a critical observation to consider when transitioning to new, more-sustainable production practices.

In part because catfish is a new product, many advances in the industry have come about from close ties between producers and scientists. The introduction of innovations and new technologies to improve production are substantial. Growers strongly support research efforts to continue these advances, and a portion of their proceeds goes to support catfish research. Technological advances include genetic stock, feed formulations, pond construction, aeration, and harvesting techniques. Focused research has improved the flavor and texture of the fish, as well as the weight per fish. Producers see the potential for improved technologies to increase the amount of fish per acre that can be produced. With ever-increasing costs of production for feed and energy, producers realize the need for changes to keep unit production costs low, to compete with foreign markets, and improve sustainability of production.

15.5   ENVIRONMENT

Environmental processes individually and in concert influence agriculture systems and shape agroecosystems (Hendrickson et al. 2008c). The environment determines where, how, and when agricultural production can occur. The diversity of climate, soil, water, and other natural resources within the United States has resulted in the development of an array of agroecological niches. Innovations directed to overcoming environmental limitations in the past considered individual system components and not the production system in its entirety. Advances in our appreciation and understanding of the complexity and interconnectedness of the agroecosystem have facilitated development of new technologies aimed at promoting environmentally sustainable production systems, most notably conservation production systems.

One important factor that impedes implementation of conservation practices is the high percentage of rented land. Cash renters have been shown to be less likely to adopt conservation practices than are landowners or share renters (Soule et al. 2000). One farmer recognized the benefits of using conservation practices, including crop rotations and cover crops. However, because of competition between local farmers for leased land, if he implements conservation practices he incurs the expense of improving the land but risks losing the lease and the long-term economic benefits of his conservation investment. Now, he specifically uses rental land for crops that deplete the soil.

A second factor that is important to the production systems is the geographic distribution of farms, particularly of leased parcels. Several farmers were limited in production practices because of the spread of acreage over greater distances. These increased driving distances were particularly critical when moving large equipment and for making timely management decisions. Geographic distribution also limited the range that poultry litter could be profitably sold, and increased the costs of feed because of added transportation costs. This may be a factor contributing to the consolidation of poultry and other animal operations in discrete geographic regions. This will become even more important with increasing fuel prices.

Environmental issues are an important concern in agricultural production. In poultry production, while the industry controls many of the inputs to the system, the farmers are responsible for all environmental impacts from the confinement buildings. Waste and dead chicken disposal are serious problems, and farmers are required to follow strict CAFO (confined animal feeding operations) regulations. Producers can sell the manure, or spread the manure on pastureland used for grazing cattle. Requirements for litter disposal are a key driving factor increasing the diversity of the farms. The catfish industry has taken a proactive approach in certifying best management practices for catfish production that actually benefit the environment by supplying cleaner water coming out of the ponds than that going in. This work has led the Environment Protection Agency (EPA) to approve best management practices for catfish production, and other organizations to certify catfish as environmentally friendly (Delta Farm Press 2007).

15.6   FUTURE CHALLENGES IN AGRICULTURE

Agriculture has changed greatly over the past few decades. Our greatest challenges lie ahead specifically, "Can an agricultural system, developed in response to forces of the twentieth century be sustainable in the twenty-first century?" (Hanson et al. 2008). In speaking with agricultural producers, one never hears that they want to be "unsustainable." In general, producers want to treat the land resource properly, yet they must operate a system that is economically viable. However, there are different understandings of what is sustainable, and different visions on how to achieve it.

Most definitions of sustainability include an economic dimension, an environmental dimension, and a social and community dimension. Here, we define agricultural sustainability as an approach to producing food and fiber which is profitable, uses on-farm resources efficiently to minimize adverse effects on the environment and people, preserves the natural productivity and quality of land and water, and sustains vibrant rural communities (Hendrickson et al. 2008a). Other aspects that must be considered for sustainable agricultural systems include (1) they must be holistic, (2) they must be scientifically acceptable, (3) they must be ecologically based, and (4) they must be designed for the long term.

Most producers want to practice sustainable agriculture. Such practices provide several long-lasting benefits to agriculture as a whole. First, they provide long-term viability and resilience of farm economics. Second, they lead to conservation and enhancement of the natural resource base. Third, sustainable systems minimize offsite environmental impacts. Fourth, producers practicing sustainable agriculture improve their farm level management skills. Finally, the socioeconomic viability of rural communities is enhanced when sustainable practices are applied.

From discussions with producers, it is apparent that the biggest question of future sustainability is economic—how do farmers stay in business? Consumers are placing increasing demands on environmental sustainability. If society truly wants to move to sustainable agricultural production, consumers must be willing to pay the full expenses of production. Today in the United States, about 11 percent of the nation's disposable income is spent on food purchases, as compared to about 23 percent in 1929 (ERS 2006). Today's consumers are demanding a wider variety of products and have greater concerns for food safety. Consumers are also concerned about how and where products are produced, and these concerns are reaching even to farm programs. The array of food items has grown astronomically. The typical supermarket carries 40,000 food items (Harris et al. 2002). However, globalization and low profit margins continue to be important issues affecting agriculture at the farm level (Archer et al. 2008).

15.6.1 ALTERNATIVE PRODUCTION SYSTEMS

Integrated agriculture has the potential to fully or partially address many of the problems that confront agriculture. Integrated agriculture can help increase agricultural diversity from the field to the farm scale, which may increase system stability. Integrated agriculture may also be the best framework to use for developing sustainable agricultural systems. In Europe, an integrated approach to crop production called "integrated farming systems" has been advocated as a sustainable approach to agriculture that can maintain farmer income and safeguard the environment (Morris and Winter 1999). Integrated agricultural systems are complex systems and developing and analyzing them can be difficult. The development of a set of principles underlying complex agricultural systems can assist in the difficult task of developing integrated agricultural systems. Principles are defined in Webster's *New World Dictionary* as "the ultimate source or cause of something"; "advice, guidelines, prescriptions, condition–action statements, and rules" (Armstrong 2001); or "guidelines or prescriptions for how to use intentions" (Morwitz 2001). These definitions range from highly definitive laws to loosely applied guidelines. We chose to define principles as a set of concepts or ideas that help to explain how systems operate. When looked at in this manner, even the simplest system relies on principles. For example, a wheat–fallow system is based on the principle that harvesting water and controlling weeds during the fallow year will enhance yields during the subsequent cropping year.

Hendrickson et al. (2008a) suggested the use of dynamic-integrated agricultural systems as a horizontally integrated system which could potentially meet sustainability and adaptability goals. Dynamic-integrated agricultural production systems are agricultural production systems with multiple enterprises managed in a dynamic manner that interact in space or time and these interactions result in a synergistic resource transfer among enterprises (Hendrickson et al. 2008b). This system uses annual and intra-annual decision making to decide what to grow based on the producers goals, management concerns, and exogenous factors. The dynamic aspect of this concept is a management philosophy that requires management decisions not be predetermined but rather made at the most opportune time with the best available information (Hendrickson et al. 2008b). Its use of multiple enterprises and tactical decision making will maintain producer flexibility in a rapidly changing environment. Because of its emphasis on producers' goals and management concerns, producers can modify it to reflect their current labor and management abilities. A bidirectional flow of information from producers to researchers allows producers to use the best possible information in making management decisions. As technology, and in particular, information technologies, grow, these can be used by producers to ease the management burdens of dynamic systems. These technologies will allow producers faster and more complete and accurate access to information on marketing and management. The challenge is the strategic design of these systems to allow them to respond to changes in an external driver to gain benefits but still maintain sufficient stability (Archer, 2005; Hendrickson et al. 2008b).

15.7 CONCLUSIONS

An examination of Southeast production systems demonstrated the importance of internal social drivers in production choices. Additional pressure came from external social and political pressures, including environmental concerns, lack of skilled labor, and changes in the Farm Bill. While farmers were supportive of new technologies, they were much more likely to incorporate these technologies if they had ready access to an expert in the area. Other factors that were not apparent in the initial analysis of production systems included the geographic distribution limitations and the importance of farmers' knowledge to implementing new procedures.

The farmers interviewed showed a great desire to remain in farming and enjoyed the lifestyle that agriculture offered. However, the lack of freedom to make management decisions, especially among the poultry producers, was a major source of dissatisfaction. The ability of producers to change the undesirable situation was hampered by debt accumulation to buy and update expensive technology that may not be profitable. Debt accumulation and lack of available funds also affected decisions regarding sustainability. Although participants expressed a desire to achieve sustainability, their ability to do so was compromised by financial concerns.

All the producer participants had multiple enterprises. They indicated multiple enterprises provided income stability and decreased financial risk. Dynamic-integrated agricultural systems expand on the income stability and risk reduction gained by integration by providing producers with increased flexibility. This flexibility can provide producers with the ability to adjust to unknown future conditions. However, the reluctance of producers to try unfamiliar new enterprises indicates the importance of knowledge transfer in these complex systems.

ACKNOWLEDGMENTS

The authors express our appreciation for the producers and scientists involved in this workgroup, and especially recognize the efforts of Drs. Randy Raper and Andrew Price at the USDA-ARS National Soil Dynamics Lab for hosting the conference.

REFERENCES

Archer, D. 2005. Weeding out economic impacts of farm decisions. In *The Farmer's Decision: Balancing Economic Successful Agriculture Production with Environmental Quality*, Hatfield, J. L., Ed. Soil and Water Conservation Society, Ankeny, IA, 63–75.

Archer, D. W. et al. 2008. Social and political influences on agricultural systems. *Renewable Agriculture and Food Systems*, 23, 272–284.

Armstrong, J. S. 2001. Introduction. In *Principles of Forecasting: A Handbook for Researchers and Practitioners*, Armstrong, J. S., Ed. Kluwer Academic, Norwell, MA, 3.

Delta Farm Press. 2007. U.S. farm-raised catfish environment-friendly. Jan 12, 2007. http://deltafarmpress.com/mag/farming_us_farmraised_catfish/index.html (accessed June 7, 2008).

Dimitri, C., Effland, A., and Conklin, N. 2005. The 20th century transformation of US agriculture and farm policy. Economic Information Bulletin EIB-3. U.S. Department of Agriculture, Economic Research Service, Washington, D.C.

Drabenstott, M. 2000. A new structure for agriculture: a revolution for rural America. *Journal of Agribusiness*, 18, 61–70.

ERS (Economic Research Service). 2006. Food CPI, prices and expenditures: Food expenditures by families and individuals as a share of disposable personal money income. U.S. Department of Agriculture, Economic Research Service, Washington, D.C. http://www.ers.usda.gov/Briefing/CPIFoodAndExpenditures/Data/table8.htm. Updated June 9, 2006 (accessed June 7, 2008).

Halloran, J. F. and Archer, D. W. 2008. External economic drivers and integrated agricultural systems. *Renewable Agriculture and Food Systems*, 23, 296–303.

Hanson, J. D., Hendrickson, J. R., and Archer, D. W. 2008. Challenges for maintaining sustainable agricultural systems. *Renewable Agriculture and Food Systems*, 23, 324–334.

Harris, J. M. et al. 2002. The U.S. food marketing system: Competition, coordination, and technological innovations in the 21st century. Agricultural Economic Report 811. U.S. Department of Agriculture, Economic Research Service, Washington, D.C.

Hendrickson, J. R. et al. 2008a. Principles of integrated agricultural systems: Introduction to processes and definition. *Renewable Agriculture and Food Systems*, 23, 265–271.

Hendrickson, J. R. et al. 2008b. Interactions in integrated agricultural systems: The past, present, and future. *Renewable Agriculture and Food Systems*, 23, 314–324.

Hendrickson, J.R., Liebig, M.A., Sassenrath, G.F. 2008c. Environment and integrated agricultural systems. *Renewable Agriculture and Food Systems*, 23, 304–313.

Hoppe, R. A. and Banker, D. E. 2006. Structure and finances of U.S. farms: 2005 family farm report. Economic Information Bulletin Number 12. U.S. Department of Agriculture, Economic Research Service, Washington, D.C.

Martinez, S. W. 2000. Price and quality of pork and broiler products: What's the role of vertical coordination? Current issues in economics of food markets. Agriculture Information Bulletin No. 747-02. U.S. Department of Agriculture, Economic Research Service, Washington, D.C.

Martinez, S. and Stewart, H. 2003. From supply push to demand pull: Agribusiness strategies for today's consumers. *Amber Waves*, 1(5), 22–29.

Morris, C. and Winter, M. 1999. Integrated farming systems: The third way for European agriculture? *Land Use Policy*, 16, 193–205.

Morwitz, V. G. 2001. Methods for forecasting from intentions data. In *Principles of Forecasting: A Handbook for Researchers and Practitioners*, Armstrong, J. S., Ed. Kluwer Academic, Norwell, MA, 33–56.

NASS. 2007. Quick stats: Agricultural statistics data base. Department of Agriculture, National Agricultural Statistics Service, Washington, D.C. http://www.nass.usda.gov/QuickStats/ (accessed June 7, 2008).

Ruttan, V. W. 1999. The transition to agricultural sustainability. *Proceedings of the National Academy of Sciences*, 96, 5960–5967.

Sassenrath, G. F. et al. 2008. Technology, complexity, and change in agricultural production systems. *Renewable Agriculture and Food Systems*, 23, 285–295.

Smith, R. 2007. Sunbelt growers planting much less cotton. Delta Farm Press. May 28. http://deltafarmpress.com/cotton/070528-sunbelt-growers/index.html (accessed June 7, 2008).

Soule, M. J., Tegene, A., and Wiebe, K. D. 2000. Land tenure and the adoption of conservation practices. *American Journal of Agricultural Economics*, 82, 993–1005.

Trewavas, A. 2002. Malthus foiled again and again. *Nature*, 418, 668–670.

U.S. Census Bureau. 2008. U.S. and World Population Clocks. http://www.census.gov/main/www/popclock.html.

16 Participatory Approaches and Stakeholder Involvement in Sustainable Agriculture Research

Karin Eksvärd, Lennart Salomonsson, Charles Francis,
Nadarajah Sriskandarajah, Karin Svanäng, Geir Lieblein,
Johanna Björklund, and Ulrika Geber

CONTENTS

16.1 INTRODUCTION

Nordic experiences of applying participatory approaches in research and development of sustainable agriculture have been highly successful. The farmers involved were participating in certification programs for sustainable agriculture systems, most of them in the organic farming certification scheme. The findings from the research reflect the current European Common Agricultural Policy (CAP) that is changing from a production focus to a multifunctional focus, and especially to deal with increasing environmental concerns. The results also reflect current difficulties over how to create incentive structures for sustainability, rules that support creativity in handling increasing local and global environmental problems, instead of forcing farmers with state regulations to comply with top-down-created regulations and subsidies that are inflexible and not site specific.

This chapter is based on the authors' collective agronomy experiences, knowledge, and many years of discussions and reflections on how agriculture, as well as the global society, is facing new challenges. To address the magnitude and complexity of these challenges will require cooperation among people with divergent interests and abilities to handle goal conflicts. During the discussion that follows we provide evidence and illustrate the potential for participatory approaches in agriculture research. Creative and systemic approaches are appropriate to handle complex challenges, by applying a different worldview and base for learning and new knowledge, as a platform, and in going from a perspective of "How can we reach them (farmers)" to a perspective of "Everybody can learn about and do research within their own farm situation." In other words, we advocate a perspective that improvements in complex messy situations can best be made through a learning process that includes collaboration, and that all participating actors have something to contribute to learning as well as the research process.

In a real sense, sustainability cannot be developed without individual action, even if we often focus on the institutional frameworks as the important means or driving forces. In that respect we believe that a key element of sustainability is about individual responsible choices, with those coming together into joint actions, within a diverse social context. In the process of inquiring and researching for better futures, participation by all appropriate and responsible individuals can ensure the mutual learning and reciprocity in action, necessary for creating agreed-upon future wanted situations.

16.2   BACKGROUND

Farmers in the Nordic region and elsewhere face increasing challenges today due to fragility of support programs, high costs of inputs, competition in the global food system, and uncertain global resource and environmental conditions. There is special concern about the effects of global warming and consequences of peak-oil and future availability of fossil fuel. Farmers have to cope with a higher risk level than ever before. They need research that is relevant to their immediate needs and applicable in each specific farm situation on one hand, and that contributes to long-term food system sustainability on the other. Present agricultural research in universities and national institutes is often driven by sources of funding, interests of specialized scientists, and need for technical publications. Too often, the needs of farmers and the needs of researchers are different. We describe in this chapter a process that involves setting research priorities not only to create new knowledge, but also to stimulate learning and change through shared planning and close collaboration among multiple stakeholders in the agriculture and food sector. This research process is based on a theoretical and practical approach different from conventional disciplinary natural science research normally used in agriculture, including broadening aspects of who has the right to ask questions, recognizing that everyone has the capacity to learn and research his or her own situation, and that contributions from several actors are necessary to understand uniqueness of place and the problems within that situation. The participatory approach is described in the context of today's agriculture as well as through projections for the future. Specific examples are provided where farmers, researchers, extensionists, and employees from farmer organizations share ownership of the research agenda. We have found that such a model in the Nordic region leads to research that is on one hand relevant to the systems being studied and results being used for situation improvement and creation of desirable futures by the actors, as well as spread to others, and at the same time creates new perspectives and knowledge for the university academic environment.

In the participatory approach model discussed in this chapter, the research process is based on social learning among people with a common interest. This raises key questions about the research process: who takes leadership and responsibility, and who benefits from the results? Participation emphasizes learning related to the issues being studied, and the impacts of the learning and changes going on, as well as about influence on the surrounding systems and interactions among people. Participation by stakeholders in research could be through cooperation with researchers who are planning and managing a project and using farmers' fields for the trials, but where conducting research is still in the hands of the researchers. But this is not what we mean by participatory research in this chapter. Rather we discuss participation when the comanagement of all aspects of a project are essential, and when the participating stakeholders have ownership in the process from the first step of identification of key problems to interpretation and application of results.

16.3 PARTICIPATORY RESEARCH IN AGRICULTURE

"We can't solve problems by using the same kind of thinking we used when we created them."

—**Albert Einstein**

Participatory research has a long history in other areas than agriculture as well, mainly in the social sciences. An impressive stream of social research has been conducted under the label of participatory research, a subset of the action research tradition, in the arenas of community development, social change, and education. Within farming systems contexts, where there is a need for a holistic approach, participatory research has been used in order to understand and research complex situations. These approaches have been evolving over the past 40 years, have been firmly based on ideas of hard and soft systems thinking (Checkland 1981; Checkland and Scholes 1990), and later developed and applied by many others (e.g., Bawden 2003). As such, the ideas around participatory research go beyond the initial ideas of liberating action by Freire (1972, 1974), emphasizing the reality of each participant as a starting point for learning, developing new knowledge and practical change.

The interest in participation in agricultural research came at about the same time as this methodology became useful in health research. Cornwall and Jewkes (1995) found: "Research approaches which emphasize participation are gaining greater respectability and attention within mainstream health research in developed and developing countries."

In practice, the term participation itself can include all things ranging from "being informed" on the one end to collegially "exercising actual control" on the other (Elden and Taylor 1983). Here, we discuss participation more in correspondence with what Pretty (1995) calls interactive participation and self-mobilization, and what Johnson (2004) calls collegial level participation. Several authors have commented on the divide between "real" and "pseudo"-participation (Okali et al. 1994; Martin and Sherington 1997) were "pseudo" is an expression for situations where participants are led to believe they have influence but actually have not. These issues of power have formed the basis for different typologies of participation often used (Arnstein 1969; Pretty 1995).

The promotion of participatory approaches to rural development began in the late 1970s by the work of people like Robert Chambers (1994), and even earlier by those pursuing the action research tradition (Hall et al. 1982; Rahman 1984). A parallel interest in participatory research within agriculture also emerged in the context of developing countries, particularly through the Farming Systems Research movement (Farrington and Martin 1988; Martin and Sherington 1997; Collinson 2000; Flora and Francis 2000). Recognition of the relevance of these experiences from developing countries in dealing with similar issues in the agriculture of industrialized countries emerged in the 1990s. For instance, a landmark conference held at University of Illinois in 1991 explored the rationale for a broader participation of stakeholders in setting the research agenda and conducting research on farms (Gerber 1991). An example of application of this strategy in the Northwest Area Foundation program was reported by Rusmore (1995), who described how close farmer involvement promoted focus on immediate problems on the farm, encouraged innovation and better adoption of results, and developed leadership capacity in the farmer community.

We view three models as important aspects of participatory research: (1) a *participatory* approach assumes that "people affected by a problem should be instrumental in solving it" (from Freire 1972), a reason for starting participatory research in the first place, (2) a *transdisciplinary* approach that includes joint activities of scientists, extensionists, farmers, and private sector (Rhoades and Booth 1992) as knowledge and experience from different disciplines and professions are needed, and (3) a *systemic* approach (Ison 2008) that treats farming as a system including the farmer, the social system aspects, and the natural system aspects. The importance of promoting information exchange among farmers and cooperative extension specialists was described by Francis et al. (1990). Successful results of grassroots research and development projects, invariably involving stakeholders, were summarized by Pretty (2002).

In participatory approaches on collegial level in agricultural research, it is useful to clarify the participatory research process as a learning cycle where answers give new questions and unthought-of areas for learning may emerge. In the application here used this learning cycle includes six main steps:

1. Situation analysis
2. Defining areas/questions of interest
3. Selection of quantitative and/or qualitative methods
4. Research and development work
5. Analyzing and concluding
6. Presentation of findings

The learning is continued as new knowledge and situations are created. It is important that learning goes on in each and every one of these steps, and is not designated only to step 5. This stresses the importance of a continuous monitoring and evaluation of the whole research process, which could be seen as a parallel seventh "step."

The interest in participatory research and development of new models in agricultural sciences has grown out of concerns about how research was approached conventionally with narrow emphasis on production. In the light of these experiences with participatory research approaches, that involve farmers and other stakeholders in the food system, what can be said about what differs from disciplinary research and the difference on outputs and outcomes of interest to farmers and the relevance to their problems?

16.4   CRITIQUE OF CONVENTIONAL RESEARCH

Although there is a wealth of useful information on crop varieties, soil fertility, and weed management recommendations that come from our agronomy research programs, often this does not reach farmers or does not fit the real-world context and complexity faced by those who must make decisions on these issues every day. One complaint is that university information is specialized and fragmented, and that the results are valid under the conditions of the experiment station but not on the farm. If the experiments under controlled conditions, designed to isolate one or two factors, are close enough to the conditions on nearby farms, there is little reason the results should not be useful. Yet the suspicion persists that results generated by the *underlying presumption* that knowledge *would be produced* through objective study *of* defined variables of the university (Hildebrand and Russel 1996) have limited applicability in farm systems. Examples of approaches to overcome this difficulty include the use of long, drive-through plots that more closely represent commercial agriculture (Rzewnicki et al. 1988), or an invitation to farmer teams to sit with university researchers to negotiate a research agenda and help interpret results that will be of value to both groups, or to do collegial level participatory research (Johnson et al. 2004) as later exemplified.

One limitation of the dominant reductionist research methodology used in disciplinary natural science research is the quest to isolate one or a few discrete factors to study while holding the rest of the system constant. Although this provides data on whether the several maize hybrids, or fertilizer rates, or pest management options are significantly different, or helps the researcher to understand narrow mechanisms of system function, the results apply in a strict sense only to the conditions and the year when the experiment was conducted. When we use conventional field plot methods and statistical analyses to compare different farming systems, it is possible to describe with confidence which system produced the greatest dry matter, protein, or monetary return. But when systems differ in a number of factors—varieties, planting dates, fertility strategy, interest of farmer—it is much more difficult to determine which components of the system were those that contributed to one or another system's success. The consequence of this quandary in research design is that university specialists work on a few key factors without the originating questions being on the complex situations but on problems unconnected of the context, and the results may or may not be of interest to farmers. This while farmers have questions about alternative systems and may have no valid way to assess them from a statistical point of view. In participatory research field experiments and statistical analysis may well be used, *after* having analyzed the needs of the situation and from there defined what subproblems to explore.

Last, the most common university research agenda necessarily is focused on those issues that can be studied under controlled conditions, with limited available budgets, and that most likely will result in technical publications and future grants for more research. Recognition and promotion in the research establishment is based primarily on research publications, even for those people with a large teaching or outreach responsibility.

In addition to the problems resulting from most research being narrowed by the traditions of the respective disciplines, we can also see a conflict between university research and farmer needs. This could be due to differences in (1) perspectives on the state of agriculture and what priority problems should be studied, and (2) opinions on what sustainable development means. Both influence the choices of research topics and methods by researchers, and amplify the differences between them and farmer clients. Table 16.1 highlights the challenges by describing two generalized, contrasting perspectives, which will have implementations on what kind of research that will be done. Our experience is that it is not unusual to express a worldview that is more related to the right column in Table 16.1, when talking generally on world problems, but use the research approach described in the left column, not being aware of other options or modifications.

Table 16.1 highlights the importance of both scientists' own perspectives of the surrounding world and their opinions/reflections on what sustainability is or should be about. These two components have great effect on how the problematic area is described, how the research question will be formulated, what to research, and with whom and with what kind of methods and from what theoretical bases. When we introduce the table in workshops with scientists, we often get two kind of common comments: "I personal do not fully agree with all statements in any of the columns, but would like to put my own perspectives on a zigzag scale in between"; "My personal perspectives is more on the right side, but I have been trained to use the methods and theories that have their bases in the more left side column."

Disciplinary research is often about key problems, but may lack a wider vision of the farming system and the relative importance of their problems compared with others that constrain the system. There may or may not be close communication with clients in the field. In the right column are generalizations about transdisciplinary research, and especially that research founded on continuous collaboration with farmers who experience problems every day in their operations. This latter participatory approach differs from the conventional in the overall worldview of agriculture and food systems used, on what will be sustainable for the long term, and the types of research and methods used for agricultural investigation.

16.5 APPLICATIONS OF PARTICIPATORY APPROACHES IN RESEARCH ON SUSTAINABLE AGRICULTURE: EXAMPLES FROM NORDIC AGRICULTURAL RESEARCH

The Centre for Sustainable Agriculture at the Swedish University of Agricultural Sciences in Uppsala contributes tools, knowledge, and professional expertise concerning participatory research and facilitation of the process. Most studies involve farmers, extension personnel, and researchers through the whole research project. The starting-point in most cases has been common interest among all actors for a farm product or production system on each farm. Goals decided by the groups can be to achieve improvements in production practices, farm management and design, and on-farm and landscape biodiversity. Three examples of ongoing projects are described here.

TABLE 16.1
A Generalized Picture of Different Evaluations and Opinions Often Connected to the Use of Different Research Approaches When Working with Issues of Sustainable Development

Scientists Make *Different* Evaluations of the Surrounding World, and Priorities of Problems

Often Connected with Disciplinary Research	Often Connected with Participatory and Transdisciplinary Systemic Research
Focus on overpopulation	Focus on an equitable distribution of resources
Solve individual environmental problems case by case	Concern on the health of the global ecosystem
Optimism on the means in techniques (to "solve" the problems)	Skepticism on the means in techniques (to "solve" the problems)
Knowledge increases the predictability	Knowledge to handle the unpredictable
Good supply of renewable energy	The energy will be a commodity in short supply

Scientists Have *Different Opinions* on What Sustainable Development Means

Often Connected with Disciplinary Research	Often Connected with Participatory and Transdisciplinary Systemic Research
An issue of resource efficient and recycling techniques	Dynamic, dependent on the context
To uncouple economic growth from natural recourse use	An issue of adaptability and learning
	An ongoing discussion and negotiation

And They Use *Different Research Approaches* and Choices of Methodologies, with Different Focus

Disciplinary Research	Participatory and Transdisciplinary Systemic Research
Parts will generate knowledge of the system	Holistic perspectives and interdisciplinarity
Controllable system models	Participatory research processes
Technological solutions	Solutions by mimic ecosystems functions and structures
Biotechnology as a frontier	Local prerequisite and limitations
Increased precision to reach high net production	Ecological and social systems adaptations by learning
Environmental adaptation by high precision and efficiency	Diversity as tools and insurance
Specialization as a tool	
Advantage in big size	
Energy efficiency	

16.5.1 ORGANIC TOMATO PRODUCTION GROUP

Members of a group interested in organic greenhouse production of tomatoes first met in February 1999. The growers are located in an area with a maximum distance between farms of 400 km. Over the years of the project, members have moved in and out of the group but a majority has provided continuity. The production areas are small scale, ranging from 240 to 1000 m². On three of the farms, both husband and wife are engaged in production. Swedish tomato production is modest in comparison with commercial operations in other European countries, and especially so with organic production. Most farms with organic tomato production also differ from specialized conventional farms in that they rarely have tomatoes as sole income and commonly sell their produce through local channels. Possibilities for formal research and education have accordingly been very small to date.

The tomato production group has focused on improvement of production and its sustainability as part of their overall enterprise mix. Farming is often a large part of their lives, and boundaries between work and other activities are diffuse. The cyclic process of experiential learning going on in the group has revealed several different priority areas for research, and clearly showed how nested the questions were. For example, a question on a tomato root disease led to new learning on nutrients, which raised questions of tomato taste, which led to questions on product differentiation and selling. During this process, the group has gone from asking questions, such as "Are we doing things right?" (single loop learning), to "Are we doing the right things?" (double loop learning), and on to ask "How do we decide what is right?" (triple loop learning) (Mason 2005). Examples of learning on the third loop level can be seen in the discussions on how they would define sustainability, using methodological pluralism and stretching their considerations across both natural and social science questions (Björklund et al. 2005).

Among the results that have been important to the group are new knowledge, changes in production practices, potential impacts on their farms and social situations, and learning how to run these processes themselves. From the start, the group established a goal to work on a collegial level of participatory research. Today, the group is responsible for facilitating its own meetings, contacting the required specialists, and seeking new competencies they need to add to the group experience in order to solve certain problems or questions.

On their own homepage (www.ekotomat.se in Swedish), the farmers promote the critical value of teamwork for enabling collaboration. They speak of the importance of taking time for new members to become part of the group and of getting to meet more informally overnight in the low work season. They also talk about why the group needs to accept differences within the group and how the wildest whims may turn out to be the best ideas to pursue. The importance of giving each other space to contribute, and to share their knowledge and experiences, is also featured as one foundation for the group's successful dynamics.

These factors in group dynamics have been as important as the technical competence in the group, and members recognize the value of a semistructured cooperation within a systemic, open process. This means that the group has worked to improve situations through first finding the important questions, then to plan and do the work needed, to analyze and evaluate, to agree on results and to present their findings as well as document, and to monitor their work, and, finally, has done all this in a constantly developing process. The members have also been able to successfully evaluate their collaboration, reflect on their learning, and work on process improvements. The group continues to meet and can be considered a successful example of participatory research in practice.

16.5.2 Crop Production Group

The group was formed by eight farmers, one crop scientist, an organic farming extension specialist, and a facilitator. Some of the farmers are also pioneers in organic crop and seed production in Sweden. The farms are situated in the middle of Sweden and have a long history of both cereal and organic production. Overall goals for the farmers are to measure and to improve management, increase sustainability, and develop useful tools for on-farm decisions.

When the group started, the expectations about the participatory research process were that collective experiences from farmers, researchers, and the extension personnel should contribute to deciding what relevant on-farm experiments should take place. The farmers also expected that the knowledge in the group should increase and be disseminated, and that research should be better focused on areas that they find important and problematic. The project has been published in a Swedish report from Centre for Sustainable Agriculture (CUL), at the Swedish University of Agricultural Sciences (SLU) (Svanäng et al. 2002).

Discussing participatory research, it is useful to highlight what happens in such a group with well formulated and focused intentions leading to solutions to day-to-day cropping problems of a technical kind. Already in the first phase, some overall questions were generated: How can we solve the long-term problem with phosphorus supply? How can we reduce the energy consumption on the farm? What is the impact of my farming on the landscape? How can we use local or "internal resources" from the on-farm ecosystems, instead of imported inputs and technologies?

During the participatory process the focus changed to the wider questions related to complex farming situations. The new questions which arose were, for example: What is quality in organic farming, in a broader sense, and what does that mean in our crop production? How shall we secure quality and assess our efforts on the farm? How big is the external energy demand in our production system? The farmers have documented how they manage their crops and also kept records to have an idea how much time, energy, and machinery they use. The results have been used as decision tools and as indicators in designing more sustainable farming practices on the farms.

Ongoing research will be focused on type and level of indicators for applied management. The type of management farmers employ may have a big functional impact, for example, to create wetlands, but also generate technical/practical consequences on the farm. Experience from the group shows the importance of a participatory approach in the work for sustainable management on farm level, in order to effectively handle the complex and goal conflict issues in this process.

The group has designed and implemented on-farm experiments on some of the initial key questions, such as: when and how to plow a grass/clover ley; measure the effect of commercial organic fertilizers; test the effects of different mechanical soil processing on couch-grass and thistle. The farmers, the scientists, and the extension people's reflections on the different processes in the project, during the period 1998 to 2001, are described in the project report (Svanäng et al. 2002).

The description of reflections from the three different groups of actors show that the farmers and the extensionists seem to have gained the most, as well as responded best to the potentials of the project. The farmers say they learned much about organic crop production, and have adapted and changed some of their practices as the project has developed new knowledge and experiences. The extension participants seem to have benefited, adapted, and implemented much of the shared knowledge and experiences from researchers and farmers, but also from the new jointly developed experiences and knowledge generated by the project.

The least impact of the project seems to have been on the researchers' professional activities and decisions. They say they learned a lot by observing the farmers practices, and from the discussions within the group on the "overall picture" of an organic farming system. They appreciated the systemic perspective of farming, the insight of the farmers' perspective and practical problems, and the personal learning process to tie theoretical knowledge to practical experiences. But they seem to have difficulties making substantial application of these new findings in their own professional research roles. The researchers struggle with how to design and interpret research on the farming system level. They grapple with experiments without replicates, uncontrolled and complex interactions on a huge number of different factors, and look on the on-farm experiences as interesting pilot trials that can guide them on what relevant problem area the researchers can explore with scientific basic research. They could also see the on-farm experiences as applicable "test experiments" for scientific research results, applied to real farming practice. The report shows that the group has not succeeded in influencing the established research agenda and they would like to establish tighter contacts with researchers from different disciplines. But they also recognize that this would demand more working involvement from other researchers in transdisciplinary teams, something that the university system does not support.

The research has provided results that are relevant at different levels of spatial scale:

1. At the field level the grass/clover trials show best results when plowing clay soil rather early in the autumn, in order to induce enough mineralization of organic matter for the following winter crop. This result is contradictory to the common rule in Sweden, where the advice is to plow a ley as late as possible in the autumn.
2. At the farm level the nutrient balances for the whole crop rotation show a shortage of phosphorus on many farms. As a consequence the farmers have begun to use commercial organic fertilizers with phosphorus.

3. On a personal level the contacts and experiences in the group have a high impact on the daily work. The different ways of documentations on the farms show the range of solutions applied among the farmers. The shortage in time available for this type of project is one issue for the future, a concern of all the stakeholders.

16.5.3 QUALITATIVE INDICATORS FOR BIODIVERSITY AT THE FARM LEVEL

Some of the members in the crop production group described in Section 16.5.2 decided to focus on how to increase biodiversity in the farming system. To get funding, the group agreed to extend the collaboration to include conventional farmers participating in an agrienvironmental scheme (Swedish Seal: http://www.svensktsigill.com/). The objective was to develop key ratios and indicators for biodiversity on farms with organic and integrated production. To start the process, the group worked out criteria for choosing the indicator they found most relevant to use. The indicators have to be easy to communicate and function well as an internal aid for improvements on the farm, and must be adaptable to each unique farm context.

Biodiversity is an area where different farmers have similar goals for their production systems, but often have different contexts in terms of local climate and ecology as well as production conditions. In the European Union, farmers are now expected to respect certain basic environmental standards without any financial compensation, and the "polluter-pays" principle is being applied by the CAP (Commission of European Communities 2004). At the same time farming above the reference level of Good Farming Practice (GFP) offers farmers payments for environmental commitments. Through this strategy CAP increasingly aims at reducing the risks of environmental degradation, while encouraging farmers to continue to play a positive role in the maintenance of the countryside and the environment by targeted rural development measures.

Farmers in the group had experiences that different agricultural production systems, as well as specific management strategies have effects on biodiversity and may enhance or reduce its levels. These experiences are also substantiated in scientific literature (Bengtsson et al. 2005; Belfrage et al. 2005, Ahnstrom et al. 2007). Furthermore, the farmers found it urgent to communicate the importance of farming for biodiversity in general as well as its importance in specific management practices. Therefore, an obvious task was to search out tools for enhancing communication with consumers and other actors in the food system. Indices for biodiversity already existing and used by some of the farmers, were seen as neither communicable nor useful tools for farm management. Furthermore, they were specific to conditions on the plains, and did not make sense in areas with a mosaic of field and forest.

In reflecting on the process, and by applying techniques for continuous monitoring and evaluating as a group, we asked ourselves whether we had detected the expected changes and if the project had been effective. By asking if this was what we wanted to achieve and whether the right methods were used, the group was able to identify and highlight several important findings:

- That in the process of elaborating quality (not only quantity) it was important to search for what types of indicators, for example, flowers or birds, are special for each region and farm.
- Each farm has its own context and there are difficulties to find a *common* indicator or a *single* species to measure biodiversity for all the farms. For example, at some farms the length of ditches was a relevant measure, which did not make any sense at some other farms where the amount of open field instead contributed more.
- That farmers themselves and their vision for the farm was a means of quantifying and understanding the level of biodiversity.
- That it was fruitful for farmers from the different agricultural/environmental schemes and for researchers and extensionists to work together.

An important lesson learned in the process of work was the need for reflections and evaluations at all stages. To reflect was especially important for the researchers and extension personnel, who were traditionally accustomed to taking large responsibility. They needed to reflect on their own actions in the group to be able to contribute without controlling. Obvious examples of this were to include all participants in the setting of dates for meetings, to distribute responsibilities, and to assure that all important decisions explicitly had to be made by the group. From the continuous monitoring, the group found that the obligation for each person to help make decisions in the group also made the individual members more responsible and patient.

16.6   KEY FINDINGS FROM THE EXAMPLES

From the examples presented here, we can see that the groups are learning from and dealing with complex tasks that often, at first sight, include goal conflicts. Through working with the problem situations, new learning often leads to new understanding and the reasons goals appear to be in conflict with each other. The participatory process can be experienced as frustrating when full understanding or clarity has not been reached on the power issue, and how the collaboration is conducted will affect the learning and the results of the work. As a deeper learning and understanding of problems happen, there is a tendency for choices to be easier, if there is a level of mutual trust and respect in the group. This is in accordance with an extensive review of literature by Cornwall and Jewkes (1995), who found: "Whilst conventional health research tends to generate knowledge for understanding which may be independent of its use in planning and implementation, most participatory research focuses on knowledge for action." In turn, knowledge to act for change takes understanding of the problem, its affinity in the context, and how to improve the situation. To reach this level of learning participants have to understand their own parts and responsibilities, as well as those of others, in relation to the problem at hand. Unless everyone in the group dares to speak up, it is difficult to mobilize the entire group's creativity to solve problems and improve situations.

Even if farmers have the ability to solve problems *within the context,* they are usually not asked by authorities. Disciplinary reductionistic research aims to get a general answer to a specific problem irrespective of context and local environment, and as a consequence the research may be less interesting in the farmer's perspective.

Thoughtful and creative farmers would better like to use their abilities to manage sustainable agriculture farms, than spend time and effort following the authority regulation within different production systems.

Trust is a key issue. This means trust in oneself, one's colleagues, in the possibilities of the group, and in the approach to research and the process that unfolds during meetings and activities. This trust can develop with time, when everyone is experiencing and learning the potentials of collaboration and ways of working effectively. Traits that are unquestionably required here are also equality and clarity. Participatory research collaboration is facilitated by clarity on who has the right to make what decisions and how, and equality in acting and deciding from knowing that everyone's contribution is of the same value. This includes not falling back in traditional roles of who is to ask, who is supposed to answer, who is to lead the meetings, and what power comes from position or prior involvement in similar activities.

Also important when starting to work with new and different research approaches in agriculture is to broaden the common natural science notion of what research is and what it could be. This means being very open toward finding the methodologies that can be used to deal with problems and questions included in the improvement of a situation, and then choosing appropriate methods. This takes new capabilities of the practitioners, not sticking to one common or comfortable area of methods and an individual branch of natural or social science. What will always remain is the importance of the continuing monitoring, documenting, and evaluating the group's progress.

16.7　CONCLUSIONS

The demand on farming to deliver new services and products, for example, bioenergy, is increasing. At the same time the environmental impact concerns and animal welfare concerns of farmers and the public are now significant in the European Union. These demands and concerns are implemented in the form of norms, laws, or regulations, but sometimes also as promotions such as subsidies/payments. This is already a practice in the European Union CAP, both as regulations and subsidies/payments. But because these institutional frameworks are set up as top-down processes that are researched at universities and implemented centrally from Brussels, farmers do not own the problem the regulations are designed to correct. The farmers' knowledge and creativity are instead focused on how to deal with the regulations in order to maximize the subsidies, instead of focus on the problems they are best able to solve in the same landscape structures and locations that the plan is designed to support.

We believe that participatory research could be better accepted and could lead to bigger impact on development if there were better integration with the mainstream institutional research frameworks, leading to some of the issues of conventional research identified earlier being addressed. Too often, university research is conducted under conditions and assumptions that are far from what clients are thinking about. If we can integrate our efforts in a transdisciplinary research framework with farmers as full members of the team, there is a much greater probability of working on questions relevant for the situation and reaching solutions that will be used. If we are to successfully handle the complex environmental problems that we are now facing, we need creative and local solutions that will promote a sustainable agriculture, and we need to integrate production and conservation. This is especially important as we approach natural limits of key resources such as fossil fuels, and it will require all of our ingenuity as teams to learn from natural systems and design our own agrifood systems for long-term sustainability.

REFERENCES

Ahnström, J., Höckert, J., Bergeå, H. L., Francis, C. A., Skelton, P., and Hallgren, L. 2008. Farmers and nature conservation—What is known about attitudes, context factors, and actions affecting conservation? *Renewable Agricultural and Food System,* 23, in press.

Arnstein, S. R. 1969. A ladder of citizen participation. *American Institute of Planners Journal*, 35, 216–224.

Bawden, R. 2003. Systemic discourse, development, and the engaged academy. Paper presented at the 47th meeting of the International Society for the Systems Sciences (ISSS), July 7–11, Heraklion, Crete, Greece.

Belfrage, K., Björklund, J., and Salomonsson, L. 2005. The effects of farm size and organic farming on diversity of birds, pollinators, and plants in a Swedish landscape. *Ambio*, 34, 582–588.

Bengtsson, J., Ahnström, J., and Weibull, A. C. 2005. The effects of organic agriculture on biodiversity and abundance: A meta-analysis. *Journal of Applied Ecology*, 42, 261–269.

Björklund, J. et al. 2005. Vad kan egentligen kallas ekologiska tomater? *Ekologiskt lantbruk,* 42. Swedish University of Agricultural Sciences. http://www.cul.slu.se/publikationer/publikationslista. asp#Rapportserien (accessed June 26, 2008).

Chambers, R. 1994. Participatory Rural Appraisal (PRA): An analysis of experience. *World Development*, 22, 1253–1453.

Checkland, P. 1981. *Systems Thinking, Systems Practice.* John Wiley, Chichester, U.K.

Checkland, P. and Scholes, J. 1990. *Soft Systems Methodology in Action.* John Wiley, Chichester, U.K.

Collinson, M., Ed. 2000. *A History of Farming Systems Research.* FAO and CABI Publishing, Wallingford.

Commission of European Communities. 2004. The Common Agricultural Policy explained. http://ec.europa. eu/agriculture/publi/capexplained/cap_en.pdf (accessed June 24, 2008).

Cornwall, A. and Jewkes, R. 1995. What is participatory research? *Social Science & Medicine*, 41, 1667–1676.

Elden, M. and Taylor, J. C. 1983. Participatory research at work: an introduction. *Journal of Occupational Behaviour*, 4(1), 1–8.

Flora, C. B. and Francis, C. 2000. Farming systems research in extension and policy formulation. In *A History of Farming Systems Research*, Collinson, M., Ed. CABI Publishing, Wallingford, 129–160.

Francis, C. et al. 1990. Participatory strategies for information exchange. *American Journal of Alternative Agriculture*, 5, 153–162.

Farrington, J. and Martin, A. 1988. Farmer participation in agricultural research: a review of concepts and recent fieldwork. *Agricultural Administration and Extension*, 29, 247–264.

Freire, P. 1972. *Pedagogy of the Oppressed*. Continuum Press, New York.

Freire, P. 1974. Research methods. In *Literacy Discussion*. International Institute for Adult Literacy Methods, Teheran, 142–153.

Gerber, J. 1991. Participatory research and education for agricultural sustainability. In *Participatory On-Farm Research Concepts and Implications*. Agro-Ecology Program, University of Illinois, Urbana, paper 91-15.

Hall, B. 1975. Participatory research: An approach for change. *Convergence*, 8(2), 24–32.

Hall, B., Gillette, A., and Tandon, R. 1982. Creating knowledge, a monopoly? Participatory research in development. ICAE, Toronto.

Hildebrand, P. and Russel, J. 1996. *Adaptability Analysis: A Method for the Design, Analysis, and Interpretation of On-Farm Research-Extension*. Iowa State University Press, Ames.

Ison, R., 2008. *Systems thinking and practice for action research. In Handbook of Action Research: Participative Inquiry and Practice, Reason, P. and Bradbury, H., Eds. Sage, London.*

Johnson, N. et al. 2004. The practice of participatory research and gender analysis in natural resource management. *Natural Resource Forum*, 28, 189–200.

Martin, A. and Sherington, J. 1997. Participatory research methods—Implementation, effectiveness, and institutional context. *Agricultural Systems*, 55, 195–216.

Mason, H. 2005. Levels of learning. *Berkana exchange*. Berkana Institute. http://berkana.tomoye.com/file_download.php?location=S_U&filename=11225772391Triple_Loop_Learning.doc (accessed May 29, 2008).

Moore, R. and Stinner, D. 2006. Ecological integration of the social and natural sciences in the Sugar Creek method. Paper presented at ASA-CSSA-SSSA International Annual Meetings, Indianapolis, November 12–16, 2006. http://crops.confex.com/crops/2006am/techprogram/P21654.HTM (accessed May 29, 2008).

Okali, C., Sumberg, J., and Farrington, J. 1994. *Farmer Participatory Research Rhetoric and Reality*. Intermediate Technology Publications, London.

Pretty, J. 1995. Participatory learning for sustainable agriculture. *World Development*, 23, 1247–1263.

Pretty, J. 2002. *Agri-Culture: Reconnecting People, Land, and Nature*. Earthscan, London.

Rahman, M. A. 1984. *Grassroots Participation and Self Reliance: Experience in South and Southeast Asia*. Oxford and IBH, New Delhi.

Reason, P. and Bradbury, H. 2001. *Handbook of Action Research: Participative Inquiry and Practice*. Sage, London.

Rhoades, R. and Booth, R. 1992. Farmer-back-to-farmer: Ten years later. In *A Conference on Participatory On-Farm Research and Education for Agricultural Sustainability*. University of Illinois, Urbana-Champaign, 17–27.

Rusmore, B. R. 1995. Use of participatory research in agriculture. In *Planting the Future: Developing an Agriculture That Sustains Land and Community*, Center for Rural Affairs, Walthill, 179–190.

Rzewnicki, P. E. et al. 1988. On-farm experiment designs and implications for locating research sites. *American Journal of Alternative Agriculture*, 3,168–173.

Svanäng, K. et al. 2002. Deltagardriven forskning—Växtodlingsgruppen. Resultat och utvärdering av arbetet under 1998 till 2001. *Ekologiskt lantbruk, 35*. Swedish University of Agricultural Sciences. http://www.cul.slu.se/publikationer/publikationslista.asp#Rapportserien (accessed June 26, 2008).

Weaver, M. and Moore, R. 2004. Generating and sustaining collaborative decision-making in watershed groups. Paper presented at the 67th annual meeting of the Rural Sociological Society, Sacramento, August 11–15, 2004. http://www.ruralsociology.org/annual-meeting/2004/Weaver,Moore.pdf (accessed May 29, 2008).

17 Retrofitting Suburban Landscapes with Sustainable Agroecosystems

Gar House

CONTENTS

17.1 THE EXISTING SUBURBAN LANDSCAPE

The suburban landscape in the United States and other parts of the developed world is composed of a repeating mix of shopping centers, parking lots, single- and multiple-family homes, lawns, asphalt streets, parks, golf courses, and office and industrial buildings. Except for the occasional backyard garden, agricultural land use is conspicuously absent in the suburban landscape. Municipal governments generate and maintain a highly discrete suburban land use arrangement through zoning ordinances. These segregated land use zones are connected by roads constructed exclusively for convenient automobile and truck conveyance. Movement and transport within and throughout the American suburban landscape is highly dependent on gasoline and diesel fuel. Without liquid fuel the current suburban system with its abject reliance on automobile and truck transport would cease to function. American suburban dwellers commute an average of 32 miles per day (Langer 2005), compounding their dependence on automobile transport. Such long daily automobile commutes to and from work will likely become increasingly difficult due to unabated traffic congestion, the rising cost of liquid fuel, and impending climate change legislation to mitigate greenhouse gas emissions. Walking as a mode of transport has been so heavily discounted that suburban communities are often constructed entirely without sidewalks.

Yet despite the fact that the suburban landscape in the United States faces serious challenges (Kunstler 2006; Bradford 2007), the actual physical structure holds potential for transitioning to sustainability, defined here as any activity or process for which the rate of use does not exceed the regeneration rate (Heinberg 2007). Many critical structural and physical components in the suburban landscape already exist: housing, water, electrical, and sewer infrastructure are all in place. Lawns and turf comprise a significant open land use percentage, estimated to cover an average of 23 percent of the total suburban surface area (Robbins and Birkenholtz 2003). Lawns can be transformed into gardens and other agricultural land use with minimal modification, demonstrating that alternative applications for the existing suburban landscape are indeed possible (Holmgren 2008). In other areas of the world, especially Asia, significant amounts of food are produced within the urban/suburban boundary, indicating the potential for this type of production in the United States. For example, Hanoi obtains 80 percent of its produce, 50 percent of its fish and meat, and 40 percent of its eggs from farmers within the city or on its fringes, and Shanghai receives more than half its produce and meat from farms in or around the city (Roberts 2008).

Given the existing physical components, what set of scalable suburban landscape modifications will restore ecological functioning to promote sustainable processes and practices? Stated in a different way, working within the existing suburban landscape, can remodeling or retrofitting our existing communities support and maintain sustainability? The solution offered here is found in using the ecosystem concept as a unifying solution framework.

17.2   FUNDAMENTAL COMPONENTS OF A SUSTAINABLE COMMUNITY

For healthy ecosystem operation, sustainable communities require significant areas of land devoted to both natural and agricultural use (Odum 1983). Recognizing ourselves as working members within the web of life rather than separate or immune from biophysical processes is a critical first step toward sustainability. Humans are definitely capable of successfully integrating into the natural environment as all other life. Ecologically aware cultures such as Native Californians (nineteenth century and earlier) interacted with nature through a process of limited, selective harvesting of their ecosystem's flora and fauna. Such cultures, in effect, "tended the wild" (Anderson 2006) and avoided permanent ecosystem degradation by treating the natural environment as their garden. These societies possessed complex and comprehensive knowledge of the location and seasonal utility of naturally occurring plants and animals in their ecosystems. Their harvesting and other activities influenced species population size as well as biodiversity. These ecologically aware cultures demonstrate that humans can indeed have a significant, positive effect on ecosystem processes and community composition. Thus, human interaction with nature has the potential to function in an analogous manner to what is known ecologically as a "keystone" predator, defined here as a species whose impact on its community or ecosystem is disproportionately large relative to its abundance. Native Californians acted as keystone predator species by significantly influencing the varieties and population densities of other species in the community (Mills et al. 1993).

When we humans think of ourselves as stewards of nature, it becomes easier to embrace the three fundamental land uses that necessarily compose a sustainable community landscape (Figure 17.1): (1) the natural ecosystem providing critical life-support processes, that is, ecosystem services including nutrient cycling, soil formation, air and water purification, and flood control via ecologically active watersheds; (2) the agricultural production system providing local food security and self-sufficiency; and (3) the built environment supporting human habitation, commerce, industry, and entertainment.

Adoption of the ecosystem as a design template offers economic, social, and physical benefits for building and maintaining sustainable communities. Natural ecosystems, critical to life-support, function quietly in the background similarly to our own vital organs, which allow us to breathe, circulate oxygenated blood, digest food, eliminate waste, and so forth. Biotic processes are familiar to us in the fermentation of grapes for making wine and in the decomposition of organic matter to make compost. Ecosystem services are simply these same or similar biological processes operating on a large scale.

Eugene Odum defined an ecosystem as "'any area of nature that includes living organisms and non-living substances that interact to produce an exchange of materials between the living and non-living parts" (Odum 1959). The ecosystem concept is a unique contribution to our understanding of our living world in that it not only includes both biotic and inert components, but also shows how all parts interact cooperatively to function as a sustainable unit (Odum 1959). Ecosystems provide a convergent, inclusive, durable, yet flexible framework for building sustainable communities. Ecosystems

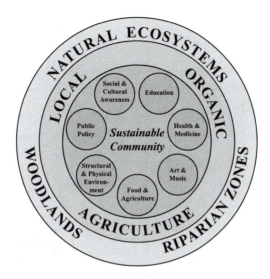

FIGURE 17.1 Three fundamental land uses necessarily compose a sustainable community landscape: (1) the natural ecosystem, which provides critical life-support processes, i.e., ecosystem services including nutrient cycling, soil formation, air and water purification, and flood control via ecologically active watersheds; (2) the agricultural production system, which provides food energy for both man and animals; and (3) the built environment supporting human industry, commerce, and habitation. (Illustration by Karen Holmgren.)

have been operational in nature from the beginning of life on Earth. So by their very existence they are a highly successful and fundamental way of organization for living communities (Odum 1969).

Emulation of ecosystem structure and processes in organic or ecologically based agroecosystems (Pimentel 1984) has proved highly successful (Balfour 1942; Howard 1947; Gliessman 2006). A substantial body of applied methodology based on ecological principles exists and is currently practiced worldwide, for example, organic or ecological agriculture (Bradley and Ellis 1993; Altieri 1995), permaculture (Holmgren 2002), biodynamic (Steiner 2006), and biointensive (Jeavons 2002).

The task at hand is to adopt and apply analogous ecological practices to the suburban landscape. Figure 17.2 shows diagrammatically how internal ecosystem control is highest in natural systems and absent in industrial agricultural and urban systems. Sustainable, ecological practices return biological processes to communities and agriculture. The goal is straightforward: let nature do its work, rather than building and relying on complex, expensive, energy consumptive technologies for cleaning air, water, and soil.

Local, organic, that is, ecologically grown, agricultural production is thus central to a sustainable community, as it returns internal control and, importantly, a margin of food security. Organic farming practices also play an important role in local carbon sequestration. Under ecological farming methods large quantities of composted animal and plant materials are routinely applied to soil as fertilizer, a practice that has the side benefit of turning these soils into sinks or areas for long-term carbon storage (Montgomery 2007; Favioine and Hogg 2008). Food security and safety alone are compelling reasons for a community to embrace urban agriculture. Proactive municipal governments have the authority to influence public policy to preserve their own natural ecosystem services as well as promote and secure a safe, local, agricultural production base. Such a prudent civic stance by local authorities demonstrates genuine public responsibility, land and resource stewardship, as well as a positive vision for their community.

17.2.1 CHANGING OUR ATTITUDE TOWARD NATURE

A healthy environment is supported by robust ecosystem activity and functionality. Our environment, and the ecosystems included therein, must be recognized as more than a luxury or amenity,

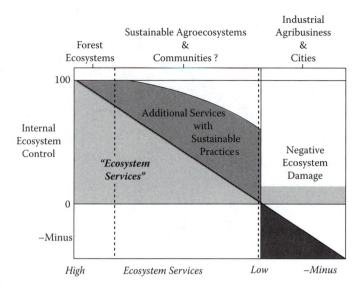

FIGURE 17.2   Internal ecosystem control is highest in natural system and essentially absent in industrial, agricultural, and urban systems. Sustainable ecological practices return biological processes or ecosystem services to communities and agriculture. The goal is to assist nature in accomplishing its work, rather than building and relying on complex, expensive, energy consumptive technologies for cleaning air, water, and soil. Importantly, by applying ecological principles to common issues, e.g., composting organic material, a community can improve its level of sustainability (move toward the left side of the diagram).

but instead as fundamental to our existence. "Ecological footprint" measurements provide one useful method for measuring our impact on the world's ecosystems (Rees 1992; Wackernagel et al. 1994). Americans are largely unaware of the importance of ecosystem services and often exhibit what might be called "nature deficit disorder" as a consequence of having little contact or personal experience with natural ecosystems. Thus, a primary step in raising environmental awareness is enhancing ecosystem literacy through understanding how ecosystems operate. At the very least, the average citizen should appreciate nature's fundamental role in providing life support to us and all biota.

Because the three critical land use components—natural ecosystems, agricultural systems, and the human-built environment—are often separated by significant geographic distances, local management, environmental responsibility, and even basic awareness are often problematic. To build and maintain a sustainable community all three land use components must be present, and, importantly, managed locally using ecological principles. Because of their close proximity to human habitation, urban agricultural ecosystems should be pesticide-free zones for health and safety reasons. Obviously, most suburban communities are a long way from accepting this scenario. Yet the goals of self-sufficiency and sustainability cannot be achieved without these three land use components existing within a reasonable distance from each other (Odum 1983; Vail 2008). The bottom line is that these three components are the physical foundation of any sustainable or regenerative community design (Pearson 2007). So the sooner we can quantify component land use percentages required to support and maintain sustainability, the better position we will be in to successfully estimate the land allocation requirement to build these communities (Bradford 2008).

Finally, the seven descriptive items in the innermost circle of Figure 17.1 list traditional human activities and organizational pursuits typically found in a vibrant community. Maintaining a viable community requires much more than physically altering the landscape. Cooperative engagement among social, cultural, political, and especially educational leaders and stakeholders is essential to build and maintain continuity. A number of contemporary writers, including Kurt Cobb, James Michael Greer, James Howard Kunstler, Richard Heinberg, and Alex Steffen, have addressed and continue to investigate such critical issues, particularly the inherent cultural impediments to embracing a sustainable lifestyle (visit Energy Bulletin at http://www.energybulletin.net/ for articles by these and many other ecologically conscious writers). Although the issue is complex, there are indications that a significant number of the population, especially within the younger generation, have begun to embrace the growing need for a comprehensive shift away from our current energy intensive lifestyle. Many environmentally focused web sites including http://www.grist.org/ and http://www.worldchanging.com/ support and promote an ethic of planetary stewardship.

17.2.2   What Would a Sustainable Community Look Like?

Figure 17.3 offers an idealized, self-sufficient, "walkable" community. The landscape has been substantially altered to accommodate local agricultural land use, and a large portion of the population has relocated to high density centers close to public transit. Others in the community have become farmers and horticulturalists, tending the agricultural areas. These modern land stewards provide food security and employment for the more densely populated centers. Distances between farms and dense living areas are close, no more than a half mile at most.

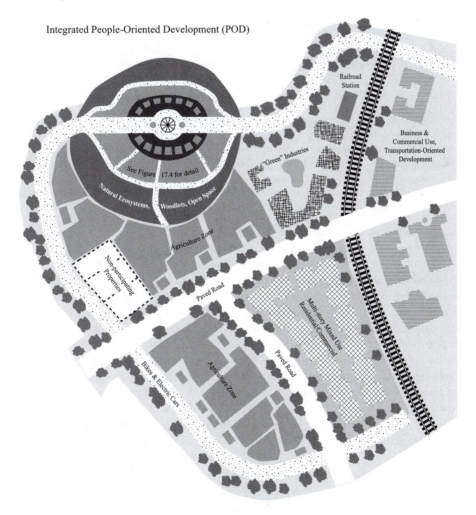

FIGURE 17.3   Hypothetical sustainable community incorporating several local land uses, including agriculture, transit-oriented-development housing and commerce, and natural or "green" space. All basic goods and services are available with a walking distance of a half mile. (Illustration by Karen Holmgren.)

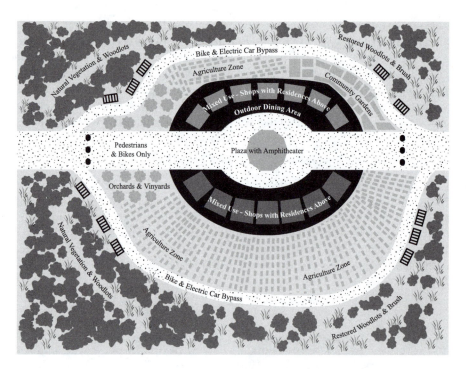

FIGURE 17.4   Diagram showing high resolution of village center components, a modified mixed-use green belt supporting dwellings, green enterprises (shops and businesses), and public amenities such as theaters and plazas. Village centers can be viewed as an American version of traditional clustered living style, which evolved in Europe and Asia. Mimicking this tradition, many families voluntarily relocate to the community's new green belt village center to become the incipient merchant class, owning shops and businesses. Essentially, village life with optimal structure and social interaction is the potential goal. Better quality of life would occur for all citizens with the revitalization of local food production and services becoming a core activity of the community. (Illustration by Karen Holmgren.)

A major objective of this suburban redesign is to provide walking access to basic necessities and amenities, all within a half-mile radius. (To maintain personal health a daily one-half mile walk—the approximate distance covered in a 30 minute walk—is recommended by the American Heart Association; AHA 2008). Land use devoted to natural ecosystems is substantially expanded, preserved, and restored. A village center (see Figure 17.4 for a detailed view) provides commerce, cultural enrichment, community identity, entertainment, and opportunities for human engagement in an environment without automobile interference. This design is sustainable to the extent that not only are community environmental by-products such as gray water and human manure processed and recycled with the aid of adjacent natural ecosystems, but a significant percentage of the community's food is grown and consumed locally, providing healthy food, a degree of self-sufficiency, and food security currently unknown within the average American suburban community. Natural ecosystem cycles and processes are thus harnessed, returning internal control and a significant measure of sustainability to the community (Figure 17.5). Rather than remaining completely dependent on food being trucked in from distance sources, a substantial amount of food is grown and consumed locally.

A critical guiding principle is to redesign the suburban environment for "walkability," meaning that all essential goods and services as well as many jobs are available within walking distance or approximately a one-half mile radius. Local merchants own and operate small stores to provide these goods and services. Land use projects that employ mixed-use zoning to improve walkability and reduction in automobile traffic are not only a welcome relief to the monotony of the classical suburban, single-family dwelling development, but also in increasing demand (Lloyd 2008).

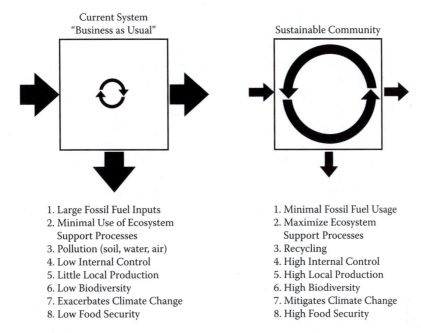

1. Large Fossil Fuel Inputs
2. Minimal Use of Ecosystem
   Support Processes
3. Pollution (soil, water, air)
4. Low Internal Control
5. Little Local Production
6. Low Biodiversity
7. Exacerbates Climate Change
8. Low Food Security

1. Minimal Fossil Fuel Usage
2. Maximize Ecosystem
   Support Processes
3. Recycling
4. High Internal Control
5. High Local Production
6. High Biodiversity
7. Mitigates Climate Change
8. High Food Security

FIGURE 17.5 When natural ecosystem cycles are harnessed, internal system control returns adding a measure of sustainability to the community (represented in the diagram by the large thick circular arrow). For example, rather than remaining dependent on food being trucked in from distance sources, food is grown and consumed within or nearby the community.

The characteristic benefits of this suburban retrofitting design include (1) reduced energy consumption, (2) less dependence on automobiles and trucks, (3) clean soil, air, and water, (4) increased use of renewable energy, (5) less pollution, especially greenhouse gas emissions, (6) a healthier citizenry and environment, (7) safe, secure food produced via organic farming methods, and (8) preservation of natural biodiversity. Walking becomes an enjoyable, even fashionable, activity, because the village center destination is readily accessible and socially compelling. High-density housing also results in significant energy savings and promotes a sense of community rather than isolated, gated fiefdoms. All these benefits are fundamental goals of a sustainable community.

The land use modifications shown in Figure 17.3 would, admittedly, require a fundamental shift in the mixed-use concept. Zoning ordinances would necessarily experience substantial alteration to include natural and, especially, agricultural land use within city limits, not a concept at present familiar to the majority of city council members. The cooperation of a number of organizations would be necessary to accomplish such land alterations, requiring the formation of a redevelopment or similarly authorized agency with active stakeholder support and participation. For optimal ecological control and management, clusters of these sustainable communities would form the basic operational components within the larger watershed to which they are geographically bound.

Figure 17.4 provides greater resolution of village center components, a modified mixed-use green belt supporting dwellings, green enterprises, for example, shops and businesses, and public amenities including theaters and plazas. Merchants owning businesses and shops would form an important core group relocating to the community's village center. A primary goal is for the structural design of the village center to promote and enhance social interaction and community engagement (Register 2006; Farr 2008). Improved health and a better quality of life are anticipated as local food production and related services grow to become a fundamental activity of the community.

Each community's population demographics would differ, but the general trend would be to coalesce into high-density centers around dependable public transit centers, particularly those supported by passenger trains. In response to these shifts, railroads are beginning to build additional lines for both passengers as well as freight (Richards 2008). A trend toward transit-oriented development is well under way in many California communities (Parker et al. 2002). Even though transit-oriented developments do not at present include an urban agriculture component, they are nevertheless a first positive step in the sustainability transition process (Lerch 2007; Hopkins 2008).

Future economic and fossil fuel energy constraints will accelerate such population consolidation, providing opportunities for agricultural enterprises. Business partnerships, families, individuals, or grower cooperatives might purchase available properties from owners or banks and begin the conversion to blocks of farmland. Thus, substantial suburban acreage has the potential to be converted to agricultural use, particularly after rezoning for urban agricultural use becomes established municipal code. Urban agricultural zoning regulations would differ substantially from industrial or conventional agricultural zoning, for example, organic farming methods would be the norm and pesticides prohibited.

In this scenario a significant number of the original tract houses would be removed and replaced with large vegetable gardens, orchards, pastures, and vineyards, especially on more productive soils. In recent years suburban development has increasingly been constructed over excellent agricultural soils, so that such retrofitting would return this land to its highest and best use (Montgomery 2007). Much of the suburban "hardscape" is structurally ephemeral. For example, asphalt and stucco are relatively easy to remove (Register 2006). So that certain activities required for urban retrofitting are "low-tech" processes, and could be performed with hand tools by dedicated crews of men and women. Under directed leadership such crews would be able to transform significant sections of a suburban environment in a relatively efficient manner and period of time. Unemployment could be ameliorated by guilds of men and women performing "green" jobs to transform their communities to sustainable landscapes and initiate local, organic practices. Families and individuals that wish to remain in the agriculturally zoned landscape would find increasing opportunity for green employment within the new organic agriculture landscape. Positions for horticulturalists, vitaculturalists, small and large animal farmers, and other highly specialized growers would all be in demand to feed the local community.

Local, organic vegetable crops, fruits, and other foods are typically grown with substantially less fossil fuel energy consumption (Pimentel et al. 2005), but require more human labor as well as a high degree of farm knowledge and ecological literacy. Industrial globalized agriculture is completely dependent on fossil fuels for fertilizers, transport, storage, and processing, leading to centralization of farms, environmentally deleterious practices, especially monoculture, and nutrient-compromised food. With its emphasis on whole biological systems, organic farming as a career requires significant field experience as well as formal and applied education. Internships and apprentice programs with experienced farmers would provide the community a stable population of career farmers, educated in the particular nuances of food production in that area and climate. With increasing demand for organic food, organic farmers are gaining respect in their community for their ability to supply a safe, secure, and diverse food supply. The California Certified Organic Farmers organization recently reached a landmark with the certification of over 500,000 acres (Central Valley Business Times 2008). Other organizations such as Roots of Change, a collaborative of diverse leaders and institutions, are unifying in common pursuit of achieving a sustainable food system in California by 2030 (http://www.rocfund.org/).

Microclimatic benefits also accrue from these scenarios. In our remodeled landscape, streets are lined with trees, cooling the area underneath by as much as 25°F, providing direct energy-use reduction benefits to the community (Lancaster 2006). With careful varietal selection, these trees could easily provide food as well as shade, for example, walnuts and olives. The original street is also reduced to one or two asphalt-maintained lanes in anticipation of significantly reduced automobile traffic, again providing significant maintenance and energy savings for the community.

Under new zoning ordinances promoting this expanded mixed use, those houses that remain in the new agricultural zones, especially those with attached garages, may be converted to small workshops and "cottage" industries supporting the community as well as the individual family owners. Retrofitting the suburban landscape will require embracing new ideas and a flexible approach. It is much easier to build new structures and developments to green standards and codes than it is to retrofit existing ones.

Such a sustainable landscape might be 20 years, perhaps 30 years in the future. By this time, individual automobiles may be an infrequent means of conveyance, and those automobiles in active use may rely to a greater degree on electricity than liquid fuels. In 30 years, fossil fuel sources, especially oil and natural gas, may be considered too valuable for burning in individual transport vehicles. Fossil fuels may be sparingly used both because of their relative scarcity and the deleterious effects recognized by the release of carbon dioxide and other greenhouse gases they emit into the atmosphere via the internal combustion engine (Hansen 2008). Sustainable communities supporting productive, well-managed pastures will have the unique option of using horses and other draft animals for personal conveyance and light freight delivery, returning aesthetically pleasant aspects of rural life to the community.

As fossil fuel energy grows more expensive and automobiles are used less frequently, parking lots and economically unviable strip malls could become prime candidates for future local farms, parks, greenbelts, and village centers. Sales, lease, and other arrangements between the city government and owners might be negotiated to provide the land for alternative uses, especially natural "life support" ecosystems, local organic farms, community gardens, pastures for grazing animals, and managed woodlots.

Citizens who remain in these sustainable communities may embrace and create lifestyles substantially altered from the high-energy-consumption level we recognize today. Attempts to maintain the energy intensive "car culture" way of life will probably attenuate. Sustainable communities and their incipient culture could evolve from the interaction and activities of these adaptable individuals and their families. Cooperation and partnerships among individuals and groups may become a more preferred and accepted method of engagement among community members. Synergy among individuals and groups will likely occur, generating opportunity for unique local lifestyles. Biological and cultural diversity and its attendant benefits may thus be given an opportunity to flourish. Organic local agriculture would form the core activity for a significant portion of the population (Heinberg 2006). If the community is proactive, fortunate, and provident of its resources, it will organize and partner to generate yields and goods of value for economic commerce with other communities. Communities most likely to succeed will be those which actively implement innovation and experimentation, share knowledge and responsibility, and, perhaps most importantly, learn how to collaborate.

A number of municipalities throughout the United States are already beginning to experience a population movement and gravitation around the original urban core. This general trend may expand to be one of mixed land use of transit-oriented developments, village centers, and agricultural ecosystems coalescing over a wide area around a central urban core (Figure 17.6). The suburban environments will thus transition to a semirural landscape with its primary economy being agricultural production and related support for the urban core.

As long as our current cultural beliefs hold that unlimited growth, inexhaustible natural resources, and materialism are immutable, transition to sustainability will be difficult. It will take great political will and cultural realignment to initiate the changes proposed here. Business-as-usual and keeping our present energy-intensive system in place seems an unwise path to the future, as energy will continue to cost more and deliver less as time goes on (Hall 2008).

Therefore, viable visions of the future are important. Humans are adaptable and clever. Our situation can be thought of as analogous to a test—those who studied generally fared better than those who did not. Communities that support pilot projects will, at the very least, have experienced the value in examining several sustainability options when responding to uncertainty.

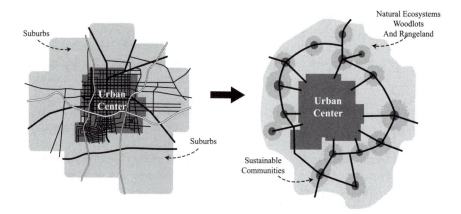

FIGURE 17.6   Diagram showing the general trend of mixed land use of transit-oriented developments, village centers, and agricultural ecosystems coalescing over a wide area around a central urban core. In this scenario, suburban environments transition to a semirural landscape with their primary economy being agricultural production and related support for the urban core. (Illustration by Karen Holmgren.)

As citizens adapt their behavior to accommodate energy constraints (e.g., conservation) and the challenges of climate change (Lovelock 2006; Brown 2006; Hansen 2008), they will, out of necessity, begin viewing existing suburban structures and landscapes in a different way. Physical changes will be required, but, as noted earlier, the suite of existing suburban resources is robust and can be adapted to generate local, sustainable commerce and agriculture. Locally focused activity is likely to become the accepted and fashionable social norm. Proactive communities will remain flexible regarding energy consumption, seeking ways to adapt to conserve energy and promoting sustainable technologies. Improvements in digital technology will undoubtedly assist in the delivery of energy on a "networked" rather than "command and control" basis.

17.3   PUTTING IT ALL TOGETHER: LEARNING HOW TO VIEW AND ADAPT YOUR COMMUNITY USING ECOLOGICAL PRINCIPLES

How does a community integrate its various parts into a working, functional ecosystem?

The following six steps are designed to provide a first approximation of a community's sustainability rating as well as specific ways to increase a community's sustainability. The methodology also reveals the strengths and weaknesses of a community.

Step 1. Form a "Sustainable Community Task Force" comprising informed individuals in the community. Ideally, select one individual for each of the seven community sectors (see Figure 17.1). Members mindful of the local ecosystem or watershed unity concept will be a prerequisite. Foster an atmosphere where intercity, interagency, and interdistrict cooperation is the norm. Next form subgroups utilizing the talent of your core leadership group. Appoint local community experts to new posts, which will have the effect of energizing the intellectual base of your community. Request input from decision makers and have them look at the future in light of ecological principles and current global issues. Acknowledge and honor their work and stress their important role as policy makers to make a successful transition to sustainability. The goal is to engage your community and begin members on a path toward sustainability (Hopkins 2008).

Step 2. Make an inventory of your community's social, environmental, and economic assets (resources) and liabilities (lack of resources). Generate a list and rank entries from 1 to 10 for strength and size. This is a critical step and should be led by city managers and planners as well as key business leaders. Consider a new office of urban agriculture with an ecological strategist/planner at the helm with strong support from the highest level of city and regional government. Start and maintain a list of appropriate questions: What can be grown in your area? What sources of information are available? What resources? Historically what was grown in your region? The object here is to make sure that all resources, including those from local businesses, fit together to work as a functional unit. The idea is to leverage your community's particular strengths. Connect existing businesses in new ways within your community. For example, ecotourism and agricultural tourism turn existing resources into destinations to observe how food is grown organically. Experiment with the ecosystem template by placing each community asset into its proper ecological grouping and describe why it fits in that ecological role. For example, reduce pollution and cut back on your community's waste stream by composting organic matter. The resulting composted organic matter can be used as organic fertilizer by schools, golf courses, and suburban gardeners.

Step 3. Economize, conserve, and preserve. Identify those parts of your community that are wasteful of resources and can use improvement. Rank your list according to level of difficulty on a scale of 1 to 10. Estimate the acres required to satisfy a particular percentage of your food supply. This will allow you to gauge your community's level of preparation regarding food security. Make an inventory of your community's physical structure and land use: percent area in roads, percent in buildings, percent homes, percent open space, and percent parks. Use this information to generate geographical maps of your watershed and consider new zoning options. Ask what can be improved and what areas can be better utilized.

Step 4. Use your land and resource inventory to assist as a baseline to develop a plan for local sustainability and food security. Begin thinking about a percentage of land to be devoted to agriculture, light manufacturing, living, and entertainment. Use existing zoning as a place to start. Examine the transport system of your community and attempt to build or find alternatives to fossil fuel–dependent trucks for delivery of foods and goods. Ask what will happen to your community if the big box stores decide to close. Advocate policies to reestablish your community's merchant class. Generate your community's long-term green, sustainable plan emulating the interconnectedness of ecosystems. Turn problems into solutions by closing loops and thinking holistically. Fundamentally, this means seeing the interconnectedness of parts. Nature has no waste and neither should a sustainable community. Design in harmony with nature to let ecosystem processes supply the energy and do the work instead of using expensive, greenhouse gas–emitting, fossil fuel energy.

Step 5. Create a chronology of events (Figure 17.7). Lay out your timeline based on realistic expectations and existing resources within your community. For example, you might consider starting by initiating a program to compost the green organic matter that is generated in your community. The state of California has mandated through AB 939 that all green garden organic matter be excluded from landfill deposition by 2012. Under California Assembly Bill 939, Integrated Waste Management Act of 1989, jurisdictions are required to meet timed diversion goals. AB 939 also established an integrated framework for program implementation, solid waste planning, and solid waste facility and landfill compliance. Therefore, such a project is a logical place to start. Turn your waste disposal problem into a solution, that is, composted organic fertilizer for use in your community's parks, golf courses, and open green areas.

Step 6. Select projects or pilot projects for your community and implement them based on careful examination of your available resources. Use your asset inventory as your guide. Examples include sewage to algae, tree planting in parks, and urban agricultural zoning and designation. Pilot projects become attractive as an undertaking by a community when they have the potential to generate revenue and provide local "green" employment. Many businesses will be new and unique. For example, developing ecotourism and ecoeducation are both possible scenarios for a pilot project. Each local community should not only leverage its strengths but also celebrate its particular cultural uniqueness and diversity. The many thriving ethnic communities across the United States confirm the economic worth and viability of diversity in our culture.

Perhaps most importantly, pilot projects provide an opportunity for proactivity in establishing sustainable procedures and designing ecologically friendly landscapes. Building such working prototypes and pilots will place communities in a better position to scale up rapidly when a crisis

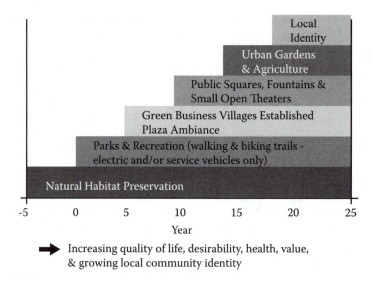

FIGURE 17.7   Exemplary chronology of events showing a timeline based on realistic expectations and existing resources within a community. Natural habitat preservation is shown as already in place—hence the negative years. Local identity is a critical developmental step in that it returns regenerative ownership to a community. (Illustration by Karen Holmgren.)

occurs, for example, climate change and energy depletion. It is disaster preparation of the most fundamental kind.

17.4   VISUALIZING A POSITIVE FUTURE

No two sustainable communities as envisioned here will appear or function identically. Each community will be a unique product of its own history and existing structure for transportation, water, energy, and diversity. Thus each community will face its own distinct opportunities and challenges for retrofitting its existing suburban resources. However, using the ecosystem as a unifying tool and solution framework provides a community with a set of basic design and operation principles. Implementing urban agriculture, restoring natural habitats with the watershed, changing zoning to accommodate mixed use, relocating to transit-oriented development areas, and so forth are all issues which can be successfully included within the ecosystem paradigm.

Implicit within sustainable communities are new opportunities for positive lifestyle changes. Humans are social by nature. We like to meet, talk, play, and compete. Sustainable communities, such as the hypothetical design offered here, have the potential to enhance social engagement, possibly discovering deeper satisfaction through an increase in those intangible, but very human, values of contentment and meaning.

ACKNOWLEDGMENTS

I would like to thank Karen Holmgren for her diagrams, attention to detail, and dedication. Also many thanks to Tracy Micka, Building Sustainable Communities secretary, for proofreading and editing the manuscript.

REFERENCES

Altieri, M. A. 1995. *Agroecology: The Science of Sustainable Agriculture.* Westview Press, Boulder, CO.
American Heart Association. 2008. Physical Activity in Your Daily Life. http://www.americanheart.org/presenter.jhtml?identifier=2155.

Anderson, K. 2006. *Tending the Wild: Native American Knowledge and the Management of California's Natural Resources.* University of California Press, Berkeley.

Balfour, E. B. 1942. *The Living Soil.* Faber & Faber, London.

Bradford, J. 2007. Relocalization: A Strategic Response to Climate Change and Peak Oil. The Oil Drum. http://www.theoildrum.com/node/2598.

Bradford, J. 2008. Can My County Feed Itself? Part 3. The Available Land-base. Energy Farm. http://www.energyfarms.net/node/1491.

Bradley, F. M. and Ellis, B. W. Eds. 1993. *Rodale's All-New Encyclopedia of Organic Gardening: The Indispensable Resource for Every Gardener.* Rodale Books, New York.

Brown, L. 2006. *Plan Bm2.0: Rescuing a Planet under Stress and a Civilization in Trouble.* W. W. Norton, New York.

Central Valley Business Times. 2008. California organic acreage reaches milestone http://www.centralvalleybusinesstimes.com/stories/001/?ID=8295 (accessed July 14, 2008).

Farr, D. 2008. *Sustainable Urbanism: Urban Design with Nature.* John Wiley & Sons, New York.

Favoino, E. and Hogg, D. 2008. The potential role of compost in reducing greenhouse gases. *Waste Management & Research,* 26(1), 61–69.

Gliessman, S. 2006. *Agroecology: The Ecology of Sustainable Food Systems,* 2nd ed. CRC Press, Boca Raton, FL.

Hall, C. 2008. Why EROI Matters. The Oil Drum. http://www.theoildrum.com/node/3786.

Hansen, J. 2008. Climate target is not radical enough. Guardian.co.uk. http://www.guardian.co.uk/environment/2008/apr/07/climatechange.carbonemissions (accessed July 14, 2008).

Heinberg, R. 2006. 50 Million Farmers. Energy Bulletin. http://www.energybulletin.net/22584.html (accessed July 14, 2008).

Heinberg, R. 2007. Five axioms of sustainability. http://www.richardheinberg.com/museletter/178 (accessed July 14, 2008).

Holmgren, D. 2002. Permaculture: Principles and Pathways beyond Sustainability. Holmgren Design Services.

Holmgren, D. 2008. Backyard answer to energy crisis. Sydney Morning Herald. http://www.smh.com.au/news/environment/backyard-answer-to-energy-crisis/2008/03/18/1205602385256.html (accessed July 14, 2008).

Hopkins, R. 2008. *The Transition Handbook: From Oil Dependency to Local Resilience.* Green Books, Devon, UK.

Howard, A. 1947. *The Soil and Health: A Study of Organic Agriculture.* Faber & Faber, London.

Jeavons, J. 2002. *How to Grow More Vegetables Than You Ever Thought Possible on Less Land Than You Can Imagine.* Ten Speed Press, Berkeley, CA.

Kunstler, J. H. 2006. *The Long Emergency: Surviving the End of Oil, Climate Change, and Other Converging Catastrophes of the Twenty-First Century.* Grove Press, London.

Lancaster, B. 2006. *Rainwater Harvesting for Drylands and Beyond.* Vol. 1: *Guiding Principles to Welcome Rain into Your Life and Landscape.* Rainsource Press, Tucson, AZ.

Langer, G. 2005. Traffic in the United States: A Look Under the Hood of a Nation on Wheels. ABC News, http://abcnews.go.com/print?id=485098.

Lerch, D. 2008. *Post Carbon Cities: Planning for Energy and Climate Uncertainty.* Post Carbon Press, Sebastopol, CA.

Lloyd, C. 2008. Is Suburbia Turning into Slumburbia? SFGate.com. http://www.sfgate.com/cgi-bin/article.cgi?f=/g/a/2008/03/14/carollloyd.DTL.

Lovelock, J. 2006. *The Revenge of Gaia: Earth's Climate Crisis and the Fate of Humanity.* Basic Books, New York.

Mills, L. S. et al. 1993. The keystone-species concept in ecology and conservation. *BioScience,* 43, 219.

Montgomery, D. 2007. *Dirt: The Erosion of Civilizations.* University of California Press, Berkeley.

Odum, E. P. 1959. *Fundamentals of Ecology.* W. B. Saunders, Philadelphia.

Odum, E.P. 1969. The strategy of ecosystem development. *Science,* 164, 262–270.

Odum, H. T. 1983. *Systems Ecology: An Introduction.* Wiley-Interscience, New York.

Parker, T. et al. 2002. Statewide Transit-Oriented Development Study Factors for Success in California. http://transitorienteddevelopment.dot.ca.gov/miscellaneous/StatewideTOD.htm.

Pearson, C. J. 2007. Regenerative, semiclosed systems: A priority for twenty-first-century agriculture. *BioScience,* 57, 409–418.

Pimentel, D. 1984. Energy flow in agroecosystems. In *Agricultural Ecosystems*, R. Lowrance, B. R. Stinner, and G. J. House, Eds. John Wiley & Sons, New York.

Pimentel, D., Hepperly, P., Hanson, J., Douds, D., and Seidel, R. 2005. Environmental, energetic, and economic comparisons of organic and conventional farming systems. *BioScience,* 55, 573–582.

Rees, W. E. 1992. Ecological footprints and appropriated carrying capacity: What urban economics leaves out. *Environment and Urbanisation,* 4, 121–130.

Register, R. 2006. *Ecocities, Rebuilding Cities in Balance with Nature.* New Society Publishers, Gabriola Island, Canada.

Richards, G. 2008. Railroads are expanding at a record clip. The Virginian-Pilot. http://hamptonroads.com/2008/04/railroads-are-expanding-record-clip.

Robbins, P. and Birkenholtz, T. 2003. Turfgrass revolution: Measuring the expansion of the American lawn. *Land Use Policy,* 20(2), 181–194.

Roberts, P. 2008. *The End of Food.* Houghton Mifflin, Boston.

Steiner, R. 2005. *What Is Biodynamics?: A Way to Heal and Revitalize the Earth : Seven Lectures.* Steiner Books, Great Barrington, MA.

Vail, J. 2008. Rhizome & Central Place Theory. jeffvail.net. http://www.jeffvail.net/2006/04/rhizome-central-place-theory.html.

Wackernagel, M. 1994. Ecological Footprint and Appropriated Carrying Capacity: A Tool for Planning Toward Sustainability. Ph.D. thesis, School of Community and Regional Planning, University of British Columbia, Vancouver, Canada.

Index

A